A Guinea Pig's History of Biology

JIM ENDERSBY

A Guinea Pig's History of Biology

Harvard University Press
Cambridge, Massachusetts
2007

First published in the United Kingdom in 2007 by
William Heinemann

Printed in the United States of America

Library of Congress Cataloging-in-Publication Data

Endersby, Jim.
A guinea pig's history of biology / Jim Endersby.
p. cm.
Originally published: London : William Heineman, 2007.
Includes bibliographical references.
ISBN-13: 978-0-674-02713-8 (alk. paper)
ISBN-10: 0-674-02713-2 (alk. paper)
1. Heredity – History. 2. Genetics – History. 3. Biology – History. I. Title.
QH431.E615 2007
576.509 – dc22
2007020824

Illustrations by Alison Kendall

For
Daniel
Walter
Sophie
Max
and Katya

The F_2 generation

Contents

Preface and acknowledgements

In the past 200 years we have gone from complete ignorance as to why some diseases run in families, to having simple genetic tests that can be bought on the internet. Over the same period, we have gone from Darwinism being pronounced dead to seeing the modern theory of evolution triumphant. All thanks to the fruit fly, the guinea pig, corn and a handful of other organisms that have helped us unravel one of life's greatest mysteries – inheritance.

Biology is currently one of the world's most exciting sciences, promising everything from better foods and cures for common diseases, to the more alarming prospect that genetic engineering could allow us to redesign life itself, including ourselves. Examining the organisms that have made all this possible gives us a different understanding of how we got here and perhaps new ways of thinking about where we might be going. Instead of a story in which great scientists have great ideas, the story of passionflowers and hawkweeds, of zebrafish and viruses, allows us to understand the different kinds of work that make science possible. It can take months, even years, to prepare the organisms for an experiment: careful breeding to select the characteristics that will be useful to the researcher, and equally painstaking work to master or invent the techniques necessary to make the experiment work. Science is as much about fund-raising and meticulous planning as it is about having brilliant insights. And examining the work involved reminds us that very few scientists work alone; they rely – in hundreds of different ways – on their colleagues and competitors, on their teachers and students, and on their technicians and assistants. Above all, they rely on the efforts of their predecessors; every scientist is, as Newton put it, standing on the shoulders of giants.

The history of biology allows those of us who are not scientists – and have no hope of mastering the technicalities of genetics – to understand something of how science works, and thus what kinds of answers it can – and cannot – give us about the variety, complexity and nature of living things, including ourselves. In choosing the organisms for this book, I have not tried to catalogue every creature that made a contribution to this story; instead, I have tried to mix the celebrities of genetics, like the fruit fly, with organisms like the evening primrose, whose role in this history is now almost forgotten. In order to keep the book to a manageable size, I have had to omit some very important characters along the way, but those which have been included give some sense of the diversity of biological research and of how much it has changed in the last two centuries.

'Meet *Drosophila*, the friendly fruit fly' is not the kind of sentence one expects to find in an academic monograph on the history of genetics, but Robert Kohler's *Lords of the Fly: Drosophila Genetics and the Experimental Life*, is no ordinary academic monograph. Kohler builds on a range of recent work in the history of science to tell an extraordinary story in a highly original way – by making the fruit flies the heroes of his tale, rather than the geneticists who worked with them. What Kohler calls 'following the fly' allows him to examine issues that are still too often overlooked by historians.

Kohler's book is perhaps the best example of a trend among academic historians of science, of paying increasing attention to the organisms used in research as a way of broadening our historical focus. Inspired in part by the ideas of the French sociologists Bruno Latour and Michel Callon, historians such as Karen Rader and Angela Creager have examined everything from laboratory mice to tobacco mosaic virus and have brought new perspectives to bear on the history of biology.[1] This book makes use of some of these recent approaches in an effort to write a broader history of biology, particularly the history of heredity and genetics. A survey of some of the key organisms that have been used since Darwin's day illustrates how biological research itself has changed – transforming the largely amateur practices of natural history into modern, laboratory-based molecular biology. It allows us to trace the slow, uncertain path – complete with diversions and dead ends – that led us from the ancient world's understanding of inheritance to modern genetics. And the story of the organisms also reveals how much work was needed to reach our current understanding of evolution.

Like science itself, the history of science depends on the work of the historian's colleagues, contemporaries and teachers – and this book is no exception. I have relied on many other people's work; that which I made the most use of is listed in the notes. Many people gave up their time to read sections of the book, offering suggestions, corrections and improvements. I would particularly like to thank: Nathaniel Comfort, Caroline Dean, Sheila Ann Dean, Ian Furner, Beverley Glover, Chuck Kimmell, Les King, Sharon Kingsland, Robert Kohler, David Kohn, Maarten Koornneef, Emil Kugler, Kate Lewis, David Meinke, Elliot Meyerowitz, Robert Olby, William Provine, Anne Secord, Jim Secord, Chris Sommerville, Shauna Sommerville, Derek Stemple, William Summers, Will Talbot, John Waller, Garrison Wilkes, Stephen Wilson and Paul White.

Special thanks to: Katie Armstrong (Talbot), Dan Austin, Rodney Bolt, Joe Cain, Kenneth Carpenter, Simon Chaplin, Soraya de Chadarevian, Anjan Chakravartty, Bonnie Clause, Gregory Copenhaver, Gina Douglas, Judith Flanders, Tom Gerats, Geoff Gilbert, Laura Green, Eivind Kahrs, Evelyn Fox Keller, Henrika Kuklick, Martin Kusch, Nick Jardine, Peter Lipton, Tim Lewens, Valerie-Anne Lutz, John MacDougal, Sunit Maity, Tim McCabe, Craig Morfitt, Gina Murrell, Charles Nelson, William Ogren, John Parker, Sridhar Ramachandran, Rowan Routh, Simon Schaffer, Peter Straus, Charlotte Thurschwell, Hugh Thurschwell, Bob Whitman, Ian Woods and

Eric Zevenhuisen. And many thanks to the team at Heinemann, especially Alban Miles, Caroline Knight, Annie Lee and Juliet Rowley.

I would also like to acknowledge: the Syndics of the Cambridge University Library for permission to quote from the unpublished correspondence of Charles Darwin. I am grateful for permission to quote from The Haldane Papers, UCL Library Services, Special Collections. Thanks also to the American Philosophical Society, Philadelphia, for permission to quote from Sewall Wright's papers; and the Archives of King's College, London, for permission to quote from the manuscripts of Reginald Ruggles Gates. My thanks to the staffs of these libraries and archives for their help, and to the staff of the Whipple Library, Cambridge University, without whom none of my work would be possible.

The research for this book was made possible by a post-doctoral fellowship awarded by the Master and Fellows of Darwin College, Cambridge; and by the Royal Society of Literature and the Jerwood Charitable Foundation, whom I would like to thank for awarding me the inaugural Royal Society of Literature Jerwood Prize. I am most grateful to them.

Lastly, my largest debts are to Ravi Mirchandani, for the huge amount of time he spent on editing the book, massively improving it in the process, and to Pamela Thurschwell.

QUAGGA

Equus quagga quagga

EXTINCT

12 AUGUST 1883

Chapter 1

Equus quagga and Lord Morton's mare

The right honourable George Douglas, FRS, 16th Earl of Morton, was perplexed; in 1820, he had sold a chestnut-brown mare to his friend, Sir Gore Ouseley, who had crossed her with 'a very fine black Arabian horse'. Both Morton and Ouseley were astonished by the offspring of this union, two foals who were clearly Arabian horses, but 'both in their colour, and in the hair of their manes, they have a striking resemblance to the quagga'.[1]

Quaggas were a kind of zebra, but it was mainly their heads and shoulders that were striped. They belong to a small, sad club (whose members include the passenger pigeon and the thylacine, or Tasmanian tiger) of animals for whom we know the exact day on which they became extinct. The last quagga died in Amsterdam zoo on 12 August 1883. At the time, the species went unmourned, as no one realized she was the last one.

Why should the offspring of Ouseley's animals look like quaggas, when both parents were horses? Morton had a theory. A few years earlier, he had owned a male quagga which he had hoped to domesticate. But like the other zebras, quaggas are practically untameable – bad-tempered and stubborn, zebras not only bite, they refuse to let go – even today, more zookeepers are injured by zebras than by lions or tigers. The indigenous people of South Africa had never managed to ride them and thus Europeans were not met by troops of African cavalry when they started their encroachments. (The history of Africa might have been rather different if zebras were as tractable as Eurasian horses.)

Morton had been 'desirous of trying the experiment of domesticating the Quagga'. In the late eighteenth and early nineteenth centuries, Africa's diseases proved the biggest obstacles to European efforts to colonize the continent. Among the many scourges the whites faced was sleeping sickness, which attacked both Europeans and their horses. Zebras and quaggas are largely immune to the disease, and Morton seems to have been one of a number of Europeans interested in cross-breeding them with horses to produce a hybrid that would be better adapted to African conditions. In 1788, the German naturalist Johann Georg Gmelin observed that the quagga was 'thicker, and more strongly made than the Zebra, and is more tractable, having sometimes been broken to the draft'. During Morton's lifetime, a traveller to South Africa reported that his host kept 'a tame quagga . . . for the purpose of making experiments in improving his breed of horses'. But reports of the quagga's temper and tameability were mixed: a few years after Morton's experiments, a British naturalist observed that the one he had seen 'is not very docile, being much more wild than the Zebra'. While another commentator suggested that it might nevertheless be possible 'to reclaim this creature, and to subject it to the service of man'.[2]

Morton acquired a male quagga and crossed it with a chestnut mare. He described the result as a female horse 'bearing, both in her form and in her colour, very decided indications of her mixed origin'; she had evidently retained either her father's stripes or his filthy temper – probably the latter, since Morton seems not to have repeated the experiment. However, it was the same chestnut mare, the quagga's former mate, which Morton had sold on to Ouseley. And it was *this* mare that then produced further quagga-like foals – even when she was mated with a perfectly ordinary horse. Her foals had striped legs and stiff, short manes that stuck straight up like a zebra's, instead of hanging down like a horse's. Stripes do occur occasionally in breeds of domestic horses, but are very rare in Arabians; neither Morton nor Ouseley had even seen manes like these on a horse before. It appeared as though the mare's encounter with the male quagga had permanently altered her, leaving some indelible imprint of 'quagga-ness' that affected the offspring of her later matings.

This incident struck Morton as sufficiently unusual that he brought it to the attention of his friend William Hyde Wollaston, president of the Royal Society of London (of which Morton himself was a fellow). In 1821, Morton's letter was published in its journal, the *Philosophical Transactions*, under the modest title 'A Communication of a Singular Fact in Natural History'. The notion that a male could influence the female's subsequent offspring by other males was not new, but had typically been dismissed as little more than folk myth. Now, however, lent authority by the influential members of the prestigious Royal Society, the quagga effect became an established – if 'singular' – scientific fact.

Nearly fifty years after Morton published, even Charles Darwin remained convinced that 'there can be no doubt that the quagga affected the character of the offspring subsequently begot by the black Arabian horse'.[3] He referred to the case as one of the crucial puzzles he needed to solve when trying to devise a theory of biological inheritance. Darwin, like others struck by Lord Morton's mare, believed that the male quagga had made some sort of lasting impression on the female horse. By some unknown means, the character of his species had left its mark on her, but no one knew how: was it, as some thought, a mental impression, an idea of quagga-ness that the mare retained in her mind? Or was it something tangible, some particle or fluid in the male's semen or blood that had been passed on and retained? Might the male subsequently lose interest in female quaggas and pursue only mares?

This book is about how some of these mysteries were unravelled. Long before Darwin, most people knew that for most organisms, as for humans, both males and females were needed to produce offspring, but even after centuries of investigation and speculation, no one knew exactly why, nor exactly what each contributed to their progeny. Given that some creatures managed without sex, what was its purpose? And what about inheritance? Why do most children look like their parents, while some are more like one of their grandparents? How are the characteristics of animal or plant breeds passed on and preserved over generations? And, given that creatures only ever give birth to creatures of the same species, where do new types of organisms come from? Given that species remain largely constant over generations, how and why do they sometimes vary?

Lord Morton's explanation of his mare's striped foals may appear unlikely, even ridiculous, as may many of the later attempts to resolve his puzzle. To understand how and why they made sense at the time, we need to be aware of the centuries of thinking about reproduction that had preceded Morton's experiment.

The harbour of Lesbos

The Philosopher must have become a familiar figure to the fishermen of Lesbos. Each morning he would come down to the harbour to inspect the night's catches, searching for unusual or bizarre fish, which he and his students would take away to study. Almost 350 years before the birth of Christ, Aristotle was initiating the systematic study of life; observing sea creatures grow and develop, with no tools more sophisticated than the naked eye.

The observation of nature – this simple act of looking – seems obvious now; where else to begin understanding the world? But to Aristotle's teacher, Plato, this beachcombing would have amounted to heresy. Plato had taught that the mind was the only sure guide to the world; only through logic can we apprehend the Forms, the ultimate abstract truths that lie behind the deceptive world of mere appearance. For example, every student of geometry knows that however carefully we draw a figure, such as a triangle, it is never perfect: the lines (which should be infinitely fine) actually have a measurable width, and their corners do not form perfect angles. Any actual triangle is simply a sketch of the idea of a triangle; no closer to the Platonic Form of a triangle than a child's sketch of Daddy is to the man himself. Plato extended his idea of Forms to everything: the dog wandering down the street is just an imperfect image of the Dog, the pure Form of doggyness; even the blue of the sky is just a shadow of the pure Form of Blue. Examining individual dogs can never reveal the underlying Form that defined Dog, any more than comparing blue skies to blue eyes reveals the Form of Blue. Only reason can penetrate the realm of the Forms.

So, had Plato lived to observe his former pupil's activities, he would have been horrified. Not just by wet feet and the undignified spectacle of a middle-aged man paddling and listening to fishermen's tales, but by the idea that observing the world, trying to catalogue the seemingly endless kinds of fish, could ever reveal the underlying Form, the universal, absolute truth of Fishiness. Yet that is what Aristotle was trying to do. Even worse, Aristotle was not even looking for the Form, as Plato would have understood the idea. By examining and comparing many specimens of a creature, he hoped to discern their essential properties: Aristotle wanted to know what all fish have in common, which allows us to correctly classify them as fish. And what distinguishes one kind of fish from another?

For Aristotle, analysing differences was the key to understanding what caused those differences, and causes were the key to everything. It is not enough to know that fish were able to breathe in water, while animals could not; for Aristotle, the question was why? What was the *cause* of the difference? His philosophical system identified four distinct types of cause. First there is the *formal cause*: identifying the structure of a subject. Then it is necessary to understand the *material cause*: what kind of stuff is this structure made from? Then the *efficient cause*: what force shaped this particular structure out of that specific material? And last but most important of all comes the *final cause*: what is the purpose for which any new structure is being made?

Aristotle applied his search for causes to every aspect of the world, from politics and ethics, to astronomy and cosmology. The nature of living things particularly intrigued him and in order to

understand them he paid particular attention to their reproduction, especially by studying their eggs. He goes to some lengths
in his writings to justify his decision to spend so much time on such
lowly matters, but it is hard not to feel that he was simply
fascinated by the natural world. As he wrote, 'to the intellect, the
craftsmanship of nature provides extraordinary pleasures for
those who can recognize the causes of things'.[4] Marine creatures
were particularly useful to him: they laid eggs, and eggs are easier
to observe than the development of mammals, whose embryos
develop inside their bodies; moreover, sea creatures lay their eggs
in water, which means they do not need the protection of a hard
shell, so Aristotle could see what was going on inside them as they
grew. He also dissected many of the creatures the fishermen
brought him and made dozens of careful observations (his *Historia
Animalium* mentions over 500 species, of which about 120 are fish).
He discovered many extraordinary things, some of which were
not confirmed for many centuries after his death, but he was also
prepared to listen to and preserve old-fisherman's tales, and
record the most implausible claims without further investigation.

Scrutinizing fish eggs as they develop into embryos and then
into tiny free-swimming larvae helped Aristotle apply his theory of
causes to living creatures. Reproduction was, he contended, the
ultimate proof that Plato was wrong: there could be no form
without matter, no 'Blueness' or 'Dogness' without something to
be the physical bearer of such properties – an actual sky or a real
dog. The ability of two dogs to produce a new dog shows that
nature has the power to impose form on matter. Aristotle ascribed
the size, shape and behaviour of an individual dog not to the
remote, abstract ideal Form of Dog, but to the messy business of
reproduction: it is reproduction that maintains the form of a
species, renewing itself, generation after generation.

Like many of those who followed him, Aristotle was enthralled
by the creatures he studied; in part he seems to have observed
them out of simple curiosity but he also wanted to know what they
could tell him about himself and other humans. He noted that 'the
inner parts of man are for the most part unknown, and so we must

refer to the parts of other animals which those of man resemble and examine them'.[5] One reason to study animals and plants was to shed light on human reproduction and how human qualities – including our ability to think about and understand the world – were passed on from generation to generation.

Aristotle's ideas were, of course, not the only ones circulating in the ancient Greek world. Just a day's sail south of Lesbos, on the coast of what is now Turkey, lay the city of Miletus, a bustling, trading port where people from all over the eastern Mediterranean came to trade spices and dyes, timber and pottery. They also brought with them their languages and customs, their gods and their stories about the origins and nature of the world. This cosmopolitan mixture may help explain why it was in this city – a couple of centuries before Aristotle's time – that Greek philosophy began; confronted with many different accounts of how and why the world worked, the Miletian Thales became the first Greek (whom we know of) to wonder whether thinking for oneself, rather than following tradition, might not be the most reliable guide to the nature of the cosmos.

Thales and his successors came up with many different ideas to explain the essential nature of the world: one of his followers suggested that living things were formed by the heat of the sun on moisture; others that everything must ultimately be made from a single, underlying substance – otherwise how could one thing turn into another (such as ice into water and then steam)? Some proposed that water must be this underlying substance, while others believed it might be air. Gradually the idea of one, single substance seemed inadequate and the competing conceptions were combined into the doctrine of the four elements – Earth, Air, Fire and Water – out of which everything was made. Some combination of these four, each of which had its own distinct properties, made up everything around us, but it was difficult to account for living things within this system. Organisms seemed to possess unique properties that other kinds of matter did not, and of these, reproduction was the most remarkable.

The Greek doctor, Hippocrates, often called 'the father of

medicine', attempted to explain some of the unique properties of living creatures with a theory of fluids called humours – the body's equivalents of the four elements. Human beings are infused with black bile, blood, yellow bile and phlegm. These affect our character: those with an excess of black bile (associated in the Hippocratic system with the element Earth) are supposed to be melancholy, while those with too much of the hot, yellow bile (Fire) are choleric – short-tempered and angry. The language retains traces of these ideas when, for example, we describe someone as phlegmatic or bilious. Imbalances in the humours also explained why we get sick, and so, for hundreds of years after Hippocrates, doctors would treat disease by trying to get the humours back into balance – by prescribing a change of air or diet, or a course of blood-letting. But in the Hippocratic tradition, animals were also considered purposeful, integrated systems that can and do heal themselves.

However, while the theory of humours offered an explanation of how living things worked, it did not explain where they came from. The Hippocratic tradition dealt with that question in two treatises called *The Seed* and *The Nature of the Child*, which proposed that both the male and female contributed some kind of seed or fluid to their offspring. (Although the female's contribution had never been seen, the Greek doctors deduced that it must exist because females clearly enjoyed copulation as much as males did.) The male seed is contained in the seminal fluid and is a combination of the strongest parts of the four humours (which is supposedly why men are so tired after intercourse). This semen is in some way distilled from every organ of the body, and so the seed carries within it some essence of each of the body's parts. It is collected in the spinal cord and transmitted to the testicles. The doctors were much vaguer about the source of the female equivalent, but it was assumed to be comparable to the female's pleasure – more prolonged, less violent and (these were, after all, male doctors) of a lower order. When the male and female seeds combine they produce a child that resembles both its parents, but to differing degrees – the relative strengths of its mother's and father's influences determine both the resemblance and the child's sex.

The Greek doctors were aware that menstruation stopped during pregnancy and assumed that the blood was being used to nurture the growing embryo. But they were much less clear as to how the growing child took shape. How did substances as amorphous as blood or semen develop into the complex body of a child? The doctors suggested the mother's 'breath and spirit' formed the child's organs, but left the details of the process extremely vague.

The Hippocratic teachings remained influential for over 1,000 years, but Aristotle was unimpressed by them, especially by their apparent inability to explain the organization of an animal's parts. His own theory abandoned the hypothesis of a female seed and argued instead that the male and female contributions were of different kinds. The male's semen was derived only from his blood, which carried the spirit or *pneuma* of each organ and nourished it. The essence of each of these different forms of nourishment – the essential 'spirit' of the heart, eyes or brain – was then distilled into the semen and passed on to the offspring, thus passing some of the father's characteristics on to his child. The female's contribution was only the menstrual blood. Aristotle argued that it must be akin to the semen, in that it too must carry something of the form of the female's organs. Thus he explained how elements of the mother's appearance and character were passed on to the child; if females contributed only substance without form, Aristotle argued, all children would be perfect copies of their fathers. Nevertheless, he viewed the female's contribution as less significant than the male's: the mother's main contribution was sustenance, while the higher faculties – character and behaviour – were, to Aristotle, a uniquely male contribution.

According to Aristotle, it is the amount of heat in the semen which determines the sex of the child. Females are naturally cooler than males, so older parents – whose ardour has cooled a little since their youth – are more likely to produce girls, as are those who copulated when a cool, southerly breeze was blowing.

If you wanted a boy, he advised, you should wait for a hot wind. However, the mother's and father's contributions are not a simple, all-or-nothing business: like their parents, each child's mind and body is divided into faculties; during sexual intercourse the parents unknowingly compete to provide the major contribution to each of their future offspring's faculties. In some areas, the male principle will dominate, in others, the female. Hence the child might have its mother's eyes, but its father's wisdom. But sometimes the intensity of this struggle can damage a faculty, which could either lead to deformities or to the character of one of the parent's earlier ancestors coming through in the child; that was why a child might sometimes have grandma's nose or their uncle's chin.

Generation after generation

Aristotle left Lesbos to become tutor to the young Alexander the Great, son of Philip of Macedon, teaching him everything from the virtues proper to a prince, to the mating habits of frogs. Like royal families before and since, the kings of Macedon inherited their right to rule; the prince inherited his father's throne and wealth, and – like any other son – also his blood and with it his courage, his skill as a leader, and his other qualities, both good and bad. Inheritance, in the biological sense, is a concept borrowed from the law, the body of customs and traditions that have always surrounded the passing on of thrones, palaces or other more humble property. Such inheritance is usually from father to son, and the mother's subsidiary role in many biological theories of inheritance parallels her relatively minor place in such dynastic successions. Yet royal brides usually came from royal families and were expected to contribute something – some land, an alliance, and some good blood – to the new prince.

Just as a young king did not have to build himself a new kingdom from scratch, he was assumed to inherit his parents' human qualities. However, this assumption raises the question why, if a king's virtues could be handed on as easily as his title, did Philip of Macedon hire the world's greatest living philosopher to educate his son? Could princes be *taught* to embody the kingly virtues? In which case, perhaps anyone could learn to be a ruler. Was there nothing special about royal blood after all?

Those who, from time to time, successfully usurped other people's thrones were, understandably, inclined to argue that strength, not bloodline, was the vital quality a king needed. Those with royal blood could be milksops, raised in luxury, with no need to struggle; as a result, they were weaker – physically and mentally – than those who had had to fight their way to the top. The decision to educate a prince could be interpreted as lending tacit support to the notion that it was the circumstances and experiences of upbringing, not any quality in the blood, that conferred nobility. Plato's celebrated dialogue, *The Republic*, which set out his vision of a perfect state, set great store by proper breeding, arguing that people of inferior stock should be prevented from reproducing, while the strongest, fittest and most intelligent should be encouraged to have more children. But Plato also considered the education of his city's rulers carefully, recognizing that children might be born among the lower orders – the 'bronze' people as he referred to them – who had the souls of rulers and should join the 'golden' class, the republic's Guardians.[6] And Aristotle's suggestions about the influence of the prevailing wind during conception illustrate that he too believed environment played some part in shaping a child's nature.

So which ultimately had the greater influence: an animal or person's bloodline or their upbringing? The question was hard to settle for several reasons, but the biggest obstacle was that people breed too slowly and live too long. Even the wisest person could not hope to observe other human lives over many generations to see whether birth or education ultimately triumphed.

Faced with this problem, those who wanted to understand inheritance followed Aristotle's example: just as he had turned to fish to understand the unknown 'inner parts of man', people turned to various animals and plants to try and unravel such puzzles as the relationship between environment and inheritance. Long before written records began, someone had the idea of saving edible seeds and growing food instead of foraging in the wild. As this practice caught on, somebody decided to save only the seeds from the plants that produced the largest fruit and planting them, while discarding the seeds from fruits that were tough, tasteless or under-sized. We have no idea when or where, but thousands of years ago this must have been the origin of plant-breeding, the first deliberate attempts to improve nature. The same thing happened with animals when farmers first chose their best animals to breed from, while the cows who produced less milk, or the sheep who had less wool, ended up in the cooking pot. But the results of these early attempts at improvement were pretty variable; sometimes the breed improved, sometimes it did not, and no one really understood exactly why.

Many of the Greek philosophers took an interest in the farmyard and farming practices. Xenophon (another pupil of Plato's mentor, Socrates) wrote a treatise on hunting with dogs, *Cynegeticus*, which recommended that only 'good' dogs be bred. Plato used the example of dog-breeding in the *Republic* to persuade his audience of the importance of breeding humans more carefully. Both he and Xenophon were mainly concerned with the health and condition of the parents – whether human or canine – at the time of conception, rather than with the achievements of the parents or their ancestors. Young, fit, healthy parents were assumed to be the key to getting good-quality offspring; the issue of whether the parents themselves came from a long line of good hunters or benevolent rulers was seldom considered.

Aristotle also discusses animal-breeding, concentrating on horses, which were used for both sport and warfare in Hellenic times. Horses were considered high-class animals: more attention was paid to their breeding than to that of other animals; and, naturally, the most attention was given to the mounts of the aristocracy, rather than to ordinary cavalry horses. Later classical writers also discuss the breeding of horses and other animals, but there was no consensus about the relative contributions of good breeding and a favourable environment. Some emphasize the selection of breeding animals, while others concentrate on their condition at the time of mating, advocating starving cows to make them more fertile, while fattening bulls, to give them more energy for the arduous male task of procreation. Meanwhile, little attention was paid to either the Aristotelian and Hippocratic theories about reproduction, since neither seemed to offer breeders much practical guidance.

As Europeans rediscovered the classics during the second millennium, both Aristotelian and Hippocratic theories received renewed attention, even though they still seemed to lack practical implications. At first, little was added to the classical tradition; it was widely assumed that the wisdom of the ancients, like that of the Bible, was not to be questioned. So most scholars were intent on preserving what remained; it was not until the sixteenth and seventeenth centuries that this reverence for the classical tradition began to be widely challenged, as a new breed of enquirer, the natural philosopher, turned its attention to the ancient questions.

The natural philosophers once again approached nature in a more empirical fashion, relying less on the wisdom of the ancients and more on their own observations and experiments. Robert Boyle and his famous air pump ushered in a new era of *experimental* natural philosophy, which involved using experiments both to make discoveries and to demonstrate the power of the new philosophy. The Royal Society (to which Lord Morton would later write) was founded to propagandize for the new approach; its full title was the Royal Society of London for Improving Natural Knowledge and its early members looked to experimentation for such improvement. They included Isaac Newton and Christopher Wren, together with others we will come across in later chapters, including Robert Hooke and John Evelyn. Among the many investigations the Society promoted were practical questions of agricultural improvement; its first publication was Evelyn's *Sylva, or a Discourse of Forest-Trees and the propagation of Timber* (1664), aimed at persuading England's landowners to plant trees that would provide better timber for the country's growing navy. The following year, Robert Hooke published his illustrated volume *Micrographia*, revealing some of the wonderful things that could be seen with the new microscopes, which had recently been improved by the Dutchman, Anton von Leeuwenhoek.

Prior to the seventeenth century, practical animal and plant breeders had paid little attention to the theories of philosophers, and the indifference was mutual; the philosophers generally ignored the observations of breeders too. However, the new natural philosophy, with its emphasis on practical knowledge confirmed by experiment, gradually began to break down this barrier. Horse-breeders and others began to read philosophical tracts on what became known as 'generation', a word that referred both to inheritance and to the development of the embryo. But since almost nothing was known about how inheritance worked, the natural philosophers initially focused on the embryo's development.

The physician William Harvey was typical of the new style of investigation, and would undoubtedly have joined the Royal Society had he lived long enough (he died in 1657, just before it was founded). Harvey is best known for his discovery of the circulation of blood, but he also published a tract on generation, *Exercitationes de Generatione Animalium*, which applied the new experimental philosophy to the understanding of generation. He demonstrated that menstrual blood did not mix with sperm in the uterus of pregnant creatures, as prevailing theories supposed. He also dissected birds and reptiles, revealing that the very earliest stages of the egg already existed long before it was laid. Harvey assumed that animals and humans, who bore live young, must possess something similar to this primordial egg, which he christened the ovum. He speculated that the semen stimulated the ovum to develop, but since he had found no traces of physical semen in any of the uteruses he had dissected, Harvey assumed that some 'spiritual' emanation from the semen set development in motion. Form gradually developed from formlessness as the stimulated ovum grew into an animal or person.

However, many of those who read Harvey's work were doubtful about this, unable to perceive where the form of the new creature actually came from. Harvey himself had been forced to assume that God personally intervened in the development of every embryo, an idea that sat rather awkwardly with the increasingly widely held view of a mechanical cosmos governed by natural laws. If, as Newton argued, God could design the sun and planets to run like perfect clockwork, with no intervention from their creator, it seemed odd that he could not devise embryos that would grow unaided. Harvey's critics therefore proposed that the organs and structures of the body were already present in the as-yet-unobserved ovum – the sperm merely caused them to develop.

Harvey's theory became known as epigenesis, the idea that the ovum was originally formless and only gradually took definite shape. Aristotle had believed the same thing, explaining it with his doctrine of final causes – the embryo took form because its purpose was to develop into a new animal. Harvey's ideas seemed old-fashioned almost as soon as they were published; despite his criticisms of Aristotle, he was seen as a rather quaint Aristotelian by many of his mechanistic contemporaries. Critics of epigenesis argued instead for preformationism, the idea that the embryo was preformed in the ovum. This idea did away with the need for any organizing force that turned formless embryos into complete animals, and it also solved the problem of explaining when and how a soul entered the developing human embryo. The preformationists argued, in effect, that there had only ever been one act of creation: when God created Eve, she had all of humanity packed into her ovaries, one inside another like an almost endless series of Russian dolls. If that seems absurd, remember that the Earth was still generally believed to have been created just a few thousand years earlier, so there had been only a few hundred generations since Adam and Eve. Even more importantly, the use of microscopes had just revealed previously unsuspected worlds in microcosm – tiny new worlds existed within each drop of water; if God could create such miniature worlds and people them with bizarre miniature creatures, it seemed quite possible that he could pack a few hundred generations of humans into Eve's ovaries.

Within a decade of Harvey's death, the new microscopes had enabled the identification of the small sacs, or follicles, in the mammalian ovary which contain the ova. It became widely accepted that the female produced the embryo on her own; the male's role was little more than to trigger development. However, Leeuwenhoek – arguably the leading microscopist of the day – disagreed; his microscopes had revealed tiny creatures, or animalcules, swimming in the seminal fluid of various animals. The nature and role of these animalcules was disputed, but it was clear that the seminal fluid was not purely liquid: the little animals had form and were active, so it became possible to believe that they were the source of the embryo. The preformationists therefore split into two camps: spermists like Leeuwenhoek and the rival ovists (although imagining all of humanity, one inside the other, into Adam's sperm seems to have been even harder for some to accept than the ovists' equivalent imaginative challenge). As microscopes improved, embryologists were able to study the anatomy of the embryo, ovum and sperm, ever more closely, looking for evidence to support or undermine the rival theories. Meanwhile, heredity was increasingly neglected.

The evidence that the preformationists, of both camps, were mistaken was staring them in the face, literally, every time they looked in a mirror. Like everyone else, preformationist natural philosophers reflected the influence of both their parents, with their mother's ears and father's nose, or vice versa. Yet few serious attempts were made to refute preformationism using this argument; when it was invoked, preformationists would defend themselves by suggesting that perhaps the mother's imagination shaped the developing embryo, or she provided some nutrient that modified the male sperm; there would be echoes of these ideas in the rival explanations later offered as solutions to the quagga mystery.

The idea that every member of a species was created at the beginning of the world, preformed and prepacked inside its first mother or father, sat very naturally alongside the belief that each species represented a single divinely created and unchanging essence, an idea in the mind of God. So the preformationists naturally did not consider the variability of living things as a possible route to understanding generation, instead they tended to view variations as monstrous aberrations from the original, divinely created form of the animal or plant.

Crossing the breeds

For the practical breeder, however, the variability that the natural philosophers scorned was the key to improving the breed, by selecting two animals or plants with the desired features and hybridizing them. Horse breeders, for example, knew all kinds of things about heredity that the philosophers did not, and because horses were important, expensive animals, much of what the breeders knew was carefully recorded. In Tudor England, for example, decrees passed under Henry VIII obliged the gentry to keep and breed horses (the higher up the social scale you were, the more you had to maintain). These laws emphasized the size of these horses: noblemen were obliged to keep horses at least thirteen or fourteen hands high. It is clear from the breeders' treatises that height was widely accepted to be an inherited characteristic – at a time when the natural philosophers still doubted or ignored the role of heredity in other domestic animal breeds.

Horse-breeding manuals also devoted much space to the question of whether pure- or cross-bred horses were better, and the evidence presented for either claim was a complex mixture of first-hand experience, classical learning and folk myth. One Renaissance writer urged that stallions should only be allowed to mate at night, so that their foals would be more wild and spirited; another argued that the more domesticated and tidy the pasture, the tamer the foals would be. Others deplored the mating of horses in stable yards, rather than running free in a field, arguing (perhaps reasonably) that restricting the animal's movements made mating duller and so produced dull and sluggish foals. Such matings, these experts insisted, would produce too many female foals because the stallion would have cooled, losing his natural heat.

We get a sense of how widespread these ideas were from Shakespeare's *King Lear*, where Gloucester's illegitimate son Edmund asks why he should be accused of 'bastardy' and 'baseness'; asserting as a well-known fact that children like himself, born out of illicit passion, 'take More composition and fierce quality' than legitimate children, conceived in 'a dull, stale, tired bed' contributing to the creation of 'a whole tribe of fops, Got 'tween asleep and wake'. 'Edmund the base,' he confidently asserts, 'Shall top the legitimate.'[7] Many in Shakespeare's audience would have agreed, that the heated passion of Edmund's parents would indeed have ensured that he possessed the 'fierce quality' that would allow him to seize his half-brother Edgar's inheritance. Shakespeare was, however, careful not to present a usurper's charter before his royal patron, and ensured that the fierce Edmund eventually came to a bad end.

The social and religious upheavals of Shakespeare's day had shaken old certainties and led many of the old aristocracy to the scaffold; as old elites lost their heads and new ones took their places, knowing who to trust – who were the real gentlemen – became a pressing issue for the mass of people whose lives and livelihoods relied on the patronage of their supposed betters. Evidence of this uncertainty can also be found, perhaps rather surprisingly, in Elizabethan and Jacobean horse-breeding tracts, which began to place an increasing emphasis on an animal's pedigree as the only sure guide to its character. Shakespeare's contemporary Gervase Markham wrote at least five books on horses; he needed money so badly that he plagiarized himself, repeating key passages of his own books in his 'new' titles. He was eventually forced to promise the Stationers' Company (the guild that enforced early versions of copyright) not to write any more. However impecunious, Markham was a proud member of the gentry, and he drew an explicit parallel between the pedigrees of human aristocrats and those of horses, suggesting that keeping track of a horse's family tree would be the best guide to judging its character. 'An ill-bred horse,' he argued, 'may beget a colt, which may have fair colour and shape, which we call beauty, yet still his inward parts may retain a secret wildness of disposition, which may be insufferable in breeding.'[8] Not surprisingly, the first stud-books and horse-breeding records also come from this period.

Markham's emphasis on good breeding was shared by another early-seventeenth-century horse expert, Nicholas Morgan, who, in books like *The Perfection of Horsemanship* and *The Horseman's Honour*, decried the fashionable Italian school of horsemanship, which stressed environment and training over heredity (although not as exclusively as English critics tended to claim). According to Morgan, 'the art of breeding of horses' had to be founded on the selection 'of stallions and mares, which are most fit and proper for that purpose'. He warned his readers that 'the least carelessness therein is the utter ruine of all the whole work', since the principle of 'like engendering like' meant that 'any evil choice' of parents would ensure 'an evill product must remaine' in the horse's eventual character.[9] According to Morgan, the superiority of animals from regions known for their fine horses was evidence of their breeding, not – as some claimed – the air, food and water on which they had been raised.

Knowing good horses from bad became crucial as horse-racing became big business. People had, of course, raced horses for as long as they had ridden them, but the sport became increasingly formalized in England following the restoration of the monarchy. Races were sponsored by the new king, Charles II, and were intended to encourage the breeding of faster war horses with superior stamina. They usually involved mature animals (six or more years old) being raced over long distances (six to eight miles) while they carried considerable weight (up to 12 stone, 76kg). Over time, racing became more codified, strict rules were laid down and the prizes became more substantial. But within a century of the Restoration, the pretence that the aristocracy's passion for racing was aimed at improving war horses had been largely dropped. Shorter races for younger horses became the fashion (boosted by the fact that first turnpike roads and then the railways made it easier to get the horses to the racecourse without riding them long distances); all the classics of the British racing calendar – the St Leger, the Oaks and the Derby – began in the 1770s. They were all for three-year-olds running over less than two miles and carrying up to 8 stone (50kg).[10]

With the new style of racing came a new style of horse, the English thoroughbred, which still dominates the racing world. During the seventeenth century, it had been widely believed that the speed and stamina of horses like Arabians (from which racehorses were often bred) were a product of their hot blood, the heat having been absorbed from the sun in the hot countries where they had been born. To protect the breed's precious heat from the damp British climate, English breeders would regularly import fresh Arabians, at great expense. But during the eighteenth century, a number of horse-breeders in Yorkshire and Lincolnshire broke with this received wisdom and began to concentrate instead on keeping their stocks pure. As their horses began to win races, it seemed that the breeders' intuition had been confirmed; it was good breeding, not a hot environment, that gave the Arabian its fiery quality. Some dissented, of course, but an increasing number of breeders held to the view that thoroughbreds were not merely a matter of speed or stamina, they were not defined by colour or the horse's physical form, not created by the time of day or the location where their parents had conceived them, nor by the heat of the stallion at the time of conception. Like an English gentleman, it was breeding alone – recorded in meticulously kept pedigrees – that made a thoroughbred.

Much of the eighteenth-century horse-breeding literature was taken up with trying to persuade breeders to pay as much attention to the choice of mare as they did to the stallion; the quantity of ink spilt on this subject suggests this was a far from universal practice. Those who held that both parents contributed to the foal's quality inevitably returned to the old question of how much each contributed. Richard Wall's *Dissertation on the Breeding of Horses upon Philosophical and Experimental Principles* (*c.*1758) argued that sire and dam should be well-matched. There was, he argued, no point in crossing a thoroughbred stallion with a slow country mare; the result would be a foal that would run faster than the mother, but be much slower than the father. This commonsense notion that cross-breeding blends the qualities of the parents was – as we'll see – one that was to have a long history, becoming known in scientific circles as the theory of blending inheritance.

Let us come back to Lord Morton, who was described by a contemporary as 'a constant attendant at the Royal Society'. He would have been familiar with the latest theories on breeding and inheritance; while he was experimenting with quaggas, the Society's president, Sir Joseph Banks and numerous other philosophically inclined gentlemen were busy observing the mating of sheep. The painter Thomas Weaver recorded them doing so in *Ram Letting from Robert Bakewell's Breed at Dishley, near Loughborough, Leicester* (1810). Under the watchful eye of Banks and others, Bakewell, one of the country's most successful animal breeders, demonstrates his method for improving sheep: careful selection of the parents, especially of the ram. He had demonstrated its efficacy by creating a new and profitable breed, the Leicester, which fattened rapidly. Weaver's painting illustrates the Georgian ideal of 'improvement', promoted by King George III himself, whose enthusiasm for agricultural innovation gave him such nicknames as 'Farmer George'. Why were he and Banks so concerned with such matters? Essentially because fatter sheep meant fewer discontented men to turn their eyes to the revolutionary events in France and America, and to dangerous notions like democracy.[11]

However, the very measures that Banks and his fellow land-owners advocated often aggravated the discontent they hoped to reduce. Half of Bakewell's home county of Leicestershire and a third of Banks's Lincolnshire had been transformed from public to private land in the preceding decades. The process had been under way for centuries, but between 1760 and 1830 it accelerated rapidly as Parliament passed a series of Enclosure Acts, which divided up common land and obliged farmers to fence in their fields. Enclosure led to the loss of vast areas of commons and, despite the compensation that was paid, often resulted in great hardship for the communities who had depended on them. As an anonymous rhyme from the period complained:

> The law locks up the man or woman
> Who steals the goose from off the common;
> But lets the greater felon loose
> Who steals the common from the goose.

Banks and his fellow improvers were nevertheless convinced that private profits – their profits – served the public good. Common land had been under-used, no one manured or drained it, no one concerned themselves with improving its productivity; *that* was why people went hungry. Once an individual owned it, and could derive a profit from it, there was an incentive for improvement, perhaps by paying Robert Bakewell's stud fees to raise the quality of their sheep. In the long run, there would be fatter sheep, fatter landlords and better-fed people, none of which, however, offered much consolation for those who were already hungry.

Enclosure was to transform the landscape of much of Britain: walls and fences began to snake across what had once been open fields and fells, barriers that attested to deep social change. Unable to scratch a living from the commons, many country people moved to the rapidly growing cities, where a new power – steam – was now running the factories, advertising for new workers with columns of belching smoke that blackened the people and the land. How were these crowds of factory and mill workers to be fed, once their direct connection with the land had been broken? Men like Banks and Bakewell thought they had the answer; they hoped to apply improvement to the whole world, 'enclosing' those parts of it – such as India, Africa and Australia – whose present inhabitants were clearly not making full use of their land. Morton's attempts to create an Africanized horse were part of a much wider European attempt, by the British in particular, to turn the world into neatly cultivated, profitable colonies.[12]

As aristocrats like Banks and Morton struggled to improve their crops, sheep and horses, many of their contemporaries were also deeply concerned with the breeding of people. The old aristocracy could trace their lineages back for centuries and offered their claims to good breeding as the justification for their wealth and power. But Morton lived at a time when the authority and wealth of the aristocracy were being undermined by a new breed of industrialists, men like Darwin's father-in-law, the potter Josiah Wedgwood. The Wedgwoods and their kind were transforming Britain not only physically with their factories, but also socially – with their newly acquired wealth. Just a few generations earlier, such people, even if they were successful artisans, rarely owned land; by the early nineteenth century, the industrious men had made their fortunes and were acquiring land and grand houses, insisting their sons had access to the country's elite universities and to the learned professions. They were, in effect, claiming that traditional good breeding mattered much less than hard work; their success was proof of the vigour of their blood.

Some interpreted this energetic vigour as proof of human equality, evidence that the American and French revolutionaries were right, that once the misplaced deference to ancient pedigrees was shaken off, even the humblest people could improve themselves. These revolutions – in agriculture, industry and politics – were all closely connected to questions of breeding. Whether aimed at improving farm animals, reducing hunger, or improving the ruling class (by staving off democracy by cross-breeding the exhausted aristocracy with the hardier industrialists), knowing how inheritance worked was crucial.

In George Eliot's novel, *The Mill on the Floss*, Mrs Tulliver observes that 'It seems a pity . . . as the lad should take after the mother's side istead o' the little wench. That is the worst on't wi' the crossing o' breeds: you can never justly calkilate what'll come on't.' That, indeed, was the problem with crossing breeds, the difficulty of calculating the outcome: would the result be a striped horse or a bad-tempered one? Would breaking down class barriers create a new class that combined a sense of *noblesse oblige* with entrepreneurial virtues and willingness to work, or just a tribe of effete fops who dropped their aitches? Without knowing the answers to these questions, how should individuals, of any class, go about choosing their mates: some undesirable characteristics, such as madness, seemed to run in families; was the same true of desirable ones, and if so, which? Plato's belief that controlled breeding would improve humanity has resurfaced often in history; the desire to make people better – and the diversity of opinions about what 'better' might mean in this context – has frequently driven the desire to understand the mechanisms of inheritance.

The right tools

In explaining the origins of his 'singular fact' to the Royal Society, Morton had described himself as trying an *experiment* in domestication. The idea that experiment was the key to understanding nature was, as we have seen, one of the Royal Society's founding principles, and it was still zealously promoting the virtues of experiment in Morton's day. But during the eighteenth century, experiments were still mainly done within the physical sciences – such as optics, electricity and chemistry. Plants, animals and people were the preserve of natural history, which concentrated on describing and cataloguing nature's diversity, rather than attempting to explain what caused that diversity; partly, of course, because it was widely assumed that living things were 'caused' by God, who had created them all. That assumption began to change during the nineteenth century, in part because the drive for colonies led Europeans to realize that the world contained a staggeringly greater variety of living things than they had previously suspected. Even for those with a firm belief in God, the idea that he needed to personally design and build every species of slug or worm seemed unlikely, possibly even disrespectful. Why could God not have designed laws of nature to produce living things, just as he had done for the stars and planets?

The discipline of biology can be defined as the search for those laws, whether divinely designed or not. The word biology itself was coined in 1800, but the science only took shape slowly over the next 100 years. Experiment would gradually become one of the things that defined a biologist, distinguishing them from their predecessors, the naturalists. Instead of, for example, simply waiting for a quagga and a horse to take an interest in one another, observing what happened, and then hypothesizing about what had caused it, the new biologists tried to conduct controlled breeding experiments, with large numbers of animals, to produce precise evidence with which to investigate their theories. But they soon discovered that large animals like horses make poor experimental animals. Like humans themselves, they breed much too slowly – it would take many human lifetimes to follow the effects of controlled breeding over dozens of horse generations. Horses also eat too much and take up too much space; only the wealthy can afford to keep them. So first naturalists and then biologists began to look for alternatives, to uncover the rules of reproduction and inheritance in small, fast-breeding animals or plants, and then to see if these rules applied to the animals biologists were most interested in, including ourselves. The history of biology has, in part, been the story of finding the right animals or plants to aid the search.

So, for the last 200 years, biologists have been looking for the right tools for the job, the organisms that can answer their questions. Passionflowers were to teach Darwin about the vigour of hybrids and how plants moved, while insignificant yellow flowers called hawkweeds confused Gregor Mendel as he struggled to discover why some hybrids were stable while others always reverted to their parental types. Francis Galton discovered that people make rather bad experimental organisms (irritatingly, they often have their own ideas and motives, which may be incompatible with those of the experimenter), and Hugo de Vries was first exhilarated, then mystified, by the evening primrose. The fruit fly turned out to be more tractable, not only in revealing to Thomas Morgan how chromosomes carry hereditary information, but also in forging cooperation between American and Soviet biologists. The humble guinea pig was one of the species which enabled Sewall Wright and others to uncover how the genetics of populations works. And when the insights gained from the guinea pig were combined with those derived from the flies, biologists finally worked out how evolutionary pressure changes one species into another. Two of the best-known figures in modern biology, James Watson and Francis Crick, would never have been able to work out the structure of DNA if a virtually invisible virus called a bacteriophage had not first revealed that it was indeed DNA and not proteins that contained the instructions for making new organisms. Yet just as most biologists were focusing on the smallest and simplest of living things, such as viruses, large, complex maize plants were teaching Barbara McClintock that genes can get up to all kinds of tricks that no one had previously suspected, such as – in effect – moving themselves around.

The people in this story are fascinating, but this is not a tale of lone geniuses changing the world single-handedly, partly because even those who seem like the loneliest geniuses turn out to have relied hugely on the work of others. As we shall see, these have included not only other scientists and naturalists, but corre-spondents and critics, plant-hunters and lab technicians, pigeon-fanciers and gardeners, engineers and inventors, fruit importers and slave traders. Even the greatest theoretician needs practical results to analyse and stimulate their ideas: whether they do their own experiments and breeding trials, or concentrate on making sense of other people's, someone has to do the meticulous – and often tedious – sheer hard work that all scientific ideas are ultimately built on. In this story, the *ideas* of science come second, in every sense, to the *work* of science.

From fruit flies and guinea pigs to maize and phage, these animals and plants are some of the stars of the story of genetics – and they deserve to be as well known as the people they enlightened. Two of the celebrities of modern biology, a modest cress plant called *Arabidopsis thaliana* and the tiny zebrafish, *Danio rerio*, are currently carrying on the work of their predecessors. Arabidopsis is the botanist's answer to the fruitfly, the first plant to have all its genes mapped; it has helped make the current revo-lution in the genetic modification of plants possible. Meanwhile, the zebrafish is helping biologists unravel an even deeper mystery, how DNA actually constructs organisms; by conveniently having completely transparent eggs and embryos, zebrafish are allowing biologists to observe every step in the development of a living organism, shedding light on many human diseases while illuminating one of the great mysteries of life, how a single cell – a fertilized egg – becomes a complete, complex organism.

Today we take the answers to many of the puzzles of inheritance for granted; we learn about genes and DNA at school, and it is almost impossible to open a newspaper without reading about the latest gene involved in obesity or cancer. However, to understand the story I'm going to tell, I need you to start by forgetting whatever you know about modern genetics. First, because hanging on to what we now believe makes it harder for us to understand how and why people once thought differently. Aristotle, Morton, Darwin and most of the other figures we shall be meeting believed many things that no one believes any more, not because they were fools, but because they made sense of the world in quite different ways. If we want to understand their work and ideas, we have to try and see the world the way they did.

More importantly, we also need to forget about today's supposedly right answers, because this is not the story about how the 'right' answers got separated from the 'wrong' ones. The history of science strongly suggests that today's right answer is almost always tomorrow's wrong one; science does not trade in eternal truths, but in temporary approximations. For a while – whether for a few days or many centuries – some sketch of how the world works will provide answers, but, far more frequently, will also propose new questions, new experiments. And, nine times out of ten, the new research reveals the old concept to be wrong in some way. Which leads to a new answer – and new questions. It is almost inevitable that much of what scientists know today will eventually be discarded, refined or re-described in ways that we cannot imagine and, as a result, our grandchildren will probably think that at least some of the theories we currently 'know' to be right are as implausible as the notion that cool breezes engender more girls, or that a quagga can pass on its stripes to offspring it did not actually father.

The plants and animals I have described here are just a handful of those that have enabled humans to understand so much of the machinery of life. Regardless of Lord Morton's view, we now know that males do not impress females (at least, not in a biological sense – and often not in any other); the stripes and other quagga-like features of the foals were an atavism or reversion, the reappearance of one of the traits of the ancestor of modern horses. They are even more interesting than Morton or Darwin guessed, not least because they help shed light on the evolutionary history of horses, giving us a glimpse of what the striped ancestors of all modern horses may have looked like. But understanding genetics allows us to do more than unravel evolution – it now allows us to intervene in it as well. As we shall see, finding the right tools for the job has often involved making them, by laboriously cross-breeding hundreds of peas or fruit flies to eventually produce a pure-breeding strain with the desired characteristics. One product of our current knowledge of genetics is that it is now possible to eliminate much of this tiresome work and simply engineer organisms, like OncoMouse®, a trademarked animal that has been designed to develop cancers. Its creators believe it will bring great benefits to humans, but like many aspects of modern biology, its development raises grave ethical problems.

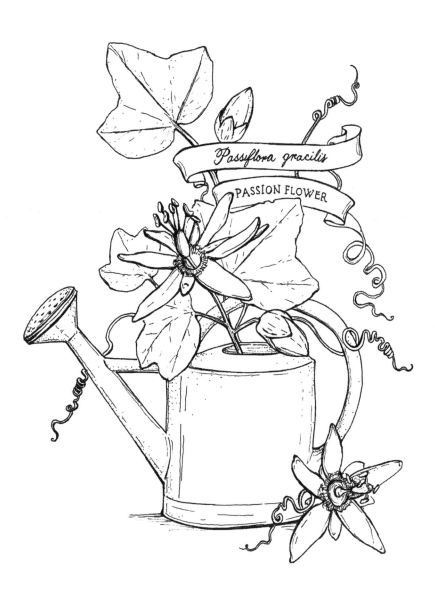

Passiflora gracilis

PASSION FLOWER

Chapter 2
Passiflora gracilis: Inside Darwin's greenhouse

A rainforest ought to be paradise for a plant: but like all desirable neighbourhoods, it can get overcrowded. Although the forest is wet and lush, a newly germinated seed usually finds light is in short supply. One way to reach it is to grow into a gigantic tree, with a thick trunk hundreds of feet high, supporting massive branches that expose the plant's leaves to the sun. But it takes time to grow such a trunk, and for every seedling that makes it, thousands fall by the wayside. An alternative strategy is to be a vine, producing a thin, rapidly growing stem that clambers up someone else's trunk – thus getting the leaves into the light with a minimal expenditure of time and energy.

Passionflowers are typical rainforest climbers: they have specialized tendrils that will cling to almost anything; they can sense light and grow rapidly towards it; and, once there, they devote themselves to the ultimate end of any organism's life – reproduction. The delicate, showy flowers from which they derive their name attract various pollinators – bees, bats, butterflies, moths or birds – that fly from flower to flower, spreading the plant's pollen as they go. Once fertilized, the ovaries swell to form passionfruit, whose bright colours and sweet taste help attract animals, especially birds and primates, which eat the fruit and so spread the plant's seeds.

Like many South American primates, brown capuchin monkeys (*Cebus apella*) are fond of passionfruit. Capuchins get their name from their brown bodies and the white fur around their faces, which reminded European travellers of the brown and white robes of Capuchin monks, who wore a distinctive white hood or *capucize* over their brown robes (cappuccino coffee gets its name from the same visual analogy). Unlike the monks, Capuchin monkeys live in sexually active social groups in rainforest canopies and eat more or less anything they can get their hands on. Brown capuchins are noisy, destructive feeders: they move from tree to tree, tearing apart fruit, cracking nuts, and leaving half-eaten plants, seeds and droppings wherever they go. The droppings provide an ideal fertilizer for germinating passionflower seeds, allowing them to begin a new scramble towards the light.

In 1553, the passionflower's strategy of attracting primates to eat its fruit and spread its seeds paid off handsomely when a member of the primate species *Homo sapiens*, known as Pedro de Cieza de León, published the first account of these exotic flowers and fruits. His story helped persuade other members of his species to spread the passionflower all over the world; the sour-sweet, egg-shaped passionfruit in your local supermarket are evidence of how well the passionflowers' primate-based strategy has worked.

León's *La Chronica del Perú* included a description of a fruit 'that is very delicious and fragrant called granadilla' (meaning a small pomegranate or *granada*). However, it was not the fruit that attracted his readers, so much as the flowers; a later Spanish writer described the flower as 'very similar to the flower of a white rose and it appears to have been carefully made to show the representation of the passion of Jesus Christ'.[1]

In 1609, Giacomo Bosio, a member of the Knights of Malta, was gathering stories about Christ's cross and the legends and miracles associated with it. Knowing of his interest, a Mexican-born friar showed him drawings of a 'stupendously marvellous' flower from the Americas that seemed to bear the symbols of Christ's Passion. Bosio was initially struck by the flower's outer ring of corona filaments, typical of most passionflowers. He interpreted this as symbolizing Christ's crown of thorns, because the species he saw depicted had seventy-two filaments – the traditional number of thorns in Christ's crown. He took the five stamens (the male parts of a flower) to represent the scourges used to beat Christ, and the five blood-red spots at the base of the specimen Bosio saw gave the plant its Spanish name, *La Flor de las cinco Llagas* ('the flower of the five wounds', i.e. Christ's four stigmata and the spear-wound). The flower also had three carpels, the female reproductive organs on the tops of which the pollen is deposited. Bosio interpreted these as representing the three nails used to nail Christ to the cross. And finally, the pollen-receiving female parts grew from a single column, which was seen as symbolizing the pillar that Christ was tied to during the flagellation.[2] The fame of the plant spread, usually in pictures that simplified its botanical features in order to make their religious significance clearer. The miraculous passionflower was soon growing in gardens across continental Europe.

In August 1612, just three years after Bosio's account had appeared, passionflowers were blooming in Paris, and at about the same time, Captain John Smith, President of Virginia and Admiral of New England – best known for his relationship with the Indian 'Princess' Pocahontas – recorded in his diary that 'The Indians plant also Maracocks, a wild fruit like a lemmon' ('Maracock' was the Algonquian Indian word for the passionfruit). It was from the British colony in Virginia that passionflowers made their way to Britain; they are mentioned in John Parkinson's *Paradisi in sole Paradisus terrestris* (1629), the first English gardening book. Parkinson called the plant 'the Virginian climer', or the 'Jesuites Maracoc', but – perhaps because of the charged religious atmosphere of the eve of the Civil War – he found it politic to reject a specifically Catholic symbolism for the plant.[3]

Nevertheless, Charles I was sometimes referred to as 'the passionflower' by English Catholics after his execution, or – as they saw it – his martyrdom. This association may have prompted John Tradescant the younger to popularize the plant to British gardeners, since Tradescant, like his father, was head gardener to the King after the Restoration. He also travelled, collected and sold exotic plants. He made several trips to Virginia and his catalogue of new plants (published in 1656) included a passionflower, which he called *Amaracock* or *Clematis Virginiana* ('clematis' simply means 'vine-branch', and was then used to describe any climbing plant). As a result of Tradescant's work, passionflowers were soon being cultivated in Britain; the Duchess of Beaufort's gardeners were growing the blue passionflower in 1699.

The great Swedish naturalist, Carl von Linné, or Linnaeus, gave the passionflower its modern Latin name: he simply translated 'passionflower' into Latin and called it Passiflora. Famously, he also standardized the way plants and animals were named, by founding a system that is still in use. Each species has a two-part name, or binomial: the second name is that of the species itself; the first is that of the group of closely related species, called a genus (genera is the plural), to which it belongs. The annual or crinkled passionflower is *Passiflora gracilis*, the blue is *Passiflora caerulea*, and so on. Modern botanists classify the genus Passiflora into a family called Passifloraceae, which consists of several other genera of tropical climbers, including Basananthe and Adenia, all of which are forms of passionflowers.

Linnaeus's botanical works were soon translated into English and his naming system became especially popular in Britain, but although the name Passiflora was soon well known, passion-flowers themselves were not widely grown in Britain until the nineteenth century. And while they owed much of their popularity to their striking colours and shapes, their progress was also aided by a change in the tax laws, by the impacts of imperialism and industrialization and, in particular, by the effects of industrial pollution.

Steam, smoke and glass

An 1845 editorial in the weekly newspaper the *Gardeners' Chronicle* celebrated the Prime Minister Sir Robert Peel's decision to abolish the tax on glass. 'All duties on glass are to be extinguished,' the paper's editor proclaimed, 'and every man who has a greenhouse, a Cucumber frame, or even a window, owes [Peel] a debt of gratitude.' The editorial claimed that 'There was not, in the whole list of excisable materials, an impost so oppressive as the glass duty.' The paper was particularly outraged because the tax had converted 'A beautiful substance . . . costing little except labour and skill . . . into a material which could only be enjoyed, even by those in easy circumstances, on the most indispensable occasions'.[4]

With Chartist agitators marching through the streets of London, starvation in Ireland and revolution brewing across Europe, the cost of a cucumber frame might not appear momentous, but the abolition of the glass tax had implications for Britain's landscape that are still with us.

At the time the tax was repealed, window glass was still hand-blown: a balloon of molten glass was pressed flat on one side, like the flat bottom of a hand-blown glass vase. As it cooled, sheets of glass could be cut from this flattened surface. The process was not only highly skilled, time-consuming and expensive, it also could not produce sheets larger than around two by three feet (60 × 90cm). But according to another popular middle-class publication, *Chambers's Edinburgh Journal*, the end of the glass tax promised to change all this: 'it is impossible to foresee to what useful purposes glass will be put', their editorial prophesied, nor 'how cheaply it will be possible to obtain it'. The journal's writer believed that abolishing the tax would give 'an elasticity . . . to the trade' because 'new enterprisers will embark in it'; and sure enough, in 1847, the abolition inspired a young man from Sunderland, James Hartley, to patent a system for making rolled plate glass.[5] Hartley machines made glass sheets that were bigger, stronger and cheaper than anything that preceded them; he became Britain's largest glassmaker and his glass was exported all over the empire. Without it, the great Victorian railway stations, with their vaulted iron and glass roofs, could not have been built, nor could the huge shop windows that displayed the dizzying pro-liferation of consumer goods that Britain's burgeoning industries were producing. But from the passionflowers' perspective, the most important thing about Hartley's invention was that it made possible large, cheap greenhouses.

Greenhouses had begun getting bigger before Hartley's

invention. In 1816, the Scottish-born garden designer and journalist John Claudius Loudon had patented a new kind of wrought-iron glazing bar for greenhouses. This permitted curved roofs for the first time, which let in more light; the older, wooden sashes blocked much of it. Loudon's bars made it possible to erect structures like the Great Conservatory at Chatsworth; designed by Joseph Paxton and completed in 1840, it was then the largest in the world. Mass-produced wrought iron allowed the great Palm House at the Royal Botanic Gardens, Kew, to dwarf even Chatsworth. But only governments or the very wealthy could afford to build on such a scale. In 1872, *Beeton's Book of Garden Management*, a product of the publishing empire built on the fame of Mrs Beeton's *Cookbook* and the *Englishwoman's Domestic Magazine*, commented that: 'Glass structures of even the smallest kind would, a very few years ago, have been considered a piece of great extravagance for any but the affluent.'[6]

To become affordable, greenhouses had to be mass-produced, which required increased demand. A sense of where that demand came from can be gleaned from *Chambers's Edinburgh Journal*'s celebration of the abolition of the glass tax:

> It is impossible to foresee the advantages of cheap glass which will be reaped by horticulturalists. Conservatory frames and other glazed implements of their art are so serious an item of expense that recent insurance companies have thought it worth their while to afford insurances against hail – a severe storm of which has been known to ruin many a struggling gardener.[7]

In a similar vein, the *Gardeners' Chronicle* commented that 'To men with whom glass was necessary for their purposes of their trade, the excise duty was a most grievous burthen.'[8]

Both publications were thinking primarily of the market gardeners who fed the inhabitants of Britain's rapidly expanding cities. The biggest problem they faced was smoke – every house and office was then heated by coal fires while steam-powered factories had to be near their workers, so their chimneys added to the pall. And on top of that, 6,000 miles of railways were built in Britain between 1830 and 1850, by which time coal production had hit 49 million tons a year – rising to 147 million tons over the next thirty years.

The Victorians applied their much-vaunted ingenuity to the problem of smoke, and various ideas were proposed to clean up the air, including a scheme to pump fresh air in pipes from Hampstead Heath into the heart of the City of London (which, perhaps unsurprisingly, was never implemented). It was not until 1851 that a more practical solution presented itself. In order to house the Great Exhibition, Joseph Paxton decided to out-do even Kew's Palm House by building the world's largest glass building; before it had even been finished, it had earned the affectionate nickname 'the Crystal Palace' from the satirical magazine *Punch*.

Others were less impressed. The art critic John Ruskin called it a 'cucumber frame' and *The Times* referred to it as 'a monstrous greenhouse' and campaigned against its being built, especially since its construction was to involve cutting down several of Hyde Park's beautiful old trees. To overcome such opposition, Paxton came up with an ingenious solution to the problem of the trees – he built around them. As a result, there were fully grown elms inside the building; the *Gardeners' Chronicle* observed that 'while the dirty, half-starved Elms, growing as if wild in the open park, made shoots at most a foot long on the average, the well-fed, well-cleaned, well-lodged trees under the [Crystal Palace's] transept, made shoots from 6 to 7 feet long'.[9] The 'monstrous greenhouse' had demonstrated to the public what horticulturalists already knew – greenhouses could address the problem of smoke. Glass buildings became so popular that in 1877 it was even suggested that a vast neo-Gothic-style conservatory be built over London's Albert Memorial to protect it from air pollution.

Industrialization made mass-produced greenhouses both practical and necessary; the smoke belching from the glass- and iron-making factories made it almost impossible for city-dwellers to grow plants out of doors. Because horticulturalists had to rely on glass to make a living, as Britain's cities grew, so did the demand for greenhouse glass. The editorial writer in *Chambers's Journal* had also foreseen that 'Private individuals also will be able to have conservatories.'[10] And, indeed, hundreds of thousands of Victorians left the Crystal Palace determined to have a greenhouse of their own. Paxton boasted that he had cut the cost of the Palace by building from standardized parts, manufactured in bulk. He took advantage of his new fame to launch a range of affordable, modular greenhouse kits, based on the same principles. Mass production made greenhouses cheap and popular; soon there was a size and price to suit everyone. Greenhouses varied from simple glazed but unheated structures to what was known as a 'stove', in which a furnace kept the interior hot enough to grow tender tropical plants – such as passionflowers.

For those who could not afford a full-sized greenhouse, or did not have a garden to put it in, there was the Wardian case, a miniature greenhouse that could sit on a table or windowsill. These were named after an enterprising gentleman, Nathaniel Bagshaw Ward, who claimed to have invented them but had in fact simply improved an existing design and then promoted it in a book with the characteristically imaginative Victorian title, *On the Growth of Plants in Closely Glazed Cases*. Ward had a passion for gardening and botany and gradually converted his London home into a verdant, social centre for London's botanists and their visitors from abroad. The gardening writer John Loudon visited Ward's home and described the planter boxes that were to be seen 'along the tops of all the walls of his dwelling house, of the offices behind, and of the wall round the yard, even up the gable ends and slopes of lean-tos'.[11] Ward held scientific soirées and took every opportunity to promote his glass cases: they were ideal for growing ferns and helped spark a fern craze; he also recommended them to botanists and nurserymen as an ideal way to bring living plants from across the empire to British gardens.

Before Ward's cases, transporting live plants on slow-moving sailing ships had been a risky business; if the salt spray did not kill them, the shock of being moved through several different climates often did. The new cases were not foolproof: one arrived in London that had been packed by someone whose anxiety about its safety clearly exceeded his knowledge of plants; he had 'painted the glass over, then covered it by way of protection with broad battens of wood and lastly nailed a thick piece of tarpaulin over the whole'. Not surprisingly, the plants arrived dead in their dark coffin.[12] But despite such setbacks, Wardian cases helped to make thousands of new types of plants available in Britain.

In its domestic setting, the Wardian case served as an elegant ornament, its living plants bringing a little colour and life to the often airless Victorian drawing-room, while attesting to its possessor's taste and interest in scientific matters. The shipboard cases brought samples of the empire's exotic fruits, flowers and vegetables to fill Britain's greenhouses, helping the plants to reach new habitats.

In addition to cheap greenhouses, the Victorians invented those other indispensable features of the British Sunday afternoon, the lawnmower and the garden centre, and they also gave us the ever-popular solution for what to do when the rain makes gardening unappealing – the gardening magazine. The steam that powered Victorian Britain's railways also powered its printing presses and churned out cheap, machine-made paper. The country experienced a massive boom in newspapers, magazines and books and as dozens of new publishers began chasing after new readers, they soon spotted gardeners as a potential market; the *Gardeners' Chronicle* was just one of a number of magazines launched to take advantage of it.

This new audience also helped resurrect the *Botanical Magazine*, which printed descriptions of new plants illustrated by beautiful hand-coloured plates. It had been published for nearly forty years when it was taken over by William Jackson Hooker, Regius Professor of Botany at Glasgow University, in 1826. He worked to increase its scientific content, but also took advantage of the exotic plant boom to reach a new audience, using connections around the world to get new plants to illustrate. In the 1830s he made contact with John Tweedie, a Scottish gardener who had emigrated to Buenos Aires some years before, hoping to obtain work designing and landscaping for wealthy Argentinians. Patrons were rarer than he had hoped, so Tweedie began to explore the country's interior looking for new plants to satisfy British gardeners' growing lust for novelties.

Tweedie's plants were soon being described and illustrated in the *Botanical Magazine*. In 1839, for example, Hooker gave a new passionflower the name *Passiflora nigelliflora* (meaning that its flowers resemble those of nigella, the plant that the spice black cumin comes from), telling his readers that 'it was discovered, in 1835, by Mr Tweedie, on his way from Mendoza to Tucuman'. Hooker described its botanical characteristics for the benefit of his more scientific readers, but gardeners were the major audience – Hooker explained that the new plant 'flowered in the Glasgow Garden in September, and seems to require the heat of the stove'. Among Tweedie's other discoveries were *Passiflora tucumanensis*, which was described as 'a free grower, and flowered copiously the second year in the stove of the Glasgow Botanic Garden'.[13] Tweedie was typical of the collectors who gathered plants and sent them home to England. In both Britain's colonies and her trading partners, gardeners, missionaries and colonial officials, naval officers and soldiers, convict supervisors and paid plant collectors were all at work trying to satisfy both commercial demand and the curiosity of men of science. Some worked for money, or for the glory of gardening, God or empire – but many were simply eager to contribute to botanical science.

The garden and greenhouse boom turned plants into big business. In the 1840s, new species of passionflower were being sold through gardening magazines for a guinea (£1.05) each. At that time, a domestic servant earned between £20 and £60 a year, so these were clearly not plants for the masses. However, by the 1850s, prices had fallen to 16s. 6d. for the showier types, down to just 2s. 6d. for the more common varieties. These plummeting prices show how much demand had increased; passionflowers were now being grown or imported in huge numbers, making them more affordable. At the same time the railways, whose smoke continued to cause city gardeners such problems, also allowed garden materials, equipment, building materials and plants to be moved quickly and cheaply around the country, all of which fostered the gardening boom. Thanks to collectors like Tweedie, commercial nurseries and the opportunities (and difficulties) created by industrialization it became remarkably easy to obtain exotic plants in Britain.

Among the *Botanical Magazine*'s rivals was the *Botanical Register*, edited by John Lindley (also editor of the *Gardeners' Chronicle*), Professor of Botany at University College London. In one issue he described a new flower he called *Passiflora onychina*, or 'Lieut. Sulivan's Passion flower' because it had first been collected by the naval lieutenant Bartholomew Sulivan, who had 'procured the seeds, with others, from the Botanic Garden at Rio de Janeiro, in 1827'.[14] A few year's after his trip to Rio, Lieutenant Sulivan was reassigned to a small surveying vessel, HMS *Beagle*, commanded by Robert FitzRoy. It set sail for South American waters again in December 1831. Alongside FitzRoy, Sulivan and the other officers and crew was a feckless young man of no fixed ambitions but with a burning enthusiasm for natural history collecting, Charles Darwin. He was to play an important part in the passionflower's story.

Sterility, marriage and the passionflower

The *Beagle* spent more than half its five-year voyage in South American waters, and while its crew were busy with the ship's real work, producing accurate maps for the British navy, Darwin was free to go on long inland journeys, sometimes for months at a time, exploring the pampas, the mountains and the rainforests. In his published journal, now known as *The Voyage of the Beagle*, he described seeing 'Numerous cottages . . . surrounded by vines', some of which were almost certainly passionflowers, since most passionflower species are native to South America. Darwin collected two unfamiliar species of passionflower while the ship was in the Galápagos: his dried, faded specimens are still in the University Herbarium at Cambridge. Sadly, he was there in

September and October, when the plants were not in flower, but his specimens still have their climbing equipment attached – slender, fragile tendrils that would come to fascinate Darwin. Indeed, since John Stevens Henslow, the Cambridge Professor of Botany who had taught him as an undergraduate, had urged his former pupil to concentrate on plants in flower, Darwin's Passiflora specimens suggest he may have already been interested in their climbing habits.[15]

After the *Beagle*'s return to Britain in 1836, Darwin distributed his specimens to various experts, in order to have them properly identified and named. At this stage in his career, his main expertise was in geology, not botany, so the passionflowers went with the rest of the Galápagos plants to his old friend Henslow. Unfortunately, Henslow was too busy with his university duties and his parish work (he became vicar of Hitcham, Suffolk, in 1837) to work on his former student's plants. They languished, neglected, for many years.

In 1843, seven years after the *Beagle*'s return, Darwin finally got tired of waiting for Henslow and offered the Galápagos plants to William Hooker's son, Joseph Dalton Hooker. Joseph had just returned from a remarkable voyage of his own – four years spent circumnavigating the Antarctic on another British naval vessel, the *Erebus*. It and its sister ship, the aptly named *Terror*, had braved icebergs and storms to sail further south than any ships had before, trying to determine the exact position of the southern magnetic Pole. Although the ships' hulls had been strengthened to help them resist the pack ice, no wooden sailing vessel could survive the Antarctic winter. As the days closed in and the ice began to extend its grip, the *Erebus* and *Terror* retreated north and wintered in Australia, New Zealand or South America, carrying out repairs and re-supplying. During these respites, Joseph Hooker went exploring and collected plants. Although he was only an assistant naval surgeon (while Darwin had paid his own way as a gentleman companion to the captain), having a father with a famous name in botanical circles opened doors for Joseph. He contacted his father's correspondents wherever he went.

Although his ship did not take him to Buenos Aires to meet Tweedie, Hooker met many similar men in Australia and New Zealand whose love of botany had led them to collect for his father, who had recently become director of Britain's national botanic garden at Kew.

On his return to Britain, Hooker decided to write not merely a description of his travels but a comprehensive flora of all the countries around Antarctica. By comparing and contrasting the plant life of different countries, he hoped to discover how plants had arrived in their present habitats. Apparently aware of his interest in distribution, Darwin offered the *Beagle* plants to Hooker, including the passionflowers. Darwin's published journal of the *Beagle* voyage and his publications on geological and other topics had made him a celebrated figure and Hooker was both surprised and flattered by his attentions. Once the plants had arrived, he wrote to tell Darwin that they 'are far more xtensive [*sic*] in number of species than I could have supposed' and that even though he 'was quite prepared to see the xtraordinary difference between the plants of the seperate [*sic*] Islands from your journal', he was nevertheless surprised by his observations of the actual specimens, which he realized would force him and other botanists to rethink their ideas about plant migration.[16] Three years later, Hooker described Darwin's passionflowers in the *Transactions* of London's Linnean Society; convinced they were new species, Hooker named them *Passiflora tridactylites* and *Passiflora puberula*.[17] Darwin's gift, including the passionflowers, began a lifelong friendship with Hooker; their regular letters and visits were invaluable to both men, and Hooker's observations and his knowledge of botanical matters were to shape Darwin's work over many decades.

After his return from the *Beagle* voyage, Darwin started trying to work out the implications of what he had seen. He did not, of course, come up with the idea of evolution as soon as he saw the Galápagos: it took years to develop and test his theories, and passionflowers were to play a small but vital role in that work. Among Darwin's surviving papers is a notebook marked

'Questions & Experiments', which he scribbled in during the first few years after his voyage. It is full of obscure jottings, which he made purely to jog his own memory; they are hard to read and even harder to interpret. Among them is a single, cryptic reference to passionflowers, which his father cultivated: under the heading 'Figs, flower', Darwin listed plants whose fertilization he wanted to investigate, including, 'Passion Flower. (as it is required to impregnate it artificially.)'.[18]

Darwin evidently knew that cultivated passionflowers need artificial pollination to produce fruit. He is likely to have read about the difficulties of growing them in the gardening magazines, the same ones where Paxton's do-it-yourself greenhouses were advertised and Tweedie's new passionflowers announced. Among these was the *Cottage Gardener*, which first appeared in 1848 and proudly described itself as 'a practical guide in every department of Horticulture'. Darwin not only read it, but wrote to it on many occasions, hoping its readers might help answer his botanical questions and supply him with fresh facts.

The *Cottage Gardener* featured regular columns on such topics as 'Greenhouse and Window Gardening'; in 1849 one Donald Beaton wrote an introductory guide to the plants that he felt should be in everyone's greenhouse. First on the list was the Sweet-scented Mandeville (*Mandevilla suaveolens*, commonly known as the Chilean jasmine), another of Tweedie's introductions. Beaton's article continued, 'After the Mandeville, I would recommend a passion flower', listing some of the hardier and more vigorous varieties that would grow outdoors or in an unheated greenhouse.[19]

Donald Beaton was a Scot by birth, who had worked to improve the spectacular gardens at Shrubland Park, Suffolk. He was an expert on plant hybridization and on bedding schemes, another Victorian response to industrial smoke – flowers grown outdoors in polluted cities died so quickly that gardeners developed the technique of raising masses of flowers under glass, planting them outside when they were in flower and replacing them as soon as the first crop faded. This technique also allowed the creation of the elaborate geometric and other designs that became known as carpet bedding, which can still be seen in some old-fashioned parks and gardens.

The year after Beaton promoted the passionflower in the *Cottage Gardener*, another of its writers, Robert Errington, contributed an article on the 'Culture of the Passifloras for the Dessert'. The opportunity to serve exotic fruit to dinner guests was a major reason to own a hothouse. In Jane Austen's *Northanger Abbey*, Catherine Morland is shown around the Abbey's gardens, in which 'a village of hot-houses seemed to arise'. It is clear that this 'village' had one main purpose: General Tilney, the garden's owner, 'though careless enough in most matters of eating . . . loved good fruit', yet despite 'the utmost care [he] could not always secure the most valuable fruits. The pinery had yielded only one hundred in the last year.'[20] A pinery was a hothouse devoted to producing pineapples, so that men like General Tilney could impress their neighbours with a display of conspicuous consumption. Passionfruit were grown for the same reason.

However, exotic plants would only impress if they could be persuaded to bear fruit and that – as Errington acknowledged – was not easy. In his advice to growers he noted that 'Another point too, and an all important one, must have strict attention – the flowers must be "set" by hand as they open. Without this, the crop can by no means be relied on.' The procedure was simple enough: 'take one of the anthers, when burst, about eleven o'clock in the forenoon, and merely rub the point of the stigma with it', but he was unsure why it was necessary. Perhaps, he speculated, it was because in most passionflowers, the anthers (where the pollen is produced) hang downward, so that when they ripen and burst, the pollen falls away from the receptive stigmas, which – as we have seen – are raised up on a central column above the stamens. Errington observed that the benefit of this arrangement to the flower is 'not very apparent' and suggested that his fellow columnist, Beaton, 'could doubtless throw some light on this curious economy'.[21]

Beaton responded in the following issue. Pausing only to mock 'the old story about the Spanish monks having mistaken this arrangement for emblems of the crucifixion' as being 'mere moonshine', Beaton claimed that despite the apparently impractical arrangement of the flowers, when they are fertile, the carpels bend to meet the stamens. He patronizingly commented that it was 'No wonder . . . that honest men like the Spanish monks and our friend [Errington], should be deceived and puzzled by "such fancies." However, we must give Mr Errington credit for wishing to clear away all impediments to such mutual understanding.'[22]

Beaton failed to answer the question of why the flower was arranged in this peculiar way, and the information he gave was largely wrong.[23] Most passionflowers are incapable of self-fertilization in the way that he describes; instead, they depend on an insect, a bird, a bat (or a gardener) to carry their pollen from one flower to another – without such assistance, they cannot reproduce.

The sterility of passionflowers both fascinated and alarmed Darwin because it seemed to contradict his theory of natural selection. Like the horse and dog breeders he had grown up with, Darwin knew that living things varied: children resemble their parents, but are never exactly like them. These small differences have long been referred to by naturalists and biologists as 'variations'; for any character, such as height, within a population there is a range – from the tallest to the shortest – which is referred to as the population's range of variation or simply its 'variability'. Neither Darwin nor anyone else at the time knew what caused these variations, but whatever it was, they could be inherited, as every farmer that mated a prize bull with the cow that gave the most milk understood – that was how the breed was improved. But in the years after the *Beagle*, Darwin was trying to explain how wild animals and plants might change – in the absence of any conscious direction of their breeding. The now-famous breakthrough in his thinking came as he was reading *An Essay on the Principle of Population* by the political economist Thomas Malthus. Malthus believed that without the checks of war and famine, population growth would always outstrip food production; the only way to avoid a struggle for food was the application of 'moral restraint among the poor', in other words, people should not have children they could not afford to feed.[24] In his *Autobiography*, Darwin recorded that at the time he read this, he was 'well prepared to appreciate the struggle for existence which every-where goes on from long-continued observation of the habits of animals and plants'; it therefore struck him that because of this intense struggle, 'favourable variations would tend to be pre-served, and unfavourable ones to be destroyed. The result of this would be the formation of new species. Here then I had at last got a theory by which to work.'[25]

This was the essence of natural selection: organisms varied at random – some variations helped the organism survive and reproduce, but others did not. The inability of plants and animals to exercise 'moral restraint' led to an intense struggle for food, space and mates; those organisms whose variations helped them to survive would be more likely to pass on these favourable variations to their offspring. Meanwhile, those with less favour-able variations would tend to have fewer offspring. And so, very slowly and gradually, over tens of thousands of generations, organisms would change and become better adapted to their environments. Eventually the descendents would be entirely different from their ancestors; a new species would have evolved.

Darwin was delighted with his theory and began to test it by seeing how well it explained his various observations and experiments; this was when plants like the passionflower became a problem. Plants obviously benefit from having both male and female parts in the same flower: they can save themselves the tedious – and often hopeless – business of finding a mate. Most flowers can simply self-fertilize and save their energy for pro-ducing and scattering their seeds. Yet not only did some plants not take advantage of this convenient arrangement, plants like passionflowers (and they are not unusual in this), seemed almost to be actively avoiding it: the flowers of many Passiflora are so arranged as to make self-fertilization unlikely (the point Errington had puzzled over in the *Cottage Gardener*), and many have the ultimate insurance against self-fertilization – if their *own* pollen falls on their stigmas, it fails to fertilize them; a phenomenon botanists describe as self-sterility. How and why did such an arrangement evolve? As Errington had wryly observed, the benefit to the plant was 'not very apparent'.

The passionflower's aversion to self-fertilization also puzzled Darwin. Ever since he had been an undergraduate at Edinburgh University, Darwin had shown a typical student's preoccupation with sex, but while his contemporaries pursued their researches in the city's bars and tea-shops, Darwin walked the beaches of the Firth of Forth, with his friend and teacher, the zoologist Robert Grant, where they collected tiny marine creatures and examined them under a microscope. Grant got Darwin interested in creatures such as corals, which are made up of hundreds of individual polyps. They seemed indecisively poised between the plant and animal kingdoms, hence Grant's name for them: 'zoophytes', or 'animal-plants'. As embryos they swim freely like animals, but then settle down to an entirely sedentary life, functioning as parts of a vast colony, almost as if they had become elements of a more complex organism, like the buds, leaves and branches of a tree.

Darwin remained interested in these questions when he was at Cambridge, where he met Henslow and heard him argue in his lectures that when plants such as strawberries propagated by sending out suckers, thus managing without sex, this was just one more entirely normal reproductive strategy, not an aberration as some of his contemporaries believed. Darwin took Henslow's and Grant's ideas with him on the *Beagle*, looking at the myriad ways that organisms procreate; some seemed to rely entirely on sex to reproduce, others managed with only an occasional coupling, while a few managed without sex at all. Among the organisms he collected on his travels were barnacles; back in Britain, he spent eight long years (1846 to 1854) studying and classifying them. He was particularly fascinated to discover that most barnacles, like flowers, are hermaphrodites – they have both male and female reproductive organs – but others have separate sexes. Perhaps the most astonishing were strange specimens in which the male barnacles were so tiny that they lived *inside* the females' shells, almost like parasites. The males are little more than tubes of sperm; the females protect and feed them. Staggered by this discovery, Darwin wondered if these species with what he called 'complemental males' might perhaps be intermediates between the common hermaphroditic barnacles and the ones that had wholly separate sexes. They suggested a route by which separate sexes could have evolved; Darwin assumed that all organisms descended from types which were asexual or hermaphroditic, since it seemed unlikely that two independent sexes could have evolved simultaneously. But what was the evolutionary pressure that drove a self-sufficient self-fertilizer – barnacle or passion-flower – to evolve towards the complex, and potentially childless, world of two separate sexes?

Along with the other baffling problems Darwin was facing in the summer of 1838 was the question of whether or not to get married. Being a man of science, he approached the question rationally, making a list of the pros and cons. Under 'Not Marry', he listed such advantages as 'freedom to go where one liked' and 'conversation of clever men at clubs'. Not marrying would also relieve him of 'quarrelling', 'fatness and idleness', and having 'less money for books'. But marriage had its attractions, including 'Charms of music and female chit-chat'. Children feature on both lists: as an advantage of marrying, Darwin wrote 'Children (if it please God)', adding, 'it is intolerable to think of spending one's whole life, like a neuter bee, working, working and nothing after all'. But he also worried about 'the expense and anxiety of children', especially 'if many children' he might be 'forced to gain one's bread' by working for a living. Despite these drawbacks, he opted for marriage. The clinching argument being that it would provide him with a 'constant companion, (friend in old age) who will feel interested in one, object to be beloved and played with – better than a dog anyhow'.[26]

Having weighed the matter carefully, Darwin proposed to his cousin Emma Wedgwood and they were married on 29 January

1839. There were already several cousin marriages in the Darwin and Wedgwood families; they were almost becoming a tradition. However, while the Darwins and Wedgwoods were becoming an increasingly close-knit clan, not everyone was convinced such unions were a good idea. While Darwin was contemplating matrimony, he read Alexander Walker's newly published book *Intermarriage*, which claimed to describe 'the functions and capacities which each parent, in every pair, bestows on children, in conformity with certain natural laws, and by an account of corresponding effects in the breeding of animals'.[27] Walker's claim that natural laws governed the breeding of both humans and animals must surely have caught Darwin's eye; soon after he got married, he decided to test ideas like Walker's by making detailed observations of plant-breeding. One of Walker's claims was that inbreeding could cause 'deformity, disease and insanity'; over the following years, Darwin worried that his decision to marry his cousin had weakened his children, who seemed rather sickly. In 1842, he and Emma watched their third child, Mary, die just a few weeks after her birth; nine years later, their beloved daughter Annie died, just ten years old; and, their last child, Charles Waring, lived for less than two years. Darwin's garden was not merely a refuge from these sad losses, it was also where he tried to understand their causes; he spent years crossing plants with each other, trying to understand the precise effects of inbreeding.

Although Darwin never left Britain again after the *Beagle* voyage, he conducted a vast correspondence with naturalists, gardeners and farmers from all over the world. He wrote with questions and received comments, specimens, ideas and observations. Putting these together with his garden observations, Darwin felt he could claim in *The Origin of Species* that '*close* interbreeding diminishes vigour and fertility':

> . . . these facts alone incline me to believe that it is a general law of nature (utterly ignorant though we be of the meaning of the law) that no organic being self-fertilizes itself for an eternity of

generations; but that a cross with another individual is occasionally – perhaps at very long intervals – indispensable.[28]

However, some of his critics attacked this assertion, arguing that he had not provided enough evidence for it; and neither they nor he were satisfied with his 'utter ignorance' of its significance, so Darwin went back to his garden to prove it.

To help him address this problem, Darwin decided he needed to add a hothouse to the range of unheated glasshouses he had already built; a stove would allow him to compare hardy plants with more tender, tropical ones. He wrote to tell Joseph Hooker, 'My hot-house will begin building in a week or so, & I am looking with much pleasure at catalogues to see what plants to get.' As soon as the building was done, Darwin was longing to fill it, 'just like a school-boy'.[29] He wanted to start growing tropical orchids to complement the native species he had been investigating for many years, but worried over their cost and told Hooker (now deputy-director of the Royal Botanic Gardens, Kew), 'I dare say I shall *beg* for loan of some orchids . . . I fancy orchids cost awful sums.'[30] Hooker responded, tongue-in-cheek, 'You will give me *deadly* offence if you do not send me your Catalogue of the plants you want before going to Nurserymen.'[31] A few weeks later Darwin visited Kew with a list of what he most wanted and when Hooker's gift of plants arrived, Darwin was 'fairly astounded at their number! why my hot-house is almost full!' He was especially delighted to 'see several things which I wished for, but which I did not like to ask for'.[32] A few weeks later, he was still crowing with delight, 'I have made list of plants, **165** in number!!!'; Darwin jokingly wondered whether such a raid on Kew's resources might lead to Hooker ending up in 'the Police Court?'.[33]

The first product of Darwin's investigations into plant fertility was his book on orchids, *The Various Contrivances by which British and Foreign Orchids are Fertilised by Insects* (1862). The fantastic shapes and colours of orchids had made them another widely popular hothouse plant, which must have helped the book's sales. Darwin had been interested in orchids ever since his undergraduate days and, as with the passionflowers, he had collected some during the *Beagle* voyage. He began experimenting with plants in about 1854 in order to investigate why even those that could potentially self-fertilize actually relied on insects to pollinate them. Granted the assumption he had made in the *Origin*, that perpetual self-fertilization lowered fertility, it would still seem reasonable to assume that natural selection would have favoured a 'general purpose' flower, one that any insect could fertilize; that would surely maximize the plant's chances of reproducing, ensuring that such general-purpose flowers would become common. Yet what Darwin found was that in many species of orchids the flower was adapted so that only *one* species of insect could enter and fertilize it. Not content with having, apparently rather perversely, given up on the easy option of self-fertilization, the orchids appeared to have gone one better and evolved so as to lower their chances of reproduction even further. And yet orchids form one of the world's largest plant families.

As he worked to unravel the extraordinarily complex flowering structures of orchids, Darwin realized that this 'lock-and-key' fit of insect and flower was beneficial to both. Because other species of insect could not get at the nectar, there was no competition for the one that could. And the flower benefited because its pollen would be spread only to other members of its species, not wasted on plants the pollen could not fertilize. These mutual benefits made natural selection into a force that had slowly, over tens of thousands of generations, reshaped both the orchid's anatomy and the insect's behaviour. Insects that inherited the preference for a specific orchid might – over many generations – eventually become the only ones that visited it, as they and their orchid became adapted to each other's anatomy. Such specialized insects would face less competition and thus fare slightly better in the struggle for food; all of which increased the chance of them passing on their preference. Meanwhile orchids that varied in a way that made them 'fit' the insect better would increase their chance of successful pollination, thus spreading the improved structure more widely in successive generations. Darwin's work revealed that what appeared to be intelligently designed – and inexplicably beautiful – structures were the products of natural selection. In each generation the most successful insects had been the ones who best fitted their flowers – together with the flowers that provided the best fit for their insects; for orchids and their pollinators, evolution was not so much the survival of the fittest as the survival of the best fitting. As Darwin put it, 'The more I study nature, the more I become impressed with . . . the contrivances and beautiful adaptations slowly acquired' through natural selection. Despite being produced by nothing more than random variation and a struggle for survival, such adaptations 'transcend in an incomparable degree the contrivances and adaptations which the most fertile imagination of the most imaginative man could suggest with unlimited time at his disposal'.[34]

Moving plants

Darwin wanted more than orchids in his hothouse, but told Hooker that 'I shall keep to curious and experimental plants.' These included carnivorous ones, which seemed another example – like the minute zoophytes he had studied in Edinburgh – of organisms that blurred the boundary between plants and animals. He was delighted to discover from the nurserymen's catalogues that he could buy carnivorous 'Pitcher plants for only 10s. 6d!'; as with the falling prices of passionflowers, the nursery trade made even unusual plants affordable. Alongside the carnivorous plants were ones that seemed to possess senses: 'Mimosa & all such funny things', that reacted to touch by closing their leaves.[35] Plants were traditionally defined in contrast with animals, as being unable to sense or move, so Mimosa was another potential case of blurring the boundary. However, if Darwin was to show that it had evolved its extraordinary abilities, as opposed to being divinely designed, he needed to demonstrate that all plants possessed some ability to sense and move.

Darwin began his investigation of plant motion with climbing plants, including passionflowers. Soon after the hothouse was finished, he told Hooker (with characteristic modesty) that 'I am getting very much amused by my tendrils – it is just the sort of niggling work which suits me & takes up no time.' In fact, it is clear from his book, *On the Movement and Habits of Climbing Plants* (1865), that far from 'taking no time', he spent years on painstaking experiments to discover just how and why plants climb. It was obvious that some plants have highly specialized adaptations for climbing. The American cup-and-saucer vine (*Cobaea scandens)*, for example, which Darwin called 'an admirably constructed climber', has branched tendrils that bear tiny hooks 'formed of hard, transparent, woody substance, and as sharp as the finest needle', which, Darwin recorded, 'readily catch soft wood, or gloves, or the skin of the hands'. This group, which Darwin called 'tendril-bearers', seemed to be the most specialized of climbers. As they grow, the tendrils revolve until they catch on a support, then they sense the light and start to clamber towards it; as he wrote to his son William, 'My hobby-horse at present is Tendrils; they are more sensitive to a touch than your finger; & wonderfully crafty & sagacious.'[36] As he worked at his hobby-horse, Darwin became increasingly impressed with the plants' 'craftiness', noting that tendrils responded to the weight of 'a loop of soft thread weighing 1/32nd of a grain' by growing towards it – a sensitivity that allowed them to detect anything they could potentially climb up. Yet the tendrils did not react when much heavier raindrops fell on them, nor to the wind.[37] Darwin was also impressed by the speed with which the climbers reacted; after experimenting with the crinkled passionflower, *Passiflora gracilis*, Darwin recorded, 'The movement after a touch is very rapid: I took hold of the lower part of several tendrils and then touched with a thin twig their concave tips, and watched them carefully through a lens . . . the movement was generally perceptible in half a minute after the touch, but once plainly in 25 seconds.'[38] Of all his 'crafty and sagacious' tendril-bearers, passionflowers seemed the most highly adapted; Darwin described *Passiflora gracilis* as exceeding 'all other climbing plants in the rapidity of its movements, and all tendril-bearers in the sensitiveness of its tendrils'.[39]

The remarkable mobility and responsiveness of these plants

was an evolutionary response to their environments. Darwin explained that 'Plants become climbers, in order . . . to reach the light, and to expose a large surface of leaves to its action . . . This is effected by climbers with wonderfully little expenditure of organized matter, in comparison with trees, which have to support a load of heavy branches by a massive trunk.'[40] That explained why plants climb, but Darwin still needed to explain how natural selection could have created such highly specialized plants. He noted that climbers divide into several kinds. Some, like ivy, use aerial roots to climb but there are also what he called spiral climbers, which he divided into twiners, which simply twist themselves around a support; leaf-climbers, which use their leaf stalks to attach themselves; and tendril-bearers, which have the most specialized climbing equipment.

Because the twiners wrapped their whole stems around the support, they used the most plant material to climb; Darwin unwrapped them, measured the stems (which were the thickest and thus the most costly part of the plant to grow), and discovered that the twiners had the longest stems. Leaf-climbers were shorter, but those of the tendril-bearers were shorter still. Reducing the amount of plant material needed to achieve the same goal showed the adaptive benefit of such specialization. By watching the plants develop, Darwin realized that tendrils were simply modified leaf or flower stalks and noted that the basic spiral movement of the twiners was essential to all three kinds of climbers. From these observations, he was able to reconstruct the probable evolutionary sequence: the twiners evolved first, with leaf- and tendril-climbers coming later. Tiny random variations in the ancestral twiners meant that some used a little more or less plant material in getting up into the light; Darwin noted, for example, that tendril-bearers can climb up the sunny outsides of bushes and trees, whereas simple twiners spent half their time in the shade. As a result, tendril-bearers could photosynthesize more efficiently and grow faster, all of which helped them conserve energy for making flowers and fruit, so they had more offspring who inherited whatever variation had benefited their parents, and so on. Gradually, generation after generation, natural selection could turn simple twiners into highly evolved tendril-bearers. In his later book, the *Power of Movement in Plants* (1880), which he wrote with his son Francis, Darwin showed that the basic movements which give plants the power to climb were found in some form in every kind of plant. Mimosa and *Passiflora gracilis* may have been exceptional, but neither was unique.

In summarizing his work on climbing plants, Darwin wrote that 'It has often been vaguely asserted that plants are distinguished from animals by not having the power of movement.' This was clearly wrong; it was more accurate to say 'that plants acquire and display this power only when it is of some advantage to them'. Most plants seldom need to move, or move too slowly for us to appreciate, but in response to a particular environment they prove to be more like animals than anyone had previously noted: predatory, perceptive, responsive and swift-moving. Darwin later wrote that 'it has always pleased me to exalt plants in the scale of organised beings', to show his readers that plants were not insensible, immobile or uncomplicated, and perhaps no climbing plant was ever acclaimed more loudly by Darwin than *Passiflora gracilis*.[41]

What is the use of sex?

As well as using passionflowers in his experiments on plant motion, Darwin decided to use them (along with many other species) in his long-running series of plant-breeding experiments. These were Darwin's attempt to answer the questions prompted in part by his marriage: what was the use of sex? Or, as he put the question, in the title of his book on the topic, what were the *Effects of Cross- and Self-Fertilisation in the Vegetable Kingdom* (1876)?

Given that Darwin was partly interested in the problem of human inbreeding, it might seem surprising that he did not choose an animal species for his experiments. As we shall see in the next chapter, humans themselves make rather poor experimental organisms, but another primate, or at least another mammal, might seem a more obvious alternative than a passionflower.

However, one of the great attractions of plants for a naturalist is that it is relatively easy to control which is breeding with which. Techniques such as netting flowers, to keep out insects, and then hand-pollinating with a paintbrush made it possible to be certain which pollen had got on to a particular plant. The first plant Darwin used was the common toadflax, *Linaria vulgaris*, but dozens of other species were soon involved, both outdoors and in the greenhouse. Darwin concentrated on growing cross- and self-fertilized specimens under identical conditions to try and prove that the cross-fertilized ones really had a competitive advantage compared with the self-fertilized.

It is exhausting to contemplate the work involved in these experiments. For each of the many species he examined, Darwin had to grow dozens of plants, which had to be kept separate from each other and protected from accidental wind- or insect-pollination. Then each one had to be hand-pollinated, its growth measured, and every seed counted. And all this had to be done twice, once for the self-fertilized plants, and then again for the cross-fertilized. Small wonder that Darwin later complained in a letter that 'I have worked like a slave (having counted about 9000 seeds) on Melastomas . . . yet have been shamefully beaten, & I now cry for aid'.[42]

Fortunately, aid soon arrived. Not long after the orchid book appeared, Darwin received a letter from John Scott, a young gardener at the Edinburgh botanical gardens, who explained that 'I take the liberty of addressing you for the purpose of directing your attention to an error in one of your ingenious explanations.'[43] Scott was still in his twenties, less than half Darwin's age; it must have taken some courage (and not a little self-confidence) to correct the country's most celebrated naturalist. Fortunately Darwin was not offended by this self-assured young man; he wrote back immediately to 'thank you most sincerely for your kindness in writing to me', adding that 'Your fact has surprised me greatly, & has alarmed me not a little.'[44] Darwin worried that he might well have made further mistakes. Despite his achievements he never considered himself a real botanist; as he told Scott, 'I know only odds & ends of Botany & you know far more.' He realized that Scott could be an invaluable assistant, telling the young botanist, 'I plainly see that you have the true spirit of an Experimentalist & good observer.' These compliments had a purpose; Darwin wondered 'whether you have ever made any trials on relative fertility of *varieties* of plants', adding that 'I much want information on this head,' especially about 'Lobelias & Crinum & Passiflora'.[45]

Encouraged by Darwin's well-aimed flattery, Scott was eager to help. He told Darwin, 'I have more than one season fertilized flowers of *Tacsonia pinnasistipula* [another passionflower] in the Gardens here,' but he had encountered the same problem that so many gardeners had faced: 'I have rarely succeeded in getting any fruit to set.' But with Darwin's encouragement, Scott offered to 'commence a series of experiments on those interesting questions', using Passiflora.[46]

Scott knew of Darwin's interest in passionflowers because they had been mentioned in the *Origin* as an example of a plant more easily fertilized by foreign pollen than its own.[47] Scott asked whether 'the *Passifloras* mentioned by you in Origin, [were] invariably sterile when treat[ed] with "own-pollen"; or is it a local occurrence?' and went on to discuss a couple of the species he had grown.[48] Darwin sent his evidence – inviting Scott decided to test it himself; soon the younger man was hard at work, crossing passionflowers, counting seeds, measuring growth and reporting his results to Darwin.[49] In return came praise ('What a capital observer you are!'); gifts ('if you would like to have any Book I have published . . . I sh^d. esteem it a compliment to be allowed to send it'); gentle criticism ('I suppose that you did not actually count the seeds in the hybrids in comparison with those of the parent-forms'); but most of all suggestions for ever more work ('I **very much** hope you will make a good series of comparative trials on the same plant of Tacsonia').[50] Scott was happy with the role of junior collaborator and asked Darwin to suggest possible experiments. 'If on reflection you would like to try some which interest me,' Darwin replied, 'I sh^d. be truly delighted', adding 'I could suggest experiments on Potatoes analogous with case of Passiflora,' and suggested that Scott also repeat some of his earlier passionflower experiments to check the results.[51] To encourage him further, Darwin sent Scott copies of the *Origin* and his *Journal of Researches* (*The Voyage of the Beagle*), and in reply Scott offered his 'sincere thanks, in humble and grateful acknowledgement of the entirely unmerited kindness you have done me'. He was grateful not only for the books, but for Darwin's willingness 'to recognise the observations of one entirely unknown, a young and ardent admirer of Science'.[52]

With Scott's help, Darwin was able to show that cross-bred plants were in fact invariably taller, hardier and more fertile than self-pollinated ones. This was sufficient to explain how mechanisms to avoid self-fertilization could spread. Imagine two varieties – A and B – of an ancestral passionflower species: both can be fertilized by their own pollen, but (thanks to random variation) A is a little more self-fertile than B. Because the chances of A being self-fertilized are higher, self-fertilized offspring will be the most common. But such offspring will also be less hardy (and thus less likely to reproduce successfully in the next generation) and so will eventually produce fewer descendants than their hardier, cross-fertilized siblings. By contrast, although variety B will leave fewer offspring, more of them are likely to be the tougher cross-fertilized kind. And not only will the cross-fertilized plants pass their slight tendency to self-sterility on to *their* offspring, but there will be further random variation in the next generation, so some of the descendants will be slightly *more* self-sterile than their grandparents were. Over many generations, variety B will gradually have to rely more and more on a mechanism like insect pollination to get its pollen to another flower (which is what must have happened to the ancestors of orchids). At first it seems implausible, but as variety B – the rarer, self-infertile variety – becomes more dependent on the uncertain business of cross-fertilization, it will gradually become more common.

Given these advantages, any random variation that accidentally favoured cross-pollination is likely to become more common.

That is why even hermaphroditic plants have often evolved mechanisms – such as insect pollination – that avoid self-fertilization to some extent. The orchids' lock-and-key relationship with their particular insect is just one of a range of adaptations that encouraged cross-fertilization with other plants.

A key part of Darwin's evidence for the evolution of cross-fertilization was that nature exhibits a wide range of different degrees of fertility: the passionflowers had taught him that 'with many species, flowers fertilised with their own pollen are either absolutely or in some degree sterile', but 'if fertilised with pollen from another flower on the same plant, they are sometimes, though rarely, a little more fertile'. While whenever plants of two unrelated species are crossed, 'they are sterile in all possible degrees, until utter sterility is reached'. He concluded, 'We thus have a long series with absolute sterility at the two ends': at one extreme, the plants were too similar to breed (almost as if some mechanism prevented incestuous unions), whereas at the other end they were too different, producing sterile offspring, as when a horse and a donkey are crossed and produce a mule.[53] By the time he published *Cross- and Self-Fertilisation*, Darwin felt he had made sense of the puzzling self-sterility of the passionflowers: preventing self-fertilization guaranteed cross-fertilization, which in turn guaranteed hardier, more fertile offspring. Evolving separate sexes would be another obvious way to avoid self-fertilization, which Darwin thought might explain how barnacles had evolved from the hermaphroditic form, via transitional ones with their tiny complemental males, into distinct male and female barnacles. As Darwin wrote at the end of his orchid book, 'nature abhors perpetual self-fertilisation'; avoiding it was the ultimate benefit of having two separate sexes.[54]

Favoured races?

Darwin loved plants. The passionflowers were just one of many genera he worked with, but they are particularly interesting because he returned to them in so many different contexts: in the unfinished draft of his 'big species book', *Natural Selection*; in the

Origin (1859); and in the *Variation of Animals and Plants under Domestication* (1868). They are one of the stars of *Climbing Plants* and they played an interesting, if minor, role in *Cross- and Self-Fertilisation*. Other plants were just as important, but Passiflora provides an unparalleled glimpse of the broad range of his botanical interests.

Yet despite his enthusiasm for his greenhouse and garden, for his passionflowers and orchids, Darwin was every bit as interested in the breeding of human beings. As we have seen, he worried about his own children but he was also concerned about the future of his country and its people. It sometimes seemed that while the vigorous, energetic empire-builders he admired had made Britain rich, they had also made it comfortable – perhaps too comfortable. Would the ease with which he and his countrymen could acquire every necessity of life reduce the impact of natural selection, thus eventually weakening the race and allowing other nations to dominate them?

Darwin had carefully avoided the subject of human evolution in the *Origin*, realizing that he would have more than enough controversy to deal with. But in 1871, when he finally tackled the subject in the *Descent of Man*, he restated his belief in Malthus's pessimistic philosophy, boldly asserting that 'all ought to refrain from marriage who cannot avoid abject poverty for their children'. He noted that 'Man, like every other animal, has no doubt advanced to his present high condition through a struggle for existence consequent on his rapid multiplication; and if he is to advance still higher he must remain subject to a severe struggle'; in other words, if too much were done to help the poor and other 'inferior' members of society, the vital struggle would be mitigated in such a way as to ensure that humanity would 'sink into indolence'. It was vital, in Darwin's view, to ensure that 'the more highly-gifted men' were 'more successful in the battle of life than the less gifted' – and that they passed on their gifts to more children.[55] Darwin echoed the concern of his cousin Francis Galton over the observation that the better-educated and wealthier members of society (who must, Darwin and Galton agreed, be the most talented) were having smaller families, while the feckless poor were out-breeding them. As we will see in the next chapter, Galton had strong opinions as to what should be done about this problem.

In addition to these wider concerns, Darwin was especially worried that cousin marriage, so common among 'superior' families like his own and the Wedgwoods, would further weaken the embattled middle classes. In 1870, he encouraged an attempt by John Lubbock, his neighbour, scientific ally, and local MP, to convince Parliament to include a question on cousin marriage in the 1870 census.[56] Disappointed by the failure of Lubbock's proposal, Darwin wrote in the *Descent*, the following year, that: 'Man scans with scrupulous care the character and pedigree of his horses, cattle, and dogs before he matches them; but when he comes to his own marriage he rarely, or never, takes any such care.' In an ideal world, he thought, 'Both sexes ought to refrain from marriage if in any marked degree inferior in body or mind', but he admitted that 'such hopes are Utopian and will never be even partially realised until the laws of inheritance are thoroughly known'. As a result, 'All do good service who aid towards this end.'[57] However fascinating passionflowers might be, it was the breeding of people that Darwin's contemporaries were most exercised about.

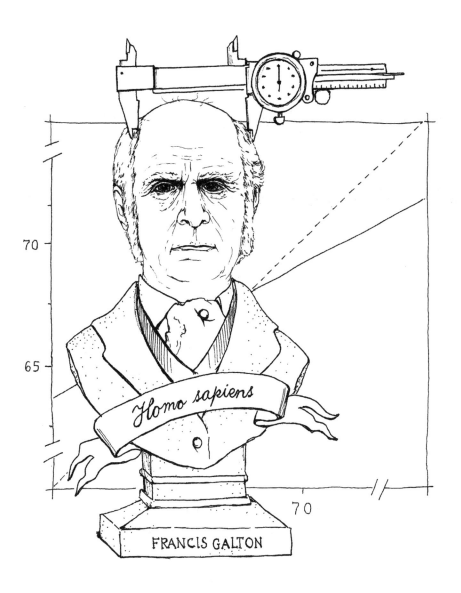

70

65

Homo sapiens

70

FRANCIS GALTON

Chapter 3

Homo sapiens: Francis Galton's fairground attraction

About 100,000 years ago, a casual observer would probably not have found the early members of *Homo sapiens* particularly remarkable. Like our ancestors, we were essentially apes with bad hair and bad posture. For some reason – much debated, yet still unclear – we'd abandoned the sensible arboreal habits of our cousins and moved out on to the African savannahs to stand more or less upright. Somewhere along the way we lost most of our attractive fur. Perhaps leaving the forests and the cover they provided prompted our ancestors to stand up, so they could spot approaching predators. Leaving the trees may also have had something to do with the hair-loss, but there is no consensus about what the connection might have been, nor is there any convincing evidence.

There was not much to suggest that these unappealing ape-like creatures would eventually colonize most of the planet, but of course we did. About 12,000 years ago we began to create permanent settlements, based on the discovery that we could grow our food instead of pursuing it. Over the next few thousand years farming became more widespread and sophisticated, allowing ever larger groups of people to live closer and closer together. A few thousand years later, *Homo sapiens* made one of its biggest discoveries: living in cities. Once urbanized, our species went on to spread across almost all the earth's continents, building larger and larger cities as it went.

Cities allowed more people to live in less space, which created two major problems. The first was getting enough for everyone to eat, and the other was disposing of the waste after everyone had eaten. On Thursday 8 May 1884, after roughly 6,000 years of coping with these closely related inconveniences, the most advanced solutions to them went on display at an International Health Exhibition in what was then the world's largest city, London. Over the following few months, over 4 million curious *Homo sapiens* came to look at displays that focused on the great problems of city life: eating and excreting.

Ingestion took up much of the available space. In the middle of the exhibition was a working dairy, where city dwellers could watch cows being milked and butter being churned – sights that were beginning to become a little unfamiliar. Around it 'are also illustrated methods of cold storage and transport of fresh meat, ice-making, the preservation of food, making bread, biscuits, &c., the manufacture of confectionary, of condiments, of cocoa and chocolate, and the production and bottling of aerated waters'.[1]

The *Illustrated London News* sent George Augustus Sala to describe the exhibition for its readers. Sala had made a name for himself writing in Charles Dickens's popular weekly, *Household Words*. He observed that if anyone could summon up the 'moral and physical courage to "do" the whole of the contents of the Health Exhibition in the course of a single day', they would at least not go hungry, since 'it is a colossal Café and Restaurant'. Visitors could if they liked 'obtain a sixpenny dinner in the restaurant of the Vegetarian Society . . . ', which planned to spend the profits from its restaurant on 'feeding the poor of London and the provinces – on strict vegetarian principles, of course – during the winter of 1884–5'. One wonders how London's poor responded to this meat-free charity, but Sala at least was impressed; 'I have partaken of the sixpenny vegetarian dinner,' he recorded, 'and found it very nice.'[2]

Those who were not bold enough to attempt the vegetarian dinner could try a '*dîner à la Duval*', in a restaurant based on 'the system so largely made use of in Paris at Duval's Restaurants'.

Great efforts were made to ensure that everyone knew what they were paying and were not over-charged, but the most striking feature of the restaurant was that 'the far end of this dining saloon is closed by plate glass windows, behind which the cooks may be seen at work preparing the various dishes'.[3]

Watching the cooks at work reassured the diners that their food was being prepared in hygienic conditions. Britain still lacked health inspectors and the purity of food was a major concern for many Londoners; the first Food Adulteration Act had been passed in 1860, but no means of enforcing it existed until the first inspectors were appointed in 1872. And it was only in 1885, after the Exhibition, that the Sale of Goods Act made sellers responsible for the goods they sold for the first time.[4] Just a few generations earlier, most people in Britain had still seen their food grown, milled, baked and sold in their own villages, but by the 1880s these were distant memories for the millions crowded into London's narrow, smoky streets. As we saw in the last chapter, London's billowing smoke made it difficult to grow fruit and vegetables close to the people who would eat them. Food now came from shops, whose weighing scales were notoriously inaccurate and whose proprietors were fond of watering the milk and bulking up the flour with alum or even more sinister compounds. The campaigning journalist James Greenwood accused dishonest shopkeepers of being thieves, adding that compared with robbers, the shopkeeper's 'is a much safer system of robbery. You simply palm off on the unwary customer burnt beans instead of coffee, and ground rice instead of arrowroot, and a mixture of lard and turmeric instead of butter. You poison the poor man's bread.'[5] The Health Exhibition's working dairy and glass-walled kitchen were responses to these anxieties, as were the displays of ingenious devices for keeping foods fresh and uncontaminated during their long journeys to the capital.

However, while the restaurants might be seen as vital to the exhibition's ostensible subject, health, there were other displays that seemed less easy to justify. The *Illustrated London News* was sure that 'the international collection of dresses and of English costumes' displayed on waxworks would be 'attractive to the fair sex', while 'persons of antiquarian tastes will be able to inspect an elaborate model of part of the City of London in the olden time'.[6] Others were less convinced; the life-size historic street was a popular attraction, as were the waxworks, illuminated fountains, tiles, pottery, iron-work, tapestries and similar decorative products, but the *Saturday Review* commented that 'those of us who have lived a few years in London will remember more than one "International" show which, opened under Royal patronage, ended up becoming a mere bazaar. There is a good deal of puffing carried on by means of them, and the more satisfactory they are to advertisers, the less they really do for the advancement of science or art.'[7] Even the more enthusiastic journalists had to confess that they could not do justice to the range of 'bronze statues and electroliers, bibles, cabbages, and parasitic pests, vile vegetable dinners [and] dairies'.[8]

The health exhibition was merely the most recent in what had become annual shows of various kinds, shows whose miscellaneous contents had been satirized in *Punch* a few years earlier. The magazine's brief guide to forthcoming 'International Exhibitions' had forecast that the 1880 show would contain 'apparatus for preventing and consuming smoke, observatories, orangeries, artificial flowers, acts of parliament, carriages-and-four, balloons, flying machines, fireworks and anything that may have been omitted in previous years'. *Punch* concluded by observing that 'fine dresses, flirtations, refreshments, season tickets, turnstiles, catalogues, military bands, crowds of people, and grumblers' would be on display every year.[9]

Many of the journalists who reported on the exhibition were content to treat it as entertainment, but the weekly magazine *Knowledge*, which prided itself on being 'an illustrated magazine of science, plainly-worded – exactly-described', was concerned to demonstrate that the show had a more serious purpose. Its publisher and editor, Richard Proctor, a well-regarded astronomer, sent one of his writers, John Ernest Ady, to cover the Health Exhibition. Ady defended its educational claims by arguing that the 'true sanitary exhibits' had been carefully mixed with 'more entertaining objects', to ensure that the casual visitor did not get bored. Meanwhile:

> The sanitary student . . . who comes intent on study is pleasurably surprised, and imperceptibly led to find how much easier his task becomes under the soothing influences of a *dîner à la Duval*, with music thereafter, and when his eye is delighted by exquisite dress, Doulton's potter's art, beautiful furniture, or quaint houses; and then to see the living cows and goats milked! Why, after that, he can go on his sanitary tour with redoubled vigour.[10]

The refreshed 'sanitary student' could then concentrate on the less salubrious aspects of the show, those that addressed the business of excreting.

By the middle of the nineteenth century, Britain was the most urbanized country the world had ever seen; the 1851 census revealed that, for the first time in the history of any country, more than half its people lived in towns. The decline of rural employment and the growth of factories forced people into cities, where life expectancy was typically half that of the countryside. Among the tightly packed city dwellers epidemic diseases spread rapidly; polluted water supplies spread cholera and typhoid, typhus was spread by lice, and warmer weather brought regular outbreaks of 'summer diarrhoea' as millions of flies feasted on the horse manure and human waste that lay in the streets, before transferring their attentions to human food. Every British city was the same; in Manchester, 'everywhere heaps of débris, refuse and offal; standing pools for gutters, and a stench which alone would make it impossible for a human being in any degree civilized to live in such a district'.[11]

London's untreated sewage flowed straight into the Thames; as a result the city stank. The need for better sanitation, especially for sewers and water treatment plants, preoccupied the Victorians, especially in the hot summer of 1858, when the Thames almost stopped flowing under its burden of refuse and the 'Great Stink' of London began. The river's stench was so overpowering that large parts of the House of Commons became uninhabitable, thus effectively focusing the MPs' minds on the urgency of sanitary reform. In less than three weeks, a bill was rushed through Parliament and money was made available to build miles of new sewers and to construct an embankment to improve the river's flow.

By the time the Health Exhibition opened, London was a little less noisome than it had been, but the problems of sanitation were still very much on Londoners' minds. The engineer and sanitary reformer Sir Douglas Strutt Galton wrote a long article about the exhibition in the *Art Journal*, condemning the terrible conditions of the poor, which resulted from ignorance of basic sanitary measures. He argued that these were everyone's concern, since 'a badly housed population is a discontented population, and legitimately so'.[12] According to the *Pall Mall Gazette*, when His Grace the Duke of Buckingham opened the exhibition he was greeted with cheers as he claimed that the opening would 'mark an era in the records of the social and domestic condition of the nation'.[13]

As if in response to the Duke's challenge, the Doulton & Company pottery (now famous as Royal Doulton) took a whole pavilion at the exhibition, where they exhibited the sewer pipes, toilet bowls and industrial ceramics upon which the company's fortunes were founded. John Ady devoted three full articles in *Knowledge* to describing the wonderful improvements in the art of waste disposal that Doulton were making.[14] Another reporter commended as 'one of the most interesting features in the Exhibition' an ingenious display that attempted to resolve the problems of eating and excreting simultaneously. Britain's fields were increasingly fertilized with guano – dried bird-droppings. Britain's guano was imported but large-scale mining in South America and the southern Pacific had already begun to deplete natural deposits, which seabirds had been diligently building for centuries; as a result, guano was becoming expensive. Britain's

Native Guano Company used the Health Exhibition to exhibit its process 'for purifying sewage, waste waters, &c., before they are allowed to fall into the rivers'. Not only did it remove 'the offensive matters dissolved in the water', but the reporter was convinced 'that vegetables grown with the manure produced by the Native Guano Company's process' were in every way superior to 'vegetables grown on sewage irrigation farms'. The Native Guano Company was awarded a Gold Medal by the organizers, but – perhaps not entirely surprisingly – their process was not to be widely adopted.[15]

Once visitors had inspected the potential improvements to their drains, the fertilizing possibilities of their own 'guano' and registered the hazards of wallpapers whose dyes contained arsenic, they could shock themselves by looking at examples of 'dress injurious to health', which included 'models, in plaster, of the liver of a healthy woman and of the same organ in a wasp-waisted votary of fashion', whose corset was laced so tightly as to displace her internal organs.[16] Also on display was a solution to this predicament, in the form of Madame Eugénie Genty's 'newly patented Health Busk, which enables the ladies, when indisposed, to unclasp their corsets instantaneously'.[17] Those with no interest in busks could turn to bees: 'the visitor who is interested in Bee Culture, will find . . . the collection of frame and straw hives' extractors, comb foundations and other appliances used in bee keeping, together with specimens of pure and adulterated honey, and of the articles used as adulterants'.[18]

At this point an exhausted visitor, well fed and slightly inebriated, dizzied by unlaced corsets and improved cisterns, could have been forgiven for giving up and going home. Which would have been a shame, because in the same corridor as the bee-keeping equipment and a collection of meteorological instruments, sandwiched between the dining-rooms and the bakeries, was a corridor 6 feet wide and 36 feet long in which one curious *Homo sapiens* had devised an experiment intended to improve his own species.

Measuring man

As the official guide to the Exhibition explained, 'Adjoining the meteorological instruments is the so-called Anthropometrical Laboratory, arranged by Mr. Francis Galton, in which visitors can have their principle physical dimensions taken, their hearing power and accuracy of eyesight ascertained, and their strength tested.'[19] Here visitors could pay 3d. to have measured everything from their 'Keenness of Sight and of Hearing' to their colour sense and 'Judgment of Eye'. Physical strength was measured too, using machines that resembled those found in Victorian fairgrounds, which estimated 'Strength of Pull and of Squeeze' and 'Force of Blow'. The laboratory's organizer, Francis Galton (no relation to Sir Douglas, the engineer), wrote that 'the ease of working the instruments that were used was so great that an applicant could be measured in all these respects, a card containing the results furnished him, and a duplicate made and kept for statistical purposes', all for the cost of the threepenny admission fee, which 'just defrayed the working expenses'.[20]

During the six months that the Health Exhibition was open, seventeen different measurements were made of over 9,000 visitors. A sergeant was on duty to superintend the crowds and a Mr Gammage, a scientific instrument maker, came in every evening 'to assist and supervise, and who maintained the instruments in efficiency'. There was also 'a doorkeeper provided by the executive', who let people in, collected their threepences, handed out and collected forms, ensured the latter were correctly filled in, and 'made himself useful in many other details'.[21] Galton was so delighted with the results that after the Health Exhibition closed, 'it seemed a pity that the Laboratory should also come to an end, so I asked for and was given a room in the Science Galleries of the South Kensington Museum. I maintained a Laboratory there for about six years.'[22] This new lab continued the earlier one's work, gathering data on nearly 4,000 people in its first three years.

To understand how all these measurements were supposed to reshape humanity, we need to know a little more about the man who made them. Even by the standards of Victorian eccentricity, Francis Galton was an extraordinary figure. His mother was a daughter of Erasmus Darwin – Charles Darwin's grandfather. Despite being a child prodigy – he was discussing Homer's *Iliad* at the age of six – he did not do well at school, and found it difficult to stick to his medical studies. In 1840, at the age of eighteen, he dropped medicine and went to Trinity College, Cambridge to read mathematics. He lived the typical life of an undergraduate: after three years of drinking, dancing, hiking and doing no work at all, he had a nervous breakdown while preparing for his finals, and left with an ordinary degree.

When Galton was twenty-two his father died, leaving him a substantial fortune. He gave up study and began to travel, in the Middle East, then Scotland, then in South-West Africa. Back in London he published his first book, *Narrative of an Explorer in Tropical South Africa* (1853). The common thread that was to link all of Galton's varied enthusiasms was already apparent in this book: measurement. During his South African travels, he wanted to determine the precise dimensions of a Hottentot woman's buttocks; since he did not speak her language he resorted to surveying her from a distance, using a theodolite (normally used to survey land); 'this being done', he recorded, 'I boldly pulled out my measuring-tape, and measured the distance from where I was to the place where she stood, and having thus obtained both base and angles, I worked out the results by trigonometry and logarithms.'[23]

Back in London, Galton turned his mathematical mind to more conventional subjects, map-making, geographical instruments and weather forecasting. He coined the term anti-cyclone and also published the first newspaper weather map (in *The Times* of 1 April 1875). Galton was elected a fellow of the Royal Society soon after his return to Britain, and some sense of the diversity of his scientific interests can be gained from the range of societies he was involved with: he was on the managing board of the Kew Observatory in 1858; he joined the Royal Statistical Society in 1860 and had also become a leading member of the Ethnological Society by that time.

In 1859, his cousin Charles Darwin published *On the Origin of Species*. Not surprisingly, given his voracious appetite for scientific novelties, Galton read it immediately and wrote to congratulate Darwin on 'your wonderful volume', which he had finished with 'a feeling that one rarely experiences after boyish days, of having been initiated into an entirely new province of knowledge'.[24] Many years later, in his autobiography, Galton recorded that he had 'devoured its contents and assimilated them as fast as they were devoured, a fact which perhaps may be ascribed to an hereditary bent of mind that both its illustrious author and myself have inherited from our common grandfather, Dr. Erasmus Darwin'. This rapid ingestion of Darwin's 'new views' encouraged Galton 'to pursue many inquiries which had long interested me, and which clustered round the central topics of Heredity and the possible improvement of the Human Race'.[25]

This was a surprising conclusion to draw from Darwin's book, which as we have seen made almost no mention of the evolution of humans, much less of their 'improvement'. However, Galton may perhaps have been inspired by the *Origin*'s famous closing passage, where Darwin wrote that:

> Thus, from the war of nature, from famine and death, the most exalted object which we are capable of conceiving, namely, the production of the higher animals, directly follows. There is grandeur in this view of life, with its several powers, having been originally breathed into a few forms or into one; and that, whilst this planet has gone cycling on according to the fixed law of gravity, from so simple a beginning endless forms most beautiful and most wonderful have been, and are being, evolved.[26]

Darwin believed he had found a natural law, akin to Newton's 'fixed law of gravity', that governed all living things, including ourselves. Some of the *Origin*'s readers were repelled by the idea that it was not God but natural selection, powered by 'famine and death', which had created them. But others – including Galton –

felt liberated from the strain of believing in a religious world-view they found increasingly implausible. He told Darwin that 'your book drove away the constraint of my old superstition as if it had been a nightmare and was the first to give me freedom of thought'.[27] To Galton and many of his contemporaries, Darwin seemed to be saying that progress was one of nature's laws, that every living thing was under the sway of the Victorian creed of self-improvement. The book's final sentence, 'endless forms most beautiful and most wonderful have been, *and are being*, evolved', could be interpreted to mean that evolution was not over. Humanity might still be improving and some future race of people would stand as distant from us as a Victorian gentleman stood from his stooped, hairy African ancestor.

Galton was also inspired by Darwin's analogy between human efforts to improve domesticated plants and animals and the way nature worked to create new species. Darwin used the familiar processes of plant- and animal-breeding, which he called artificial selection, to give his readers some sense of the changes that were possible. Humans had taken the common domestic rock pigeon, *Columba livia*, and by selective breeding over a few hundred years had produced the extraordinary array of ornamental pigeons that fascinated Victorian pigeon fanciers: the English carrier, the short-faced tumbler, the runt, the barb, the pouter and the fantail. If these bizarre birds, with their extravagant tail feathers and extraordinary shapes, were shown to an ornithologist who did not realize they were domesticated, they would, Darwin felt sure, 'be ranked by him as well-defined species', perhaps even as members of different genera. Given that artificial selection could do so much in so short a time, just imagine, Darwin suggested, what natural selection can achieve as it works away over 'the long lapse of ages'.[28]

This sense that living creatures offered almost infinite scope for improvement caught Galton's imagination. A few years after reading the *Origin*, he wrote two articles on 'Hereditary Talent and Character'. He began by noting that the 'power of man over animal life, in producing whatever varieties of form he pleases, is enormously great. It would seem as though the physical structure of future generations was almost as plastic as clay, and under the control of the breeder's will. It is my desire to show . . . that mental qualities are equally under control.' The idea that an animal's mental characteristics could be improved by breeding seemed entirely plausible; one only had to consider a gentleman's best friend and companion, his hunting dog, whose instincts to point and retrieve had been sharpened by generations of careful, artificial selection. But Galton had a rather more controversial target in mind; the mental qualities of his fellow humans. Contrary to common prejudice, Galton wrote, 'I find that talent is transmitted by inheritance in a very remarkable degree.'[29]

Galton's evidence for this claim was to examine the biographies of eminent men (women seldom attracted his consideration) to see if talent ran in families. He was convinced that it did. He and his famous cousin, two eminent scientific men who shared a grandfather, were just one of hundreds of examples that showed famous men were more likely to be related to each other than to crop up randomly among families of nonentities. Just as prize-wining horses were more likely to sire future Derby winners, so the leading judges, authors, essayists, musicians, divines, artists and scientific men were most likely to father equally successful sons. Wherever we look, Galton argued, 'the enormous power of hereditary influence is forced on our attention'. The implication, he felt, was clear: the talent that made men eminent in their fields was inherited, so if eminent men could be persuaded to breed with women 'who possessed the finest, and most suitable natures, mental, moral, and physical', they would produce outstanding offspring. And if the children of these select marriages were to be equally choosy about their own spouses, it might be possible to accelerate Darwin's process dramatically. As Galton put it:

If a twentieth part of the cost and pains were spent in measures for the improvement of the human race that is spent on the improvement of the breed of horses and cattle, what a galaxy of genius might we not create! We might introduce prophets and high priests of civilization into the world, as surely as we can propagate idiots by mating crétins. Men and Women of the present day are, to those we might hope to bring into existence, what the pariah dogs of the streets of an eastern town are to our own highly-bred varieties.[30]

Whatever the practicality, much less the morality, of such a scheme, it was a somewhat perverse reading of Darwin, who had looked to *natural* selection to make whatever improvements species might in future undergo. At the very least, the *Origin* seems superfluous to Galton's view, since he was simply promoting an extension of the well-known, longstanding practices of artificial selection to humans; no one needed to read Darwin to know that humans had been effectively breeding dogs for hundreds of years. Yet Galton felt the *Origin* was vital, because it had helped him discard his religious views.[31] Darwin had persuaded him that humans were animals like any other, subject to the same laws of inheritance and competition; if a breeder could reshape a dog's mind so that it retrieved more effectively, why could not a human mind be bred to improve its ability to paint, compose or theorize? If people could be persuaded to put aside their superstitious belief in their divinely created uniqueness, they would see themselves as capable of potentially infinite improvement.

Galton was thrilled by his own idea. He cast around for a succinct and memorable name for his theory, and initially called it viriculture, but eventually settled on the term eugenics.[32] He promoted it in magazines and eventually wrote a book, *Hereditary Genius* (1869), which was essentially a much-expanded version of his original article, bolstered with many more accounts of eminent men. Since he had no way of directly measuring a person's intelligence or other mental qualities, Galton had to use posthumous reputation as a guide, bolstered by his – largely unsubstantiated – conviction that whatever 'eminence' men attained must be predominantly a product of their innate gifts, not of their being well-connected members of privileged families.

Yet, despite the weight of his evidence, Galton's idea did not catch on. The reviews of *Hereditary Genius* were almost all hostile. The Manchester *Guardian's* reviewer focused on what remains one of the central objections to eugenics: 'Who is to decide whether a man's issue is not likely to be well fitted "to play their part as citizens?" '[33] Even more annoyingly, many reviewers were simply amused by Galton's idea; one wryly commented that if Galton's 'happy and philosophic system' were to be adopted, 'we shall no more hear of a lady "throwing herself away" upon an unworthy object'.[34]

One of the largest obstacles in Galton's path was that he had no real evidence for his controversial claim that mental qualities were inherited. Most of his contemporaries believed that humans had unique, divinely created mental qualities that set them apart from other animals. And even the less religious believed that such factors as parents' health and habits shaped their children. Alcoholism, for example, was considered a degenerative disease, whose effects were inherited; drunken parents weakened their minds and bodies by drinking, and then passed these weaknesses on to their children. By the same logic, it was assumed that if fathers and mothers were healthy they would tend to produce healthy children, regardless of their bloodline, and that improvements in their mental ability were most likely to arise from better education. Such arguments drastically curtailed the appeal of Galton's views, since they suggested that – even if the ill-effects of alcohol were inherited – people were primarily shaped by their environments; temperance campaigns and improved sewers rather than selective breeding were the most urgent task for those who wanted to improve humanity. The 1860s was a fairly prosperous decade in Britain: the country maintained its industrial lead over its competitors; it gained new colonies, and with them both cheap raw materials and new markets; and the economy continued to grow. The nation seemed to be getting steadily richer and many influential people felt they could afford to pull down some slums and build a few schools and drains.

There was nothing new about debates over the roles of inherited and environmental factors in shaping human nature;

they had been going on since at least the late eighteenth century, pitting 'hereditarians' against 'perfectibilists'. In very simplified terms, the latter argued that if the progress of civilization improved the conditions of human life, humans themselves would be improved. The hereditarians rejected this approach, insisting that good breeding was the key to improvement. Galton liked to think of himself as having reinvigorated the hereditarian argument and he brought the term 'heredity' into English (from the French *hérédité*) to symbolize his supposedly new approach. But in fact it was widely accepted that clever parents tended to have clever children, mainly because they often shared an advantageous environment but to some lesser degree because intelligence was inherited; the issue for Galton was to overturn the received view, by establishing that breeding played the dominant role in forming someone's intelligence.[35]

For most of his life Galton seems to have simply assumed the hereditary factors were dominant, and in 1874 he adopted a memorable verbal contrast to summarize his view:

> The phrase 'nature and nurture' is a convenient jingle of words, for it separates under two distinct heads the innumerable elements of which personality is composed. Nature is all that a man brings with himself into the world; nurture is every influence from without that affects him after his birth.[36]

He borrowed the phrase from Shakespeare's *Tempest*, in which Prospero describes the monstrous Caliban as 'A devil, a born Devil, on whose nature nurture can never stick'. Convinced that nature dominated nurture, Galton was sure that reformers were deluding themselves and would waste the money contributed by wealthy taxpayers like himself on their schemes for environmental and educational improvements. 'I have no patience,' he snarled, 'with the hypothesis occasionally expressed, and often implied . . . that babies are born pretty much alike, and that the sole agencies in creating differences between boy and boy, and man and man, are steady application and moral effort.' In case anyone had missed the point, he added, 'It is in the most unqualified manner that I object to pretensions of natural equality.'[37] Despite his Darwinism and contempt for traditional religion, Galton was a deeply conservative man and felt a contempt, tinged with fear, for his social inferiors. He had no patience with the filthy, immoral Calibans who swarmed through London's slums, demanding reforms they expected eminent men to pay for.

The return of Lord Morton's mare

The problem Galton faced was that he, like his contemporaries, had no idea how biological inheritance worked. Even for simple physical attributes, the patterns of inheritance were complex, but they were even more so for mental ones which were so hard to measure. As the *Guardian*'s hostile review of *Hereditary Genius* had put it: 'Do not weak men have strong children, stupid ones wise, wicked good? – while on the other hand, do we not find the weak emanating from the strong, and bad from good?'[38] Neither the reviewer, nor Galton, nor any of their contemporaries had any idea of what was actually passed from generation to generation, nor of how it was passed. As Galton later noted, at this time scientific opinions about heredity were 'vague and contradictory'; however, 'most authors agreed that all bodily and some mental qualities were inherited by brutes, but they refused to believe the same of man'.[39] Galton needed to understand how inheritance worked before he could seriously attempt to prove that it, not education and environment, determined human mental qualities.

Shortly after *Hereditary Genius* appeared, Galton received a letter from Darwin, who exclaimed enthusiastically that 'I do not think I ever in all my life read anything more interesting and original. And how well and clearly you put every point!'[40] Darwin was also contemplating the question of heredity. His lack of a theory of inheritance had been seized on by some critics of the *Origin* as the weakest link in his argument. In 1867 a Scottish engineer, Fleeming Jenkin, published a critical review of the *Origin* that Darwin himself had to acknowledge 'seems to me one of the most telling Reviews of the hostile kind, & shews much ability'.[41] Jenkin had three objections to Darwin's theory, of which the most serious focused on inheritance. Jenkin – like Galton, Darwin and most of their contemporaries – assumed that when two organisms mated, their offspring displayed a mixture of the parents' characteristics. This seemed a reasonable theory (after all, we still talk about a baby having its father's nose, but its mother's eyes) and was known as 'blending inheritance'. Jenkin thought blending was fatal to Darwin's ideas: he asked his readers to imagine what would happen if a white man were shipwrecked on an island inhabited by black people. Assume, proposed Jenkin, that he possessed 'the physical strength, energy, and ability of a dominant white race'. This superior white man is then in a position analogous to that of a new type of plant or animal, which has been thrown up by the chance variations that were crucial to Darwin's theory. Jenkin suggests that we imagine this Robinson Crusoe to have 'every advantage which we can conceive a white to possess over the native':

> Our shipwrecked hero would probably become king; he would kill a great many blacks in the struggle for existence; he would have a great many wives and children . . . Our white's qualities would certainly tend very much to preserve him to good old age, and yet he would not suffice in any number of generations to turn his subjects' descendants white.[42]

Jenkin's argument was a simple one, and his racist assumptions should not blind us to its logic. No matter how many wives our hero acquired, they would all be black and so, no matter how many children he had, they would tend to be darker-skinned than their father, and – given that there would be no white spouses available for his children to choose, the white king's offspring would also have black wives and so their grandchildren would be darker still. In a few generations, the old king would be dead and his descendants would be as black as ever.

Jenkin admitted that white skin might not, in itself, convey any advantage, but he assumed that whatever superior qualities the white races possessed – strength, vigour or courage – were traits that were inherited in exactly the same way as skin colour and so would undergo the same dilution over successive generations. The same would be true for all traits in all organisms: one fast antelope could not make its species faster any more than one freakishly tall flower could make its descendants taller.

Darwin was worried by Jenkin's argument and he made some changes to the following edition of the *Origin* (the fifth, of 1869) that he hoped would address it. He acknowledged that if a new form arose as an isolated freak of nature, or 'sport', it would not be able to alter the nature of the species, but he did not believe natural selection was invalidated by blending inheritance; instead, he simply focused his argument on the normal level of variability in a population, instead of on the sports. Suppose, he argued, that a species of bird has beaks that normally vary from being completely straight to being very curved. Imagine what would happen if a change in the birds' environment gave curved beaks an advantage: perhaps a drought made the insects that the birds normally ate scarce, so they have to dig into cracks in rock to retrieve seeds instead. If a curved beak worked better for this kind of feeding, the straight-beaked birds would do less well in the competition for food and so have fewer offspring. Meanwhile the curved beaks would be doing well and – being more numerous and better fed – breeding more successfully. Their offspring would face the same competition in the next generation and those with the most curved beaks would do best. Over many generations, beaks would become more and more curved. Such small, gradual variations (which were, in any case, much more common than sports), allowed natural selection to work just as Darwin had foreseen.

However, while Darwin felt he had seen off Jenkin, he was acutely aware that the lack of an explicit mechanism of inheritance was still a major weakness in his argument. He had been puzzling over the problem ever since his days at Edinburgh University; and had given it a great deal of thought while he worked on what he called his 'big species book'. This had absorbed him for nearly twenty years, but in 1858 he had been stunned by a letter from another naturalist, Alfred Russel Wallace, who had hit on the idea of natural selection entirely independently. Terrified that Wallace would get credit for the idea, Darwin rushed *The Origin of Species* into print, describing it as an 'abstract' of the larger book he still hoped to write.

Once the *Origin* was out, and Darwin's intellectual priority was established, he returned to the unfinished manuscript of his 'big book', eventually turning sections of it into *The Variation of Animals and Plants under Domestication* (1868), in which he focused on a collection of puzzles relating to heredity – in the hope that solving them would help him understand how inheritance worked.

To modern readers, the *Origin* can seem long-winded at times, but any reader who finds it heavy going should try reading *Variation*; had it not been for Wallace's accidental intervention, Darwin's theory (if it had ever appeared at all) would have been buried under such a mass of detailed information that it is possible no one would ever have understood its significance. *Variation* piles on example after example of breeding, from dogs and cats through to gooseberries, from pigeons to peaches, canaries to cherries, from asses to apricots. Only a very patient reader would have perceived the significance of the various phenomena that interested Darwin. Among these were 'reversion': why is it that babies sometimes have their grandmother's or great-grandmother's nose, rather than their mother's? Regeneration: why is it that if you cut the tails of some lizards, they are able to grow new ones? And Darwin was also interested in the curious case of Lord Morton's mare.

Being a country boy at heart, Darwin drew many of his examples of reversion from the farmyard, asking why breeds of sheep that for generations had been bred without horns should suddenly throw up a lamb with horns. In part, he hoped that understanding reversion might solve the problem of blending inheritance: whatever caused such characters to appear had obviously not been swamped, so perhaps this unknown cause could explain how improved variations might survive, spread and develop into species. After pondering the question, Darwin argued that whatever it was that caused horns to appear in hornless breeds, it must persist, like words 'written on paper with invisible ink', which were always present but could not be seen until some unknown factor caused them to reappear.[43]

Darwin was also fascinated by the way creatures such as salamanders could grow new limbs. This ability suggested to him that whatever the mysterious 'invisible' characters were, they need not be restricted to a creature's reproductive organs, but appeared to be diffused through its body. Finally, there was Lord Morton's mysterious mare. As we have seen, the mare had been mated with a male quagga and had produced stripy, quagga-like foals. But the pure-bred mare had also produced striped foals when she was subsequently mated with another horse. This suggested to Darwin that there was a 'direct action of the male element on the female form'; some essence of the quagga had been left in the mare.[44]

Darwin's challenge was to find a connection between his hereditary puzzles (he cited plenty of others). *Variation*'s 800 pages of detailed cases finally concluded with his answer, the 'Provisional Hypothesis of Pangenesis'. He suggested that every part of an organism must 'throw off minute granules which are dispersed throughout the whole system; that these . . . multiply by self-division, and are ultimately developed into units like those from which they were originally derived'. He christened these granules 'gemmules', and argued that 'they are collected from all parts of the system to constitute the sexual elements'.[45]

Looked at from the perspective of modern genetics, Darwin's theory looks quaint, but it is important neither to dismiss it, nor – even less appropriately – to interpret gemmules as ancestors of modern genes. Darwin's idea grew out of a long tradition of thinking about 'generation', which linked together all sorts of things – such as inheritance, development and healing – that we now see as separate. To understand why he thought pangenesis was plausible, we need to understand how it seemed to solve his various problems. In the case of reversion, for example, he argued that the ancestral gemmules for horns lie dormant even in hornless sheep, waiting to reappear. And because gemmules were supposedly dispersed throughout the organism, they explained the regeneration of missing limbs; the necessary 'leg gemmules' were circulating elsewhere in the salamander's body.

Darwin thought of his gemmules almost as tiny creatures; they multiplied within an organism and then combined to produce the offspring's characters. He argued that if 'unmodified and undeteriorated gemmules' were present in two parents, they 'would be especially apt to combine'.[46] This suggests some kind of competition among the gemmules; the 'pure' un-hybridized gemmules are described as 'undeteriorated', and so they predominate in the offspring. Darwin presumed that in such a competition, the male elements would be stronger, which explained how the quagga's gemmules had made their impact on Lord Morton's mare. But the idea of competition also seemed to explain how new, improved characteristics could survive and spread.

Gemmules had a final trick up their sleeve. Darwin – like most of his contemporaries – thought that there was good evidence that characteristics which an organism acquired during its lifetime could be inherited – and that they often were. He was particularly interested in the idea that the way an organism used – or neglected – one of its features might be passed on. This idea was an ancient one, but in Darwin's day was mainly associated with the French naturalist Jean-Baptiste Lamarck, the man who coined the word 'biology'. Lamarck, who was an evolutionist long before Darwin, argued that organisms that ran became faster because they exercised and strengthened their running muscles; he also believed that evolution, which he called transmutation, occurred because creatures could pass these advantages on to their offspring, so that gradually the whole species got faster. Although few people in Britain had ever heard of Lamarck, much less read him, the widespread belief that you could 'catch' something like alcoholism and pass it on to your children was a comparable idea. Darwin certainly believed that some acquired characteristics were inherited, but he wondered 'how can the use or disuse of a particular limb or the brain affect a small aggregate of reproductive cells, seated in a distant part of the body . . . ?' In other words, precisely what did all that running do to your eggs and sperm that allowed you to pass on your acquired talent? Pangenesis was intended to explain this too: since gemmules were produced throughout an organism's life, an altered organ would produce altered gemmules; faster legs produced 'faster' gemmules.[47]

Darwin tried to explain all these issues – from reversion to the case of Lord Morton's mare – using a persuasive analogy similar to those he had used so successfully in the *Origin*:

> Each animal and plant may be compared with a bed of soil full of seeds, some of which will soon germinate, some lie dormant for a period, whilst others perish. When we hear it said that a man carries in his constitution the seeds of an inherited disease, there is much truth in the expression. No other attempt, as far as I am aware, has been made, imperfect as this confessedly is, to connect under one point of view these several grand classes of facts. An organic being is a microcosm – a little universe, formed of a host of self-propagating organisms, inconceivably minute and numerous as the stars in heaven.[48]

Galton read *Variation* as he was putting the finishing touches to his own *Hereditary Genius*, and he added an extra chapter endorsing pangenesis. He was particularly excited by the thought that gemmules were discrete entities – thus ensuring that traits were passed on intact without blending. He even argued that with characteristics such as human skin colour, where blending seems indisputable, the intermediate colours were in fact a very fine mosaic of the two distinct parental colours. Galton also commented that because gemmules were discrete entities, 'the doctrine of Pangenesis gives excellent materials for mathematical formulae'.[49] But although pangenesis sounded exciting in theory, Galton – like many of *Variation*'s readers – remained worried by the lack of evidence for it. Neither Darwin nor anyone else had ever observed a gemmule, so Galton decided to prove that they existed.

In the blood

Darwin had referred to gemmules being 'dispersed throughout the whole system' before collecting in the sex organs, so Galton expected to find them circulating in the blood. Blood was proverbially synonymous with breeding: after all, what made a horse or an aristocrat superior? – Its bloodline. Galton decided to test his cousin's idea by seeing whether this time-honoured conceit was literally true: could blood transfusions be used to pass on hereditary characteristics? He started experimenting with rabbits, transfusing blood from black and white ones into pure-breeding silver-greys, in the hope that they would then produce some piebald offspring. He chose rabbits because they take only a few months to reach sexual maturity, so any results would show up fairly quickly. Galton consulted Darwin about which breeds to use and kept him updated on the progress, or – as it turned out – the lack of progress. All Galton could report was 'No good news', despite Darwin's 'valuable advice & so much encouragement'.[50] Galton tried various techniques and different amounts of blood, but among 124 offspring from twenty-one litters, not one 'mongrel' appeared.

Disappointed, Galton concluded that pangenesis must be wrong. He published his results in the *Proceedings of the Royal Society*, bluntly concluding that 'the doctrine of Pangenesis, pure and simple, as I have interpreted it, is incorrect'.[51] Darwin was uncharacteristically angry and claimed Galton had misinterpreted his theory. He pointed out that he had 'not said one word about the blood', adding that it should be 'obvious that the presence of gemmules in the blood can form no necessary part of my hypothesis', since he had clearly claimed that pangenesis operated in organisms like plants, which do not have blood. Galton's conclusion must therefore be considered 'a little hasty'.[52]

Galton himself might well have felt annoyed, given that Darwin had never raised this objection in all the months they had been corresponding. But if he was, he kept his irritation to himself, telling Darwin he was 'grieved beyond measure, to learn that I have misrepresented your doctrine'.[53] The two collaborated on further unsuccessful rabbit experiments for another eighteen months, but by the end of 1872 they were still getting nowhere; Galton wrote to Darwin that 'the experiments have, I quite agree, been carried on long enough'.[54]

Despite this setback, Galton retained considerable faith in pangenesis and published his own, modified version of the theory. He extended Darwin's approach to explaining the phenomenon of reversion by assuming that there were two kinds of gemmules: those that lay dormant, which Galton christened 'latent' and those that were expressed in the individual, which he called 'patent'. Each organism contained a mixture of latent and patent gemmules, derived from different generations of their ancestors. He compared this mixture to a parliament, consisting of 'representatives from various constituencies', while acknowledging that his analogy 'does not tell us how many candidates there are usually for each seat, nor whether the same person is eligible for, or may represent at the same time, more than one place'. Galton hoped that his readers would find 'no difficulty' in seeing that the particular set of characters found in any organism were 'the result of election'.[55] Yet Darwin, like most of Galton's other readers, found this analogy entirely incomprehensible.

Galton's logic becomes a little clearer once we understand his goal 'of applying these considerations to the intellectual and moral gifts of the human race'. He was not especially concerned with exactly where his parliament of gemmules was situated nor how they were transferred from generation to generation. He was more worried by the argument that children were sometimes much more (or less) intelligent than either of their parents, a fact that his critics offered as 'proof that intellectual and moral gifts are not strictly transmitted by inheritance'. If there was no correlation between parental intelligence and that of their offspring, it was a fact that killed eugenics stone dead, so Galton's analogy was intended to show that in species and varieties that had been heavily interbred, like most domesticated animals, each individual was a random selection of ancestral gemmules. He asked his readers to imagine 'an urn containing a great number of balls, marked in various ways'; these represented the traits, latent and patent, that made up an individual. When two animals mated, the result was like a handful of these balls being 'drawn out . . . at random as a sample'. He wanted people to think of children not so much as a mixture of their parents' characteristics, but as a mixture of the characteristics of their parents, grandparents and even more distant ancestors. Your own parents might not be geniuses, but if – like Galton – your grandfather had been, then the latter's characteristics could resurface, which explained your exceptional talents. Given that, as Galton put it, the human species 'is more mongrelised than that of any other domesticated animal', an adequate study of the relationships between children and their forebears would eventually 'prove that intellectual and moral gifts are as strictly matters of inheritance as any purely physical qualities'.[56]

Unlike Darwin, Galton imagined the mixtures of gemmules that made up an animal, plant or person as resulting from chance. As he had written when he first read *Variation*, pangenesis 'gives excellent materials for mathematical formulae': he was thinking in particular of the then novel mathematics of statistics and probability, in which he was an expert. Darwin, by contrast, was a mathematical lightweight, unable to make sense of even the simplest equations. The other distinction between Darwin's theory and Galton's was that Darwin hoped to explain how the acquired effect of, for example, exercising a muscle could be passed on. Galton, however, was adamant that the 'effects of the use and disuse of limbs, and those of habit, are transmitted to posterity in only a very slight degree'.[57] Again, this was essential to his eugenic argument – if improved bodies could be inherited, why not improved minds? If Lamarck and Darwin were right, better schools – and gymnasia – might do more to improve humanity than Galton's breeding schemes could.

Galton's conviction that acquired characteristics could not be inherited led him to anticipate what would become one of the central dogmas of later biological theory: in 1883, the German biologist August Weismann announced that although eggs and sperm contained the material that built an animal, the hereditary influence of the sex cells could only flow one way, into the body cells: any subsequent changes to the animal's body – whether they were brought about by injury or exercise – were not incorporated into its sperm or ova and therefore could not be transmitted to the next generation. As we will see in later chapters, this doctrine, which became known as the 'continuity of the germ-plasm' (that being Weismann's term for the heritable material) was to become vital to the way genetics developed in the twentieth century. Weismann did not know of Galton's papers when he first formulated this idea, but later acknowledged his English predecessor. It is noticeable, however, that Galton had no theory of biological inheritance – his theory rested on nothing more solid than his political conviction that nature must always dominate nurture; that in itself does not of course mean he was wrong.

Despite the bafflement that greeted Galton's announcement of his modified theory of inheritance, he remained convinced that it must be largely right. But he still faced the problem of gathering sufficient data to prove it. A few years after giving up his rabbit experiments, he tried again, this time with sweet peas (Lathyrus). These were common flowers that rarely cross-fertilized (which made it easier to keep different strains apart), and they were hardy and prolific. Nonetheless, Galton's first experimental crop failed, so he sent packets of seeds to friends and relations all over the country, including to Darwin, with elaborate instructions on exactly how to plant and harvest them. The mature plants were to be returned to Galton, who counted and weighed their seeds to demonstrate that seed weight was almost entirely a matter of heredity – the heaviest seeds produced plants with the heaviest seeds. However, these results were not as clear-cut as they first appeared, not least because Galton had sorted the seeds he had initially planted by size, not by lineage, and so had intermingled seeds from many varieties. It was therefore possible that the eventual differences in seed sizes were simply a product of the different environments in which the plants had been grown. However, none of this really bothered Galton; as he himself observed, 'It was anthropological evidence that I desired, caring only for the seeds as a means of throwing light on heredity in man.'[58]

Sweet peas seem to have proven as frustrating as rabbits as a research tool, and in any case it was never going to be easy to prove that either sweet peas or rabbits possessed 'intellectual and moral gifts', much less that these were inherited like physical characteristics. If Galton were to prove this central conviction, upon which his entire eugenic philosophy rested, he needed evidence from humans. In 1874, he successfully proposed to the Council of the Anthropological Institute that they collect height and weight data from students at a cross-section of English schools. These kinds of measurements might seem irrelevant to the issue of mental and moral faculties, but Galton was a firm believer in the adage *mens sana in corpore sano* ('a healthy mind in a healthy body'), and assumed the two were invariably linked – and linked by heredity, not by a shared healthy environment. This was a widely held view (more than one observer commented that it could well have served as the Health Exhibition's motto). Physical and mental vigour were always linked in Galton's mind; as he noted: 'A collection of living magnates in various branches of intellectual achievement is always a feast to my eyes; being, as they are, such massive, vigorous, capable-looking animals.'[59]

Unfortunately measuring schoolboys did not allow comparisons between generations, so could shed little light on heredity. So in March 1882, Galton began to call for the foundation of national human-measuring, or 'anthropometric', laboratories, 'where a man may from time to time get himself and his children weighed, measured, and rightly photographed, and have each of their bodily faculties tested, by the best methods known to modern science'. These would not only assess physical qualities, but test mental abilities such as memory and hand-eye coordination. The measurements would be recorded alongside full medical histories and photographs, and information on the 'birthplace and residence, whether in town or country, both of the person and his parents'. Galton recognized that people would need to be motivated to have themselves measured, so he proposed that the labs function as a sort of careers advice centre. Since his work had proved that nature trumps nurture: 'It follows . . . that it is highly desirable to give more attention than has been customary hitherto to investigate and define the capacities of each individual.' Such investigations would allow us 'to forecast what the man is really fit for, and what he may undertake with the least risk of disappointment'. He also argued that the labs should be welcomed by doctors who, instead of keeping anthropometric equipment in their own consulting rooms, 'could send their patients to be examined in any way they wished, whenever they thought it desirable to do so'. Galton added that 'the laboratories would be of the same convenience to them that the Kew Observatory is to physicists, who can send their delicate instruments there to have their errors ascertained'.[60] This is a revealing analogy, since it made human patients equivalent to the 'delicate instruments' used by physicists: it suggests that, for Galton, his volunteers were merely a means to an end.

Evolving the average man

Despite Galton's enthusiasm, no one seemed interested in founding the labs he was proposing, so he decided to do it himself, convinced that – whatever rabbits and sweet peas did or did not prove – it was only direct evidence from humans that would make his case. However, Galton soon found that his earlier difficulties were nothing compared with those posed by humans. 'The stupidity and wrong-headedness of many men and women' was, he decided, 'so great as to be scarcely credible'. For example, his lab included an instrument for measuring strength, which he intended to be as simple as possible: it was a tough wooden rod in a spring-loaded tube – all the subject had to do was punch it and see how far into the tube the rod went. Galton 'found no difficulty whatever in testing myself with it', but within a few weeks of opening of the Health Exhibition, 'a man had punched it so much on one side' that he broke the rod. Galton replaced it with a stronger, oak rod, 'but this too was broken, and some wrists sprained'.[61] He commented that 'Notwithstanding the simplicity of the test, a large proportion of persons bungled absurdly over it . . . and often broke the rod or hurt their knuckles.'[62]

However, broken equipment and bruised fists were only the beginning of Galton's troubles. As the crowds flowed through the lab's doors, he found it was essential 'to keep parents and their children apart', because 'the old did not like to be outdone by the young and insisted on repeated trials', which wasted his precious time. His rabbits may have not have given him the results he wanted, but at least he did not have to stop them showing off. Nor did rabbits need to have the experiment explained to them, whereas human subjects needed time-consuming explanations and demonstrations; Galton bemoaned the 'waste of the attendant's time in idly watching examinees puzzling over tests'. A partial solution was to take people through the lab in pairs, so 'that one explanation and illustration might suffice for both'. This had the added benefit that 'the promptest minded man of the two was usually the one who presented himself first, [so] the less prompt man had the advantage of seeing his companion perform the test before he was called upon to do so himself'.[63] Even so, it all took time and the lab was only able to measure around 90 people a day.

When George Sala, who had a reputation for being fond of the occasional drink, had described the Health Exhibition, he was delighted to report that alcohol was widely available, in every form from mint juleps and arrack punch to the fermented mare's milk known as *kumiss*, produced on the premises by horses and attendants from the steppes of Russia. Sala congratulated the organizers for treating the public like adults, allowing them to 'eat and drink what they like. In the gardens they make smoke. They are not turned out of the building at an unduly early hour. They may stay there, if they will, until ten o'clock at night.'[64] This liberty may have pleased Sala, but it caused Galton some annoyance, since 'on some few occasions rough persons entered the laboratory who were apparently not altogether sober'.[65]

Despite the occasional inebriate and all the other difficulties, Galton got enough results to publish several papers. In the course of analysing his measurements, he developed or improved basic statistical tools that are still in use today, including percentiles (a method of estimating the proportion of the data that should fall above and below a particular value) and correlation coefficients (a measure of the degree to which two variables are related). However, it was not how he measured but what he measured that caught the public's imagination. In the course of comparing the strength of the male and female visitors to the lab, Galton observed that 'very powerful women exist', who could squeeze with a force of 86lb – equal to that of many men. But, he added, 'happily perhaps for the repose of the other sex, such gifted women are rare', so rare that 'the population of England hardly contains enough material to form even a few regiments of efficient Amazons'.[66] These throwaway remarks caught the sharp eye of the satirical weekly *Punch*, which published a poem observing that a 'Maiden of the mighty muscles / Famous in all manly tussles', would be able to keep her husband on the straight and narrow:

> That if in the dim hereafter
> Any husband should play tricks
> You would with derisive laughter,
> Give a 'Squeeze of 86'.[67]

Ignoring such jocularity, Galton ploughed on both with the analysis of his results and with promoting the benefits and usefulness of laboratories such as his. While he recognized that the results lacked some precision, he argued that they were 'of considerable importance' because he had recorded both whether his subjects had been born in the hazardous city or the healthy country, and where they now lived, and so, he thought, afforded 'materials for testing the relations between various bodily faculties and the influences of occupation and birthplace'.[68]

However, perhaps the most obvious problem with Galton's measurements was that he had no way of assessing the very thing that interested him most, the mental qualities of his visitors. While planning the lab, he had sought advice from various experts about what kind of equipment might be used for this purpose, but no one had any useful suggestions. One option might have been to measure people's heads, as a possible indicator of the power of their brains, but as Galton explained, 'it would be troublesome to perform on most women on account of their bonnets, and the bulk of their hair, and would lead to objections and difficulties'.[69]

As with the earlier schoolboy measurements, the physical tests Galton administered were intended to measure overall health while others, such as those for sight, hearing and ability to distinguish between similar colours, were intended to provide some measure of intelligence, but not in the way one might expect. It was widely believed that animals had sharper vision and more acute hearing than people – good scores in these tests were thus a measure of how close to the animal state, how primitive or savage, the visitor was. Those possessed of the supposedly uniquely human virtue of intelligence should score badly in such trials. Yet even if these measurements did indeed shed light on intelligence, they shed little on the question of if, much less *how*, it was inherited. Galton had collected brief pedigrees of visitors to the lab, so that he could analyse the relationship between the heights of parents and children, but while the data was interesting, it provided no conclusive evidence of the inheritance of mental qualities.

Galton tried again. He attempted to persuade doctors to gather data on hereditary diseases that ran in families, even offering a £500 prize for the best analysis, but got no takers. In 1884 he made two attempts at gathering the same data directly from the public. He offered another £500 prize to whoever did the best job of completing a fifty-page questionnaire, *The Record of Family Faculties*, on their own family's heredity and health, but he only got

150 replies, for which he gave out a few small prizes. He also devised, edited and arranged to publish the *Life-History Album*, a prototype of something he hoped would eventually be presented to all new parents, so that they could keep a record of their children's development. The children themselves could then take over and complete it for their children, and so on. But even if the albums had met with a more enthusiastic response, it would have taken several generations to accumulate sufficient data for analysis.

The insoluble difficulty Galton confronted was not so much that mental abilities were so hard to measure, tough though that proved, it was that *Homo sapiens* breeds so slowly. Even a long-lived specimen like Galton – who was almost ninety when he died – could not hope to trace the breeding patterns of his own species for long enough to get reliable data. And, if that was not bad enough, humans make such recalcitrant laboratory animals; those very mental qualities he was so interested in allowed his specimens to make their own decisions: simply persuading them into the lab to be measured in the first place was hard work.

In 1890, Galton published his latest attempt at persuasion, a little pamphlet extolling the benefits of anthropometrics, which was sold for just 3d. in his second lab, at the Science Museum. The first chapter asked, 'Why do we measure humans?' and Galton offered various answers, similar to the ones he had given in his earlier article; the identification of aptitudes and talents, and perhaps spotting potential health issues that could be corrected. And, of course, he stressed the benefits to pure science. Yet anyone who bought the pamphlet, perhaps inspired by the thought that they were contributing to such noble goals, might well have been put off by its latter sections, in which Galton turned his attention to the issue of human variety. He admitted that he was really only interested in exceptional individuals, commenting that 'an average man is morally and intellectually an uninteresting being' and therefore 'of no direct help towards evolution, which appears to our dim vision to be the goal of all living existence'.[70] Galton's interest was in assisting exceptionally fine specimens of humanity, while aiming to eliminate the exceptionally poor ones. The average person played no part in this scheme, other than to be a 'sensitive instrument', the benchmark that defined who was exceptional. Small wonder that few of his contemporaries were excited by the prospect of having themselves measured.

Galton's contempt for the 'average man' may well have grown out of his frustration with the 'stupidity and wrong-headedness' of the crowds that had visited his lab, who wandered in drunk, wasted his time puzzling over simple tests or were incapable of completing them without breaking his equipment. How were these urban Calibans ever to be transformed into the 'prophets and high priests of civilization' or made over into 'massive, vigorous, capable-looking animals'? The answer, clearly, was that they were not: theirs were natures upon which nurture would never stick, so his priority was to help rid the population of such worthless specimens. More than a decade before the Anthropometric Laboratory opened, Galton wrote that he looked forward to the day when 'the non-gifted would begin to decay out of the land' just as 'inferior races always disappear before superior ones'. This shift would, he assumed, be 'effected with little severity', since the 'gifted class' would treat their inferiors 'with all kindness'. But only, he added menacingly, 'so long as they maintained celibacy. But if these continued to procreate children, inferior in moral, intellectual and physical qualities, it is easy to believe the time may come when such persons would be considered as enemies to the State, and to have forfeited all claims to kindness.'[71] This appalling prophecy was one of the few Galton made that came true: as we shall see in later chapters, the early twentieth century witnessed a revival of interest in his ideas. In countries as different as Sweden and the USA, tens of thousands of people were compulsorily sterilized 'in the name of eugenics', but the worst horror came in Nazi Germany, where Galton's ideas inspired the policy of sterilizing, and eventually of exterminating, the unfit and those from 'inferior races'.

Galton died in 1911, too early to observe the horrors his theory would inspire, so we will never know how he would have responded to them. What is certain is that his faith in his ideas remained undiminished at his death: he left £45,000 (the equivalent of over £3 million today) to found a national eugenics laboratory and endow a professorship of eugenics. He could afford to leave such a large sum because, ironically given his lifelong preoccupation with inheritance, Francis Galton died childless. His ideas were to be his only children.

Hieracium auricula

Chapter 4
Hieracium auricula:
What Mendel did next

A tiny seed, weighing only a fraction of a gram, floats in the air. It is suspended from a little parachute of fluffy bristles; the slightest puff of wind will carry it away from its parent plant to find somewhere new to grow. This is the seed of *Hieracium auricula*, commonly known as the pale hawkweed. Like most hawkweeds, *H. auricula* has small flowers, pale yellow in this species, and looks rather like its close relative, the dandelion. Both dandelions and hawkweeds are considered a nuisance by many humans, especially gardeners. They will soon take over a lawn if not dealt with; that tiny parachute, the pappus, is an adaptation that helps the seeds spread. Once they land, they germinate quickly and are soon producing new flowers and seeds, enabling them to spread still further.

Wind-blown seeds are not the only characteristics that make hawkweeds and dandelions efficient weeds. Both have flattened leaves that hug the ground, so that if a passing animal eats the flowers before they can set seed, the plant survives to produce new flowers. Happily for the flowers, these leaves also make them resistant to another major predator: gardeners armed with lawn-mowers. When a lawn is mown, the plant's leaves are flattened further but not destroyed; soon, new stalks and flowers will appear. Getting rid of hawkweeds or dandelions requires pulling them up, which reveals another of the plant's survival mechanisms – a long, strong single root, like a miniature carrot, which is called a tap-root. Unless that is pulled up, destroying the flowers and leaves will not get rid of the plant; the root stores enough energy to grow new ones.

Hawkweeds are not just a nuisance to gardeners. Although native to Europe, they have now spread across much of the United States, where they have become invasive weeds. *Hieracium auricula* is known in Montana as meadow hawkweed and is classified by the state government as a 'Category 2' noxious, or harmful, weed; in Oregon, where the same species is usually called yellow hawkweed, it is also a designated weed, and in Washington state it is a 'Class A noxious weed' – the locals hate it so much that it has become known as yellow devil hawkweed. Hawkweeds are also a problem in Canada and in New Zealand, where a closely related species, *Hieracium pilosella*, or mouse-ear hawkweed, has been so successful at invading pasture land that it forms dense mats of leaves which exclude other kinds of vegetation. Mouse-ear hawkweed is edible, at least if you are a sheep (which the majority of New Zealand's mammalian inhabitants are), but it is not as nutritious as the plants it displaces and so is reducing the productivity of the country's pastures.

Obviously, even the lightest wind-blown seeds could not have enabled hawkweeds to get from Europe to New Zealand; it was, of course, the humans who are now struggling to control them who brought the plants to these new territories in the first place.

In Britain, the medicinal properties of hawkweeds were first described in the mid-sixteenth century by William Turner, sometimes referred to as the father of English botany. His *New Herball* became famous because the plant descriptions of his earlier books were complemented by superb woodcut illustrations. Turner informed his readers that 'The nature of Hawke wede is to coule [cool] and partly to binde' (that is, to cure diarrhoea).

Turner's work was copied and improved on by later English herbalists and physicians, such as John Gerard, John Parkinson and Nicholas Culpeper (or Culpepper), over the late sixteenth and seventeenth centuries. Culpeper described hawkweeds in his book *The English Physician* (1652), better known as *Culpeper's Herbal*, which included a detailed list of Britain's native medicinal plants and the diseases they would cure. Culpeper noted that hawkweeds 'hath many large hairy leaves lying on the ground', which looked like those of a dandelion. He also described how its 'small brownish seeds' were 'blown away with the wind' as the plant 'flowreth & flies away in the Sumer Months'. Drawing on the work of the ancient Greek botanist Dioscorides, Culpeper gave the plant's 'Vertues and Use', which included the fact that the juice, if taken with a little wine, 'helpeth digestion' and thus is good for removing 'crudities abiding in the stomack'. It also 'helpeth the difficulty of making Water'. And when applied to the outside of the body 'it is singular good for all the defects and diseases of the eyes, used with some womens Milke'.[1]

It was their supposed efficacy in treating eye diseases that gave the hawkweeds their name. The Roman historian Pliny the Elder, writing nearly 2,000 years ago, recorded the plant in his *Natural History*, his fabulous compendium of fact, myth, observation and hearsay on every aspect of the natural world. Pliny mentions a kind of lettuce 'with round, short leaves' that was 'called by some *hieracion* (hawkweed), since hawks, by tearing it open and wetting their eyes with the juice, dispel poor vision when they have become conscious of it'. Following the lead of the short-sighted hawks, humans had investigated the plant and found that 'With women's milk it heals all eye-diseases'.[2] Medieval falconers used the plant to treat birds that seemed to be becoming short-sighted.

Culpeper's distinguished predecessor, John Parkinson (who first described the passionflower in English), also produced an immense herbal, which he called the *Theatrum Botanicum* (1640). Among almost 4,000 plants included in its 1,755 folio-sized pages was *Hieracium pilosella*, which he called 'mouse-ear', suggesting that you should give it to your horse before visiting the blacksmith so that it 'shall not be hurt by the smith that shooeth him'. He also observed that shepherds were careful not to let their sheep feed in pastures where the plant was growing 'lest they grow sicke and leane and die quickly after' (which perhaps underlies the New Zealanders' hostility to the plant).

The first European settlers in America brought books on herbal medicine with them or printed their own; one of the first medical books to be produced in America was an edition of *Culpeper's Herbal* (1708). But settlers soon discovered that many European medicinal plants were not to be found in the new world, so they imported the plants' seeds. Concerned, no doubt, for their horses, eyes and digestions, humans brought hawkweeds to the Americas, among dozens of foreign plants also imported for their medicinal uses. (Many more came accidentally after their seeds arrived mixed in with newly disembarked animal's fodder.) Dandelions were spread in the same way as hawkweeds: the scientific name of the common dandelion is *Taraxacum officinale*; 'officinale' comes from the Latin *officina*, meaning a shop, because dandelions were sold for medicinal purposes (many medicinal plant species have this same 'second', or specific, name for this reason). In America, Europe and the countries Europeans colonized, plants were the source of most medicines until well into the nineteenth century; in 1881, the *American Journal of Pharmacy* noted that *Hieracium venosum* (rattlesnake weed) could be used to treat tuberculosis; 'at least', the *Journal* continued, 'it seems to have a well-deserved reputation for that disease among cattle'.[3] Hieracium seeds for medicinal use can still be bought from herbal medicine sites on the internet, although most sites are responsible enough not to ship the seeds to Oregon, which has an anti-hawkweed quarantine.

With human aid, hawkweeds spread around the world, but it was in their native Europe that they first attracted the attention of botanists. As Culpeper had noted, 'there are many kinds of them', thousands of species have been described since his day, and it remains uncertain exactly how many species there are, partly because classifying living things is a complex business. In some cases there are very clear and obvious differences between species: no one could confuse a hawk with a hawkweed, any more than they could confuse one with a handsaw, but distinguishing one kind of hawkweed from another is much harder. Hawkweeds share this property of being hard to classify with plants such as brambles (the genus Rubus) and dandelions (Taraxacum). The fact that these groups are both difficult to classify and also pernicious weeds is, as we shall see, not a coincidence.

Lumpers and splitters

By the mid-nineteenth century, hawkweeds, brambles and dandelions were at the centre of a botanical war in which 'lumpers' faced off against 'splitters'. These two factions brought two very different philosophies to bear on one of the most contentious scientific subjects of the day, the classification of life. In essence, they disagreed about how many kinds of living things there were in the world, an issue that was fundamental both to important scientific and religious questions.

Beginning with the discovery of America, European knowledge of the incredible diversity of the world's animals and plants had been growing rapidly – and at an increasingly rapid pace. Initially, Europeans tried to fit the plants of the New World into the categories they had inherited from the ancient Greek authorities. For centuries, naturalists had been following in the tradition of authors like Pliny, compiling and writing commentary on ancient wisdom, but from the Renaissance onward Europeans were forced to recognize that there were more plants and animals in the world than even the wisest Greeks had dreamed of. These new plants and animals needed new names and new classifications. Within 100 years of Columbus's arrival in America, the Cambridge Professor of Botany, John Ray, observed that while the ancient Greek botanist Theophrastus had recorded just 500 species of plants, his own *Historia Plantarum Generalis* contained 17,000 species.

It was largely because of this massive expansion of knowledge that – in the century after Ray's book appeared – the Swedish naturalist Linnaeus carried out his massive reform of classification. Part of the trouble was that naturalists, botanists, farmers and florists all gave plants their own local names: *H. auricula* is not merely known as pale hawkweed, yellow hawkweed and yellow devil hawkweed, it is also known as kingdevil hawkweed and as the smooth (or in Connecticut, 'smoothish') hawkweed. And it has other names in Canada and New Zealand, as well as obviously many more in many different languages across the world. Even worse, different species may have the same common name in different countries. This confusion of common names was one of the reasons Linnaeus introduced standardized scientific names: he named the hawkweed genus Hieracium in his *Species Plantarum* ('Species of Plants', 1753) identifying and naming almost thirty different species, from *H. alpinum* to *H. venosum*. Today, the *Index Kewensis*, one of the most authoritative databases of botanical names, lists over 11,000 species of Hieracium. But in the intervening years, names have come and gone.

The almost endless details of the history of biological naming are not important to our story; suffice it to say that the proliferation of names and the renaming of species led to chaos. When naturalists wrote to each other discussing, comparing or exchanging specimens they – quite literally – did not know what they were talking about. One naturalist would decide that a particular plant was so different from those already named that it had to be considered a new species, and so it had to be given its own name. A second botanist would find the differences less significant than the similarities and decide that the plant merely represented a variety, and thus did not deserve a new name. Meanwhile a third might decide that both the similarities and differences were significant, and so classify the plant as a subspecies, which meant adding a third name to go with the other two; as recently as 1999, Linnaeus's original *H. alpinum* gained a new subspecies, *Hieracium alpinum augusti-bayeri*, discovered in northern Romania.[4] And, if that were not sufficient confusion, some twentieth-century botanists have proposed moving many species out of Hieracium into the genus Pilosella.

Back in the nineteenth century, some naturalists decided that this proliferation had gone too far: they accused those who continued to name new species of focusing too narrowly on insignificant differences between plants, of splitting hairs, and thence of being 'splitters'. The outraged splitters fought back, dubbing their opponents 'lumpers', for wanting to lump together as a single species plants that were obviously not the same. The first recorded use of these terms is in an 1857 letter of Charles Darwin's, where he told his friend Joseph Hooker that 'It is good to have hair-splitters and lumpers.'[5] Hooker would undoubtedly have disagreed, but was too busy to respond immediately: his wife Frances gave birth to their fourth child, Marie Elizabeth, the week Darwin's letter arrived. However, a couple of years earlier, Hooker had published his views on the subject, arguing that it was better to 'keep two or more doubtful species as one', and that by doing so 'we shall avoid the greater evil' – the endless proliferation of species. The 'hair-splitters' who maintained dubious species by preserving their separate names were simply causing chaos; if his readers doubted him, Hooker invited them to 'witness the state of the British Flora with regard to Willows, Brambles, and Roses'.[6] Willows and brambles, like hawkweeds, are notoriously difficult to separate into clearly defined species.

The question of exactly how many species of a particular plant there are might seem trivial, especially when the plant in question is a worthless weed, but to botanists it was an absorbing problem. Classification was much more than a matter of mere list-making: it touched on one of the biggest issues in nineteenth-century science: what were species and where did they come from? Like most of his contemporaries, Linnaeus had been convinced that species 'reckon the origin of their stock in the first instance from the veritable hand of the Almighty Creator', and that God, 'when He created Species, imposed on his creations an eternal law of reproduction and multiplication within the limits of their proper kinds'.[7] Only God could create a species, and once he had done so, it could not change; yet within twenty years of making this unambiguous assertion, Linnaeus was not so sure. His uncertainty was prompted by hybridization. His garden at Uppsala, like other botanic gardens across Europe, contained plants from all over the world. Sometimes pollen from one species landed on plants of another species and – very occasionally – these accidents created new hybrid varieties. These unplanned experiments led curious gardeners to start making deliberate ones, in an effort to produce attractive or productive new varieties. Most of these crosses failed: either the offspring were sterile or the hybrids quickly reverted to the parental type, but now and again a hybrid appeared that seemed to breed true. Confronted with examples in his own garden, Linnaeus had to admit that these stable hybrids 'if not admitted as new species, are at least permanent varieties'.[8]

In 1759, a few years after Linnaeus had made his concession, the Imperial Academy of Science in St Petersburg offered a prize of 50 ducats (well over £5,000 in modern British money) for an essay that would finally settle the old question of whether or not plants really had separate sexes. Linnaeus entered the competition and won it. His essay cited various examples of plants that did not set seed unless both male and female plants were present and argued that hybrid plants were proof of plant sexuality, since the characters of both parents were combined in the hybrid offspring. But Linnaeus clinched his argument by sending with his essay the seeds of a hybrid goatsbeard (Tragopogon), a kind of edible plant, also known as salsify, that is related to hawkweeds and dandelions. Linnaeus had crossed two kinds of goatsbeard in his garden and found that his hybrid form bred true. It therefore counted, in his view, as a new species and he named it *Tragopogon hybridum*.

Linnaeus's prize-winning essay concluded that 'It is impossible to doubt that there are new species produced by hybrid generation' and that where there were 'many species of plants in the same Genus' they 'have arisen by this hybrid generation'.[9] But he was still convinced that God had originally created all living things, so he proposed that an original set of parent plants had been directly created and that the profusion of families, genera, species and varieties were produced by mixing together God's original types. Linnaeus did not believe new types of plants could have arisen in any other way, nor could humans make new species at will; most human-made hybrids would, he argued, prove infertile or revert to their parental types. Nonetheless, he did challenge the long-standing view that *all* hybrids were necessarily sterile.

The apparently natural production of hybrids was what made

groups like hawkweeds so complicated, and which attracted the attention of some botanists. The nightmarish difficulty of classifying them suggested that in these groups of plants, the boundaries between species and varieties were blurred; some took this messiness as evidence for theories of transmutation, or evolution, arguing that plants like hawkweeds were in the process of evolving into new species. Furthermore, studying such groups might produce evidence as to how evolution worked: what was the mechanism that changed one species into another? Others disagreed profoundly, insisting that God had created all species exactly as we see them now, and that species had not – indeed, they could not – change. To suggest otherwise was blasphemous; if humans could not classify hawkweeds, it was because they had not yet understood the perfect plan of God's creation.

As these debates raged, the hawkweed seeds drifted silently on the wind. Some came to rest in a priory garden in the town of Brünn (modern Brno), where they caught the attention of a man called Gregor Mendel.

Mendel was obscure in his lifetime, but today most people are aware that he was a simple, uneducated Austrian monk who, while playing around with pea plants in his garden, discovered the basic laws of modern genetics. Yet his breakthrough was ignored, partly because he was cut off from the scientific world of his day, but also because the one famous scientist he contacted, the botanist Carl von Nägeli, sent Mendel off on a wild goose chase to investigate the genetics of hawkweeds. Nägeli may even have done this deliberately: jealous of his younger rival's brilliance, he set poor, innocent Mendel to work on a famously intractable group of plants, confident that the monk would never be able to sort them out. Frustrated by his failure, Mendel died in heartbroken obscurity. One final ironic twist to the story is added by the fact that Mendel supposedly sent a copy of his paper on peas to Charles Darwin, who never read it. Darwin found German difficult, so Mendel's paper survives alongside Darwin's other papers in Cambridge's University Library, its pages still uncut. If Darwin had only known what it contained, he would doubtless have abandoned his pangenesis theory, adopted Mendel's, and saved the biological world from having to rediscover Mendel in the twentieth century and re-establish the science of genetics.

Unfortunately, almost every word of the previous paragraph is incorrect. Strictly speaking Mendel was not Austrian, he was a German-speaking Moravian; Moravia was then a province of the Austro-Hungarian empire and is now part of the Czech Republic. (Nor was he in fact a monk; the Augustinian order, to which Mendel belonged, are friars.) Far from being a simple, uneducated man, Mendel had studied both biology and mathematical physics at the University of Vienna, where he had been taught by some of the best-known men of science of the day. Nor did he discover anything by accident; he performed carefully planned and well-designed experiments. The scientific society he belonged to had many distinguished members and its journal was widely read, so his work was not entirely ignored – although there are good reasons why it did not make quite the impact he hoped it would. (Nor is it strictly true to say that Mendel's work was 'rediscovered' in the twentieth century, but that is a story for the next chapter.) And Mendel never sent a copy of his paper to Darwin; at least, no such copy exists and there is no record of it ever having existed. But even if he had done, Darwin is unlikely to have found it of much interest, as we shall see. Finally, the story of Nägeli and the hawkweeds is also entirely inaccurate; the actual history helps us to understand what Mendel was doing and why, and – most surprising of all – it shows us why, despite the enormous importance of his work, he neither discovered modern genetics nor invented its basic laws.

In a Moravian monastery garden

Mendel's father, Anton, was a farmer and his mother, Rosine, was a gardener's daughter. He grew up in Moravia's rich farming country, amid vineyards and sheep, with a garden that had beehives tucked in-between the fruit trees. Anton had to spend half his week working for his landlord; the Austro-Hungarian empire, like much of Europe, was still run on essentially feudal principles. Young Johann Mendel (Gregor was his name in religion, once he became a friar) learned his first lessons about plants and plant-breeding by watching people like his parents, who had been struggling for generations to improve their crops and incomes. But by the time Mendel was born in 1822, the proverbs and folklore that had once guided farming communities were being rapidly displaced by the new scientific methods. Anton worked with the town's parish priest on a project to improve the yield and hardiness of fruit trees by grafting and breeding; together they produced nearly 3,000 trees which were distributed among the local farmers. Mendel's work is perhaps best understood as an attempt to carry on his father's efforts to make their land more fruitful.

As we have seen, humans had been improving crops and animals for thousands of years before Mendel's time. By the eighteenth century, breeders like Robert Bakewell knew how to produce bigger, fatter sheep; his fellow Englishman Thomas Knight applied similar techniques to plants. Knight was the first president of Britain's Horticultural Society and was one of the first to publish information about these new techniques. The first step was establishing which two individual plants had actually been crossed: instead of leaving this vital business to the wind, birds or bees, Knight investigated how to fertilize flowers artificially – pioneering the techniques Darwin would later use on his passion-flowers. Knight's hope was that new kinds of fruit trees could be created by crossing varieties that had desirable characteristics, but trees take so long to mature and bear fruit that it would take several lifetimes to discover if an experiment had worked. So he hit on the idea of first trying out his techniques on rapidly growing annual plants; after much consideration, he chose the common pea (*Pisum sativum*) for his experiments, 'not only because I could obtain many varieties of this plant, of different forms, sizes, and colours', but also because of 'the structure of its blossom', which prevented stray wind-blown pollen from getting into the flowers – much like Darwin's orchids. If a bee failed to visit the pea flower at exactly the right time, the flower's shape ensured that it would be self-fertilized, a fact which, as Knight commented, 'has rendered its varieties remarkably permanent'.[10] By simply netting the plants to keep insects away, Knight could fertilize the plants by hand, and so know which plants had been cross-fertilized by which.

Knight began his pea experiments in 1787 and a dozen years later was able to publish his results in the Royal Society's *Philosophical Transactions*. His paper was translated into German the following year and soon became well known among continental breeders. However, the anonymous German translator of Knight's paper added a footnote, observing that the techniques of artificial fertilization it described were already well known in German-speaking Europe thanks to the work of Joseph Gottlieb Kölreuter.

Kölreuter was the first naturalist to carry out systematic experiments on hybridization, partly – as we shall see – because he was interested in Linnaeus's question as to whether hybridization could create new species. But his plant-breeding research had a more pragmatic purpose; as Kölreuter wrote, he hoped that he 'might one day be lucky enough to produce hybrids of trees, the use of whose timber might have great economic effect', especially if, as he hoped, such hybrids might be faster growing, enabling them to reach maturity 'in half the time' it took their parent species.[11] This was what attracted the interest of people like Moravia's farmers, the promise of new, improved animals and plants, rather than resolving what was or was not a new species. In Moravia, this goal was furthered by Christian Carl André, who came to Brünn at the end of the eighteenth century to promote the natural sciences in the region.

André seems to have had an enthusiasm for founding scientific societies with unfeasibly long names. He began with the 'Moravian Society for the Improvement of Agriculture, Natural Science, and Knowledge of the Country' (which later changed its name, presumably to save time, to the Agricultural Society). He was intrigued by Bakewell's techniques and founded another society – the 'Association of Friends of, Experts on, and Supporters of Sheep Breeding' – to promote and develop scientific sheep-breeding, a major concern since Brünn was then the centre of the Hapsburg empire's textile industry. And when André heard about Knight's work with fruit trees, his first thought was to found a 'Pomological [fruit-scientific] and Oenological [wine-scientific] Association'. It also eventually changed its name – to the Pomological Association – and among its members was the newly appointed abbot of the Augustinian priory in Brünn, Franz Cyrill Napp, an enlightened man, dedicated to ensuring that the friars should provide both practical and spiritual guidance to the local people. With this in mind, he set up an experimental nursery garden; Mendel was one of the young friars who worked in it.

Despite his humble background, Mendel was well educated: unlike most peasant children, he had attended a *Gymnasium* (secondary school), which had a small natural history museum attached – founded at André's suggestion – that helped foster Mendel's interest in nature study. Unfortunately, while Mendel was still in his teens, his father was injured in an accident that left him unable to work, leaving the family unable to pay young Johann's school fees. From the age of sixteen he had to support himself by tutoring other pupils. He hoped to become a school-teacher, but since his family could not afford to send him to university, he entered the Augustinian priory of St Thomas as a novice in 1843.

Despite it not having been his first choice of career, the priory suited Mendel. Napp encouraged the friars to study science, especially agricultural and horticultural subjects. This was not purely for the benefit of the local people; the priory had considerable debts when Napp took over and he hoped that modernizing its farms and fields would help pay them off – like their neighbours, the friars bred sheep and sold the wool at a profit. As he worked to get the priory's finances back in order, Napp took a close interest in Bakewell and Knight's techniques and encouraged the friars to study them as well.

After a year's probation, Mendel began to study theology. After rising at 6 a.m. and attending Mass, he studied in the priory library, which contained scientific books as well as religious ones, or worked in its garden, learning how to hand-pollinate plants to create improved varieties. Many of his fellow friars shared his interests and he found himself in the midst of a stimulating community, full of lively conversation about science, its uses and its religious implications.

Mendel began to study for the priesthood in 1848, a year when revolutions briefly convulsed much of Europe. The nature of these revolts and the demands they made varied widely, but there were widespread calls for an end to the types of semi-feudal ties that bound people like Mendel's father. Another common claim was for improved education, to allow the newly emancipated peasants to join the modern world. In Brünn, Napp was a prominent supporter of these reforming demands; he demonstrated his solidarity by publicly saying Mass for students who had been killed in the fighting. Although the new Austro-Hungarian emperor, Franz Josef I, quickly emasculated the newly created parliament, effectively ending the revolt, the abolition of feudal labour and the educational reforms survived.

As the revolution fizzled out, Mendel completed his theological studies and became a priest, but his health was often poor and it was clear that he was ill-suited to regular parish duties. With Napp's support, he was appointed to a full-time teaching post at a *Gymnasium*. However, one of the reforms that had been introduced after 1848 was a new education act, which required all teachers to take university exams. Although Mendel was a gifted – and by now, experienced – teacher, he did not have the now vital qualification, so in 1850 the school's headmaster despatched him to the University of Vienna. Mendel failed, partly because his examiner seems to have been prejudiced against members of monastic orders working as teachers, but also because no one had coached him in how to prepare appropriately for university exams.

Fortunately, the educational reforms also had a positive effect on Mendel's life. The Hapsburg government had decided that the rapidly industrializing country needed new types of schools, technical schools, which would emphasize practical studies such as science, engineering and mathematics in order to equip students for the new world. With Abbot Napp's enthusiastic support, schools of this kind were established in Brünn, which was an increasingly industrialized city, but teachers for these modern subjects were in short supply. Despite his exam failure, Mendel seemed to possess talents his country would need, so one of his examiners in Vienna suggested he be given the opportunity for further study. Napp agreed, and the priory paid for Mendel to return to Vienna to continue his studies.

Back in Vienna, Mendel studied physics with Christian Doppler (after whom the Doppler effect is named, i.e. the way the frequency and wavelength of something like a sound appears to change when its source is moving relative to the listener), who emphasized the importance of designing elegant experiments and taught the most advanced mathematics of the time, statistics and probability. Although physics was Mendel's first love, he also studied chemistry, palaeontology and plant physiology. The latter was taught by Franz Unger, who shared Doppler's interest in experimental design; this emphasis on hands-on practical work was another legacy of the post-1848 educational reforms.

Unger was probably the most influential teacher Mendel had. He introduced his young student to the latest scientific ideas, especially to the then radical new cell theory. Back in 1663, the English natural philosopher Robert Hooke had looked at cork, a tree-bark, through an early microscope. He observed the regular, empty spaces in the cork, which reminded him of the rows of tiny rooms in which monks lived, so he dubbed them 'cells'. The name stuck, but it was only in the mid-nineteenth century, as better microscopes were mass-produced, thus becoming cheaper, that proper investigation of these cells began. By the time Mendel was studying at Vienna, using a microscope was a standard part of his course.

In 1838, two German naturalists, Theodor Schwann and Matthias Schleiden, had discussed cells over coffee. Schleiden described how every single plant cell he examined possessed a dark central core, a nucleus (it was first described and given this name by the English botanist, Robert Brown, after whom Brownian motion is named, i.e. the random movements of particles suspended in a fluid). As he listened, Schwann realized that he had observed something similar in the animal cells that he studied; a year later he published a book – *Mikroskopische Untersuchungen über die Übereinstimmung in der Struktur und dem Wachstum der Tiere und Pflanzen* ('Microscopical Researches on the Similarity in the Structure and Growth of Animals and Plants', 1839) – which became the founding document for an important new biological theory (despite his failure to mention Schleiden's contribution – or anyone else's). Schwann argued that cells are the basic building blocks, the atoms, of all living things – they are not only the simple units from which every physical structure is built, they also form the cogs, gears and engines that make organisms work. Every living process, from digestion and respiration to circulation and reproduction, depends on a precise arrangement of specialized cells. And yet, despite their tight interconnections, cells remain separate entities within the bodies they belong to. The new theory presented a new picture of organisms, as colonies of separate, living machines. Initially, Schleiden and Schwann believed that cells formed like crystals, new ones coalescing within the body as an organism grew. However, experimental work in the 1850s proved them wrong, and gradually most biologists accepted that cells always came about through either the fusion or division of other cells. This became the central dogma of cell theory: every cell comes from another cell. Schleiden was one of Unger's scientific heroes and it was probably his encouragement that led Mendel to acquire a copy of Schleiden's crucial book (*Grundzüge der wissenschaftlichen Botanik*, 'Basic Principles of Scientific Botany', 1842–3), which he read closely.

Cell theory played a vital role in ending the long-running biological debate about what exactly happened during sex. Were male and female influences mixed, or did the action of sperm or pollen on ovum simply stimulate a preformed organism to develop? A belief in some kind of mixing was becoming increasingly widespread, but its proponents still had to explain how this mixing happened. Plants presented a further complication, which was that the pollen settled on the stigma, some distance from the plant's ova, which developed in the ovary at the base of the carpel. Since there was evidently some kind of male influence, how did it travel down to the ova?

As they tried to understand these questions, many naturalists assumed that male and female influences were contained in two kinds of fluid, which were blended during fertilization. Even plant fertilization, which seemed to be a pretty dry business of dust-like pollen grains settling on a flower, was assumed to involve fluids in some way; Kölreuter, for example, argued that once the pollen was ripe, the grains it contained liquefied and were squirted out on to the flower's stigma. He suggested that the ways in which male and female elements were mixed in their offspring was best understood in chemical terms, just as acids and alkalis combined to form new substances with new properties. If he was right, perhaps the male fluid simply seeped down to the ova.

Gradually new, improved microscopes allowed botanists to peer more closely into the flower's private life. In 1827, the French botanist Adolphe Brongniart observed that grains of pollen grow when they land on the stigma of a flower; they produce a tiny pipe, the pollen tube, which slowly develops until it reaches the plant's ovary. This is how the male influence is transmitted; Brongniart argued that the pollen grain appeared to contain what he called a 'spermatic granule', very similar to an animal's sperm; it was this tiny parcel, not some mysterious fluid, that carried the male's contribution to the new plant. His observations brought botanists into a controversy already well-established among the zoologists: did the sperm actually penetrate the egg, or did it merely stimulate the egg to develop?

This was the question the new cell theorists aimed to settle. Schleiden and Schwann argued that the plant and animal kingdoms were united at their most fundamental level, the cell. Every plant and animal was made of cells, all of which had nuclei, and all new cells were formed from existing cells. In which case, Schleiden argued, the pollen tube could not be transporting 'a pre-existing embryo'; instead, it must carry a single cell which fuses with the female cell to form a new cell – the first cell of a new plant body. Schleiden was convinced that he had driven the last nail into the coffin of preformationism (the idea that the embryo was preformed in either sperm or ova); not everyone agreed, but his theory proved very popular in German-speaking Europe. Unger was one of Schleiden's supporters and argued that it was now clear that both parents contributed to the character of the new plant.

With his head full of these new ideas, Mendel returned to his priory in 1853 and, thanks to his university education, was able to get a decent teaching job. He also began to experiment with plant-breeding, applying what he had learned to the old problems his parents and grandparents had addressed before him: how to produce better crops. Some of his students were also farmers' children, and they later remembered being taken to the priory and shown Mendel's garden. Gardening spilled over into the classroom too, and Mendel would sometimes demonstrate the techniques he used for his experiments, showing the students how to make little paper caps with which to cover the flowers to prevent unwanted pollen entering them. The spectacle of a celibate friar explaining the sex lives of plants inevitably led to occasional schoolboy titters, at which Mendel would exclaim crossly, 'Do not be stupid! These are natural things.'[12]

Mendel's experiments had a distinctly practical purpose; one of his first scientific papers concerned a species of weevil that was devastating pea crops in Brünn. Peas were important: like their peasant neighbours, the friars grew them to sell and to eat. They grew several varieties, some of which were easy to shell, which saved time, but were not as sweet-tasting as other varieties. Some of the sweet varieties, on the other hand, grew on very tall plants, making them harder to pick and vulnerable to storm damage. If only, Mendel thought, one could take the most useful characters of each variety and combine them. The obvious solution was to cross-breed the sweet – but tall and hard-to-pick – varieties with the bushy, easy-to-pick, short ones, but how could he be sure that the improved varieties would not revert back to one or other of the original forms after a few generations? Mendel knew, both from his parents and his teachers, that this was an old problem, so as he began his experiments, he read what other people – particularly Knight, Kölreuter and Carl Friedrich von Gärtner – had done before him. He had already read some of their work in Vienna and it is clear from his notes in the margins of Gärtner's book, which still survives, that he read them closely and thought carefully about their ideas.

Hybrid species?

As we have already seen, edible peas, *Pisum sativum*, were one of the species that Knight had used in his experiments, which would have been another reason for Mendel to re-read his work. Gärtner also performed some experiments with edible peas. Among their attractions to a researcher was that pea-growers sometimes found several different-coloured peas in the same pod; that made it easy to see at a glance whether you had a pure-breeding strain. However, Gärtner was interested in more than creating sweeter peas; the question of creating stable new varieties of hybrid peas was also a way of investigating Linnaeus's old problem of whether or not hybridization could create new species. As Mendel read and experimented, combining his practical skills with his university education, he too became intrigued by this question.

At the time when the St Petersburg Academy of Science announced its prize competition, Kölreuter was in charge of the Academy's natural history collection. He had hoped to win the prize himself, but missed the deadline and was probably some-what unhappy about Linnaeus's triumph. He planted Linnaeus's hybrid goatsbeard and found that – contrary to Linnaeus's claims – the seeds did not all come up true. Although he fully accepted Linnaeus's case for the sexuality of plants, Kölreuter refused to accept that hybridization could create new species.

The year after Linnaeus won the Academy's prize, Kölreuter

published what would have been his entry, a book describing his many experimental crosses: altogether about 500 different hybridization experiments involving over 100 species. Like Knight, Kölreuter understood that experimenting with trees was impractical, so he had spent many years crossing tobacco plants, (Nicotiana). Kölreuter was sceptical of Linnaeus's claims because in his own experiments he found that hybrids always reverted to the parental form. Especially if the original form was growing nearby; the parental pollen always seemed more potent than the hybrid pollen, so Kölreuter concluded that any hybrids that occurred in the wild could never be stable. Yet, despite his enormous hard work, he never received the recognition he deserved.

Kölreuter's problems were exacerbated by his circumstances. He was not wealthy enough to devote himself to his research full-time; he had to work for a living. His position in St Petersburg lasted only a year before he returned to Germany. Subsequently he moved around, taking whatever short-term positions he could find, until finally becoming Professor of Botany at the University of Karlsruhe. For many years Kölreuter had to perform his experiments on potted plants that he carted around with him on his travels. Even when he was settled in an institutional position, he found that the gardeners were often incompetent or deliberately unhelpful and would forget to water his plants or attend to the experiments in progress.

It would be difficult to find a greater contrast between Kölreuter and Carl Friedrich von Gärtner, who was probably the most famous expert on plant breeding in the German-speaking world during the nineteenth century. While Kölreuter's father had been a humble apothecary, Gärtner's father was Professor of Botany at St Petersburg. Carl planned to follow a medical career and began as an apprentice at the royal pharmacy in Stuttgart, going on to study medicine and chemistry at some of the nineteenth century's leading universities. Having a famous botanist for a father was just one of Gärtner's many advantages over Kölreuter. When his father died, Gärtner became wealthy enough to devote himself to botany full-time. As a young doctor, he took advantage of his father's money to travel across Europe, meeting its leading naturalists. It was soon after he returned that he first read Kölreuter's book (Gärtner's father, Joseph, had known Kölreuter). Young Carl was fascinated and decided to devote himself to plant hybridization. To assist him with his experiments, Gärtner had a large private garden and a diligent paid staff.

As we have seen, Kölreuter had accepted Linnaeus's proof that plants were indeed sexual beings and he was confident that his own experiments had proved once and for all that in plants, both parents were essential to the production of offspring (even if he was not quite sure how this mixing of parental characteristics was effected). He wrote that 'even the most stubborn of all doubters of the sexuality of plants would be completely convinced' by his work; if they were not, 'it would astonish me as greatly if I heard someone on a clear midday maintain that it was night'.[13] Yet shortly after Kölreuter's death, his work was indeed challenged by August Henschel, a German physician and botanist from Breslau who claimed that Kölreuter's results only reflected his artificial techniques. Like Knight, Kölreuter had relied on tricks such as 'castrating' plants (removing the anthers) and then dusting them with pollen from another species; such unnatural methods were, Henschel argued, bound to produce monstrosities. Henschel even regarded growing plants in pots as problematic: unnatural conditions would produce unnatural results. He argued vehemently that only natural methods could reveal nature's secrets and, as we will see, he was not the last scientist to criticize breeding experiments on these grounds.

The controversy created by Henschel's book was another factor that prompted Gärtner to focus on plant hybridization. In 1830, Gärtner was hard at work with his plants while also trying to complete the massive book on botany that his father had left unfinished when he died. As he worked, the Dutch Academy of Sciences offered a prize to anyone who could answer this question:

What does experience teach regarding the production of new species and varieties, through the artificial fertilisation of flowers of the one with the pollen of the other and what economic and ornamental plants can be produced and multiplied in this way?

It is no coincidence that the Dutch Academy should have been interested in more effective ways of creating new 'economic and ornamental plants'; then, as now, bulb-growing – especially tulips – was a major Dutch industry.

The Dutch Academy was disappointed when no one entered their competition, so they extended the closing date to 1836. Gärtner did not hear of the prize until 1835, but he submitted a hastily written summary of his experiments. The Academy gave him an extension to write them up properly, and in 1837 he was awarded the prize. His book – *Experiments and Observations on Hybridisation in the Plant Kingdom* – was initially published in Dutch, limiting its circulation, but eventually appeared in German in 1849. It contained details of 10,000 experiments on 700 species which revealed 250 hybrids; by far the largest, most comprehensive experimental study of hybridization that anyone had performed. Its sales were disappointing (Gärtner was not a scintillating writer), but Darwin owned a copy and thought it so useful that he wished it were better known. Mendel also owned a copy, which still survives at the Mendel museum in Brno; it is clear from his marginal notes and underlinings that he read it very carefully.

The ultimate conclusion of Gärtner's work was agreement with Kölreuter: Linnaeus had been wrong – hybridization could *not* produce new species. Gärtner's experiments had convinced him that only hybrids between varieties of the same species were fully fertile; any attempt to cross two different species led to sterile offspring, just as horses and donkeys produced mules. Then Mendel's old teacher, Unger, published the results of his own hybridization experiments, which discussed Gärtner and Köelreuter's work in detail. But Unger came to the opposite conclusion and decided that new species *could* be created by hybridization. He did not suggest that this was the primary means by which new species were created, but he offered it as evidence against those who insisted that species could not evolve; Unger was a transmutationist, a believer in evolution, and his views caused religious controversy. He briefly faced the threat of dismissal for his unorthodox views, so Mendel could not have been unaware of how important – and potentially dangerous – his pea plants could be.

The results of Mendel's pea experiments might seem too familiar to need retelling, but the way they are usually described obscures one important fact: Mendel did not invent modern genetics. To see why, we need to understand what he himself thought he was doing.

By carefully keeping bees away from his plants, Mendel was able to create a series of separate pure-breeding strains. When he published his results, he explained why he had chosen peas: firstly 'interference from foreign pollen cannot easily occur', but just to be sure, 'a number of the potted plants was placed in a greenhouse during the flowering period; they were to serve as controls for the main experiment in the garden against possible disturbance by insects'.[14] In addition, there was 'the ease with which this plant can be cultivated in open ground and in pots'. As Knight had found before him, Mendel thought the pea's 'relatively short growth period' was a 'further advantage worth mentioning'; there was no need to wait years for the result. And finally, although 'artificial fertilization is somewhat cumbersome', in peas 'it nearly always succeeds'. He explained his technique: before the pollen could ripen, 'each stamen is carefully extracted with forceps, after which the stigma can be dusted with foreign pollen'.[15]

Mendel spent two years testing 'a total of 34 more or less distinct varieties of peas', which he acquired from commercial seed dealers; in this, as in other aspects of his experiments, he was more careful than any of his predecessors – Doppler and Unger's lessons on good experimental design had been well learned. Eventually Mendel selected '22 varieties that showed no variation after 2 years of testing'.[16] He separated these into seven pairs of contrasting characters: some varieties always produced yellow peas, others green; some were tall, while others were short. He began with simple crosses, such as yellow with green, but while Kölreuter or Gärtner had made use of only a few plants, Mendel made use of hundreds – he remembered the statistical methods he had learned in Vienna and knew that in order to use them he would need large samples to eliminate 'a mere chance effect'.[17] When the first generation of hybrid peas flowered and bore fruit, Mendel excitedly opened their pods and what he found stunned him: every single pea in every pod was yellow. The green trait had simply disappeared.

As he counted all his yellow peas, Mendel must have wondered where the green had gone: had it gone for good, or would it reappear? For hybridizers, that was the key question. To answer it, he planted his new generation of yellow peas and waited for them to grow. As they came into flower he once again busied himself, removing their anthers and dusting them with each others' pollen, ensuring each plant only received pollen from another of the hybrids. As the pea-pods began to swell and ripen, he must have counted the days till he could open them. When he did, he had a second surprise: the green had come back, but only in some of the plants.

There must have been times when Mendel wondered whether he really needed so many plants – he recorded that his experiments had involved 'more than 10,000 carefully examined plants' – but the need for statistically meaningful results drove him on; as he carefully counted and calculated his results, his decision paid off: he found that one plant in four now produced green peas – the yellow peas occurred in a ratio of 3:1 to the green. The same ratio appeared in the other experiments with each of his seven, carefully chosen, traits.

Mendel then did a third round of experiments: crossing the plants from the second generation with the original pure-breeding strains. This revealed that the yellow peas from the second generation were not all the same: when he crossed them with the original pure-breeding yellow strain, he still got some green peas, but not as many as in the previous generation of crosses. Clearly, some of the yellow peas from his first cross must have contained the green colour, but in a hidden form. Once again, he counted and calculated and deduced that there were three types of peas; to save time, he used letters to indicate each type: pure-breeding yellow were marked with a capital '*A*', while pure-breeding green were indicated with a small '*a*'. But there was also a green-yellow mixture, which although it came out yellow could still produce green peas in the next generation. Mendel marked these as '*Aa*'. For every pure-breeding *a* or *A*, there were two of the mixed *Aa* type: his 3:1 ratio was really a 1:2:1 ratio – one *a* to two *Aa*, to one *A*. The plants that had only the *a* (green) character produced green peas, while those with only the *A* (yellow) character produced yellow peas, and the mixed (*Aa*) forms also produced yellow peas. Because the yellow colour was able to dominate the green, Mendel christened it the dominant character, while the green was recessive, because it tended to recede or hide itself.

To anyone familiar with modern genetics, these ratios and their associated letters are familiar but slightly wrong: in current notation there are always two letters – *aa*, *Aa* or *AA*. Each letter represents what geneticists now call an 'allele', one of the possible forms that a gene can have. In this case, the gene for pea colour has two forms, the dominant (*A*) and the recessive (*a*). These come in pairs because the plant inherits one from each parent; if both its parents had the dominant form, the plant will be *AA* (and produce yellow peas); if they both had the recessive form, it will be *aa* (and produce green peas). But if one of its parents was pure-breeding green and one was pure-breeding yellow, it would get one of each and be *Aa* or *aA*, both of which produce yellow peas. If you have understood that, you may as well forget it, because that is not what Mendel concluded. If you are confused, that is excellent – because so was Mendel.

The apparently trivial difference between Mendel's notation and modern notation turns out to be both fascinating and revealing because it helps us understand how Mendel thought –

and how different that was from the way we now think. We can have a sense of roughly what it was that Mendel thought was being passed on from generation to generation from the term he used to describe the mysterious entities he had indicated as *a*, *A* or *Aa*; he called them *Anlagen* (singular, *Anlage*).

Anlage is a German word with no precise English equivalent. Mendel borrowed the term from embryology, where it refers to a primordium, the earliest stage of some part of a developing creature. If we translate *Anlage* as 'rudiment' we get some flavour of Mendel's thinking: yellow plants (*A*) passed on the rudiment of yellow colour, the tiniest seed of the peas' eventual yellowness – yellowness is, after all, all they have to pass on. In the same way, green plants (*a*) passed on the rudiment of green. Hybrid plants (*Aa*) looked yellow, because yellow was dominant, 'stronger' than green in some way, but they could pass on either the rudiment of yellow (*A*) or green (*a*) to their offspring. From Mendel's perspective, it would have made no sense to have written *aa* or *AA* for the pure-breeding lines: a pure yellow plant contains only the rudimentary form of yellow; what could be meant by saying it was yellow-yellow? It is obvious from Mendel's writings that he did not know what kind of thing an *Anlage* might be; it was – in some form or another – the quality of yellowness, but he had no idea of precisely what form it took.

Mendel conducted a lengthy series of further experiments, in which he traced two characters at once, then three. Although the maths and the notation became more complicated, the results were essentially the same: in hybrid forms, any one of the available characters could be passed on, at random. That meant that only pure-breeding strains were stable; they contained only one character, which never varied. By contrast, whatever the appearance of hybrids, they contained hidden characters and so reversion could happen at any time. He was thrilled by the ratios he had found, which reminded him forcibly of basic algebraic formulae. It seemed that he had discovered the fundamental mathematical laws that governed hybridization, but he did not think that this initial law was a universal law of heredity; Mendel was careful to say it only applied to Pisum. Remember that his ratios only applied to the mixed *Aa* forms, the hybrids: they appeared to have no relevance to the majority of pure-breeding strains. Perhaps that is why Mendel's published 1865 paper on peas produced relatively little response. Most of those who read it were plant breeders, interested in the problem of creating new, stable hybrids. To them it seemed that – once you cleared away all the baffling mathematics – all Mendel was telling them was what they already knew: most hybrids revert to their parental type. (And that, incidentally, is almost certainly what Darwin would have thought if he had ever read Mendel's paper.) Some historians have even suggested that Mendel was as disappointed as his readers. He was trying to produce better hybrid varieties, and while the law governing reversion was fascinating, it was not really what he was after.

Hawkweeds are not peas

In an effort to generate interest in his mathematical laws of hybridization, Mendel sent copies of his Pisum paper to various notable men of science. Almost the only one who replied was Carl von Nägeli, Professor of Botany at Munich; while Mendel would undoubtedly have preferred to have had more responses, there was probably no one whose opinion he valued more. Nägeli had worked with Schleiden and done pioneering work on the structure and growth of plants. He was also highly respected by Mendel's teacher, Unger, who had described Nägeli as the man 'who has given us both ground plans and elevations of some plant structures in which each element is marked with the number its architect intended for it'.[18]

In his first letter to Nägeli, Mendel mentioned that he was continuing his hybridization experiments and had selected a few interesting plant groups for further work, including Hieracium. Mendel had already chosen hawkweeds before he made contact with Nägeli, yet the myth persists that Nägeli persuaded Mendel to tackle this obdurate group. This tale is so well known that the American novelist Andrea Barrett included it in a short story called *The Behaviour of the Hawkweeds*, a phrase that is a quotation from the first biography of Mendel (published in 1924), which began a long tradition of blaming Nägeli for wasting Mendel's time.[19]

It is clear from Mendel's first letter to Nägeli that he was already hard at work on hawkweeds. He described how artificial pollination in this genus was 'very difficult and unreliable because of the small size and peculiar structure of the flowers', but adds that 'Last summer I tried to combine [*Hieracium*] *Pilosella* with *pratense, praealtum,* and *Auricula;* and *H. murorum* with *umbellatum* and *pratense,* and I did obtain viable seeds; however, I fear that in spite of all precautions, self-fertilization did occur.'[20]

Mendel's choice of the intractable hawkweeds was prompted by their reputation as a complex group, hard to classify because they formed hybrids so easily in the wild. During the 1860s, several Brünn naturalists were discussing wild Hieracium hybrids and debating their relationship with the apparent parental species. And the fact that Nägeli was a well-known Hieracium specialist may well have prompted Mendel to write to him.

What interested Mendel about hawkweeds was that they seemed to have a natural ability to create the true-breeding hybrid forms he was interested in: Hieracium 'possesses such an extraordinary profusion of distinct forms that no other genus of plants can compare with it'. As a result, their classification was especially complex: 'The difficulty in the separation and delimitation of these forms has demanded the close attention of the experts.'[21]

In his letters to Nägeli, Mendel discussed each of his Hieracium crosses, describing both his anticipated and actual results; from the way he describes them, it is obvious that Mendel *did not* expect to see the Pisum results repeated; in fact, he would almost certainly have been disappointed if they had been. He argued that the nature of Hieracium hybrids was an important one and, 'we may be led into erroneous conclusions if we take rules deduced from observations of other hybrids to be Laws of hybridisation, and try to apply them to Hieracium without further consideration'.[22] In his private letters to Nägeli, Mendel mentioned that now and again, one of his Hieracium hybrids would revert to type, but he never mentioned these problems in his public reports to the Brünn Natural History Society, who were kept continually informed of the unvarying nature of the seedlings. Clearly, Mendel saw the occasional variation as the exception to the rule, probably the result of some error on his part and thus not worth reporting. The constant, unvarying hybrid progeny were what he reported.

In one letter, Mendel commented to Nägeli that 'I cannot resist remarking how striking it is that the hybrids of *Hieracium* show a behaviour exactly opposite to those of *Pisum.*' Some historians have interpreted this as an expression of exasperation at the refusal of hawkweeds to behave like peas. In fact, it is clear from the very next sentence of the letter that Mendel was far from exasperated, since he wrote: 'Evidently we are dealing here with an individual phenomena, which are the manifestation of a higher, more fundamental law.'[23] What Mendel was referring to was that when hawkweeds were crossed, the first generation varied but subsequent ones remained constant; whereas in peas, the first generation were all the same, while later ones varied. He told Nägeli how surprised he had been when he crossed the yellow hawkweed (*H. praealtum*) with a golden hawkweed (*H. aurantiacum*); as he had expected, some of the hybrids looked like a mixture of the parental species but others looked much more like the yellow hawkweed in some regards, while looking like hybrids in others. Even more remarkably, when the pale hawkweed (*H. auricula*) was crossed with meadow hawkweed (*H. pratense*), three different types appeared.

Mendel regarded the variability of the first generation Hieracium crosses and the subsequent stability of the new types as two different phenomena but, far from despairing over the unexpected result, he expanded his research to investigate it. When he published his first paper on Hieracium in 1869, he summarized his provisional findings: 'Although I have already undertaken many experiments in fertilisation between species of *Hieracium*, I have succeeded in obtaining only the following 6 hybrids, and only one to three specimens of them.' Nägeli had expected that artificial fertilization would prove completely impossible because the tiny flowers of Hieracium were so hard to work with and they self-fertilized so readily, which made it hard to cross-pollinate them deliberately. Mendel described how he had overcome this difficulty: 'in order to prevent self-fertilisation,' he wrote, 'the anther-tube must be taken out before the flower opens, and for the purpose the bud must be slit up with a fine needle.'[24] He was proud of his skill; it was difficult to produce Hieracium hybrids artificially, but he had shown it could be done. And he had shown both that at least some of these artificial hybrids were fully fertile – and that the hybrid forms were stable, as long as they were self-fertilized. Half Mendel's six hybrids involved the pale hawkweed, *H. auricula*, the most cooperative member of this unruly genus.

Mendel concluded his Hieracium paper by noting that 'the question of the origin of the numerous and constant intermediate forms has recently acquired no small interest since a famous *Hieracium* specialist has, in the spirit of the Darwinian teaching, defended the view that these forms are to be regarded as [arising] from the transmutation of lost or still-existing species'.[25] The 'famous *Hieracium* specialist' was, of course, Nägeli, who had written that 'I see no other possibility that the *Hieracia* types have originated through the transmutation of extinct or still living forms.'[26] Evolution created this confusion, but in most groups the intermediate types had become extinct, leaving a group of species with definite gaps between them, which allowed them to be straightforwardly classified. Nägeli suggested that botanists were observing hawkweeds at a slightly earlier stage in their evolution, when proliferation had begun but extinction had not yet pruned the group into distinct, comprehensible species.

Interestingly, although Mendel described Nägeli's view as being 'in the spirit of the Darwinian teaching', Nägeli was slightly sceptical about Darwin's version of evolution: he accepted that species evolved but disagreed with Darwin that natural selection was the main force driving their evolution. Like several other German biologists, Nägeli thought that living things must possess some inner drive towards perfection that created new species;

natural selection served only to eliminate unsuccessful variations. For Mendel, however, this was a minor distinction: Nägeli was a transmutationist and, despite his respect for the professor, Mendel seems to have rejected evolution altogether. His precise views are hard to ascertain, but it seems likely that he agreed with Linnaeus that hybridization could make new species by combining existing ones, but only God could create the first types of plants which subsequently hybridized to produce the full diversity of species.[27] The hawkweed experiments seem partly to have been intended to show that Nägeli was wrong about the intermediate forms; they were not new species in the process of formation, but simply an unusual kind of hybrid, one that did not revert to type. Investigating this unusual, but potentially very valuable, behaviour was Mendel's other goal – if he could work out the law governing stable hybrids, he might have found something for plant-breeders to get excited about.

In summing up his Hieracium work, Mendel commented that in Pisum the first generation of hybrids all looked the same, but their descendants were 'variable and follow a definite law in their variations'. By contrast, in Hieracium 'the exactly opposite phenomenon seems to be exhibited'; the first generation varied in unexpected ways, but then remained constant. He noted that something similar had been observed among willow trees, the genus Salix. Mendel speculated that the confusing multiplicity of intermediary species in genera like Salix and Hieracium was 'connected with the special conditions of their hybrids', which proved so unexpectedly stable, but he acknowledged that this was 'still an open question, which may well be raised but not as yet answered'.[28] Mendel wrote that in 1869, at a time when he planned to continue his work on Hieracium, but his superiors in the Augustinian order had other ideas. Abbot Napp died in 1868 and Mendel was elected as his replacement. He told Nägeli that he had a few misgivings about taking the position, since it would inevitably eat into his time, but he needed the money to help pay for the education of his two beloved nephews. His doubts were well-founded. The tiny Hieracium flowers needed to be hand-pollinated with a magnifying glass under artificial light, but by 1870, Mendel's eyesight was beginning to fail. Eventually his institutional duties intervened, his garden was neglected, and he wrote to tell Nägeli that, sadly, he had had to give up his experiments.

The hawkweeds set seed and drifted away, looking for someone else to take an interest in them. They had to wait until the twentieth century for humans to unravel Mendel's 'open question', partly because it turned out to be two questions. The first one – how do all the hybrid species of hawkweed survive – was answered in 1903, when it was revealed that not only can hawkweeds manage without sex, they almost invariably do. The ova of hawkweeds can develop into viable seeds without being fertilized – one reason why the hybrid forms that puzzled Nägeli and Mendel remained stable, even in the wild when no one was keeping foreign pollen away. Had he but known it, a major clue to this aspect of the Hieracium mystery lay under Mendel's nose: the bees in his beehives had mastered the same trick. By Mendel's day it was known that the drones, or worker-bees, in a hive developed from unfertilized eggs. This phenomenon is known as parthenogenesis in the animal kingdom and as apomixis in plants. Apomixis is common in hawkweeds, and in their cousins the dandelions (Taraxacum), as well as among brambles (Rubus). It is one of the things that makes these genera such efficient weeds; with most species, if one seed lands on a lawn, it may grow and flower, but since there is only one plant it cannot reproduce. But that single, drifting Hieracium seed has no need of a second plant to fertilize it – once it has germinated, any lawn is under threat.

Understanding apomixis would have solved one of Mendel's puzzles; why the Hieracium hybrids persisted instead of reverting to type. Modern botanists have given up on Linnaeus's puzzle as to whether these persistent intermediate forms are really true species or not, and they hedge their bets by referring to such troublesome cases as 'aggregate species' made up of many 'microspecies'. But the origins of these microspecies, of the 'extraordinary profusion of distinct forms' that first attracted Mendel's attention, has a different explanation, and it took another plant, the evening primrose, to lead us to the answer to Mendel's second question.

Oenothera lamarckiana

Chapter 5
Oenothera lamarckiana: Hugo de Vries led up the primrose path

In the 1820s, fur-trappers in North America first made contact with a tribe they called the Klamath. South-western Oregon supports relatively few fur-bearing animals, so – having no immediate need to take their land – white settlers took little further interest in the Klamath; it is a measure of how isolated they remained that some twenty or thirty years after they had first met whites, the tribe is said to have owned only a single gun. However, as North America's settler population grew, so too did the pressure on Indian land, and in 1864 the Klamath were relocated to a reservation. About thirty years later, a botanist called Frederick V. Coville passed through and, perhaps aware that Native American numbers were declining, took the time to record some of the Klamath's plant lore, including their medical knowledge and their legends about plants. Coville recorded a plant called the *wasam-chonwas* that, he explained, got its name from 'the following story, in which I have retained as nearly as possible the sentiment and sequence of the Indian narrator':

A long time ago, when the animals lived and talked like men, the coyote, or prairie wolf, who was very keen and smart, but a good deal of a sneak, just as he is today, met one day the Indian Christ, Isis, who could do anything he wanted to – could make flowers, 'grub' (i.e., food), anything. The coyote said, in a bragging way, that he, too, could do these things just as well as Isis, and Isis said: 'Very well, go ahead and make a flower.' Then the coyote, who knew that he really could not make much of anything, was greatly ashamed, but he went off in the grass a little way and vomited, and on that spot pretty soon this great, rank, yellow-flowered weed came up. And that was the best the smart coyote could do.[1]

Wasam-chonwas means 'coyote vomit plant' (*wäs* – coyote, *äm* – plant and *chón-wäs* – vomit). Undeterred by (or more likely, simply ignorant of) its associations with vomiting coyotes, early settlers took the yellow-flowered plant back to England where, in addition to various medical uses, it was cultivated for its nut-flavoured root. John Parkinson gave the plant its common English name, evening primrose, because it opens in the evening to attract the moths who are the usual pollinators. A few decades later, John Ray noted that the larger-flowered varieties were already common in English gardens, thanks to their yellow flowers, which attracted many gardeners; the Duchess of Beaufort, who – as we have already seen – was a pioneer grower of passionflowers, also grew evening primroses.

Just as Linnaeus named the passionflowers Passiflora and the hawkweeds Hieracium, he also gave the evening primrose its modern scientific name, Oenothera. As with Hieracium, Linnaeus adapted this name from one in Pliny's *Natural History*, which appeared to be derived from the Greek *oinos* (wine) and *thera* (a hunt). This slightly mystifying combination led some to suggest that this 'wine-catcher' plant was used either to encourage wine-drinking or to help wine-drinkers recover from its after-effects, but the connection with wine was almost certainly a mistake. Pliny's name was originally Onothera, from *onos*, the Greek for an ass or donkey, so the name meant 'donkey-catcher'. The plant Pliny referred to (whatever it was – no one really knows) was probably a mild narcotic, used to knock out wild asses so they could be caught and eventually tamed. But donkey-taming seems to have appealed less to botanists than wine, and it was the spurious association with wine that caught on; by 1860, the Scottish doctor William Baird was confidently assuring readers of his *Dictionary of Natural History* that 'The roots [of *Oenothera*] are eatable, and were formerly taken after dinner to flavour wine, as olives now are; hence its name *Oenothera*, or wine-trap.'[2]

So when Coville – who had studied botany at Cornell University – recorded the Klamath legend about the vomiting coyote he ignored the indigenous American name for this indigenous American plant, and realized it was the species named *Oenothera hookeri*. A name that quietly commemorates Europe's colonization of America's plants, since the genus has a Latin name with classical Greek roots, while the species name, *hookeri*, honoured Joseph Hooker (the British botanist who had been so helpful to Darwin). But *O. hookeri* had at least been given its name in America, by two American botanists, Asa Gray and John Torrey. *Oenothera lamarckiana*, like most species of this American genus, was named by Europeans in Europe, many of whom had never been near America. In 1828 the Swiss botanist Augustin-Pyramus de Candolle produced the third volume of his massive *Prodromus* ('Outline') of the vegetable kingdom, which he intended to be a definitive catalogue of every known plant. Among the thousands of species was another new Oenothera with the specific name *lamarckiana*, named by one French botanist, Nicolas-Charles Seringe, to commemorate another, the pioneering evolutionist Jean-Baptiste Lamarck.

Oenothera lamarckiana had been introduced into European gardens because it was so beautiful. In 1890, more than eighty years after de Candolle first published the name, it caught the eye of the Dutch botanist Hugo de Vries, who described it as 'a beautiful, freely branching plant, often attaining a height of five feet or more', whose flowers 'are large and of a bright yellow color, attracting immediate attention, even from a distance'.[3] He recognized the plant as:

the great evening-primrose or the primrose of Lamarck. A strain of this beautiful species is growing on an abandoned field in the vicinity of Hilversum, at a short distance from Amsterdam. Here it has escaped from a park, and multiplied.[4]

After inspecting the escaped plants more closely, de Vries must have felt like dancing for joy: he had searched for years for a plant like this, examining over 100 wild species in vain, but in *Oenothera lamarckiana* he had found what he had long looked for, the plant that would prove Darwin wrong. Why? Because Lamarck's primrose would take evolution out of the realm of speculation and bring it into the laboratory, where it would help solve two of the most pressing problems confronting evolutionary theory: the fact that natural selection probably could not work in the way Darwin had envisaged, and – even if it could – it was too slow.

Darwin's deathbed

Given the enormous prestige that surrounds Darwin's name today, it is perhaps slightly shocking to realize that by the end of the nineteenth century, many biologists believed Darwinism was as dead as its founder.

Part of the problem was Fleeming Jenkin's argument about blending: that one exceptional organism could not improve a species because its advantages would always be diluted when it was forced to interbreed with its unimproved neighbours. As we have seen, Darwin felt he had addressed this problem by arguing that it was not isolated sports of nature that drove evolution, but common, small variations within a species. The pressure of competition meant that these were slowly accumulated and the situation Jenkin imagined simply never arose. Evolution was slow and gradual or, as Darwin liked to put it, *Natura non facit saltum*, 'nature makes no leaps'. Evolution's slowness explained, among other things, why no one had ever seen a species change:

As natural selection acts solely by accumulating slight, successive, favourable variations, it can produce no great or sudden modification; it can act only by very short and slow steps. Hence the canon of '*Natura non facit saltum*,' which every fresh addition to our knowledge tends to make more strictly correct.[5]

Perhaps the most important reason for Darwin to assume that evolution was slow was the profound influence on him of the geological theories of his friend and mentor Charles Lyell. Geology was a crucial science in the nineteenth century, the key to mapping and exploiting the coal and iron on which Britain's industrial wealth was being built. Unsurprisingly, the first paid government jobs for men of science in Britain were for geologists, but for a wealthy gentleman like Lyell, being paid to do science was not really respectable; it was amateur geologists, those who pursued knowledge for its own sake, who were the intellectual leaders of the science. Lyell hoped to join this leadership by taking geology beyond the mere description and cataloguing of rocks; he wanted to give a causal account of geology, one that would transform it into a true science, or, as Lyell would have said in the terminology of his day, to make it more 'philosophical'.

However, geological theorizing was potentially controversial, since it might lead the theorist to challenge the literal truth of the Bible. So Lyell faced the dilemma of devising a theory that was as respectable and uncontroversial as possible. In his massive three-volume book, the *Principles of Geology*, he confronted the problem that the dominant philosophy of science in nineteenth-century Britain was based on induction. Scientific workers were expected (in theory, at least) to work with a mind free from theories and simply gather together facts by observation and experiment. Once they recognized the general patterns that linked their facts, they could build on these to formulate laws of nature. (Induction is, therefore, the opposite of deduction, which is the process of deriving predictions from known laws.) Naturalists claimed to be working by induction, gathering specimens, classifying and then generalizing, but how could induction be applied to the history of the Earth, since no one had been around to observe it?

Fortunately for Lyell, geologists had accumulated many facts about the past that he could work with. One of the most important was that extinction had occurred – it explained why particular fossils were only found in certain rock layers, or strata. These characteristic fossils were used by geologists to identify strata. But how were these disappearing organisms to be explained? Some geologists – the Frenchman Georges Cuvier being the most famous – had assumed that the Earth had undergone catastrophic changes, great floods, earthquakes and similar upheavals of a magnitude unlike anything that occurs today. There were good reasons for his assumption: geologists could see that massive rock formations had been tilted or even smashed. It certainly looked as if great violence had been involved. But Lyell rejected this interpretation, not least because if, as Cuvier's solution implied, the forces at work in the past were unlike those of our own time, geologists could say nothing certain about the Earth's history. So Lyell made the opposite assumption to Cuvier's: he assumed that the forces we see in operation today – such as volcanoes, earthquakes, wind and water erosion – were the only ones that had ever affected the Earth. And he also supposed that they had always operated with the same intensity as they did now. He renounced all discussion of super volcanoes that shattered continents or massive floods that drowned the entire planet.

Lyell's geological philosophy became known as uniformitarianism, because it assumed uniform forces, while the opposite assumption became known as catastrophism. The *Principles of Geology* carried the subtitle 'An Attempt to Explain the Former Changes of the Earth's Surface, by Reference to Causes *now* in Operation'. He argued that the present is the key to the past. The frontispiece to the book encapsulated his view: it showed an ancient Roman temple at Serapis near Naples (now thought to have been a market). On its pillars were the marks of damage caused by marine organisms comparable to barnacles which had bored into the stone to attach themselves to it. The marks occur in neat rings around the pillars, proving that the building had once been under water, but obviously the Romans had built it on land, which is where it now stands. So the land had sunk (or the sea had risen) in the last 2,000 years and then the process had reversed itself, leaving the pillars exposed. However, for Lyell the most important aspect was that the pillars were still standing, despite having been drowned and then raised again. That showed that the process must have been incredibly slow and gradual – otherwise the pillars would have fallen down.

Darwin modelled his theory of natural selection on Lyell's geological one: both entailed slow, gradual change accumulated over millennia. As Darwin wrote: 'I am well aware that this doctrine of natural selection . . . is open to the same objections which were at first urged against Sir Charles Lyell's noble views'; times had changed and Lyell's views were increasingly accepted by the time Darwin wrote the *Origin*. 'Natural selection can act only by the preservation and accumulation of infinitesimally small inherited modifications,' he wrote, and just as 'modern geology has almost banished such views as the excavation of a great valley by a single diluvial wave, so will natural selection, if it be a true principle, banish the belief of the continued creation of new organic beings'.[6]

Unfortunately, however, not everyone embraced Lyell's views as enthusiastically as Darwin did. By 1871 Darwin found he was being haunted by 'an odious spectre' called Sir William Thomson.[7]

Thomson (later Lord Kelvin, after whom the Kelvin scale of temperature is named) was a Scottish physicist and mathematician who in 1866 had published a short paper with the rather confident title, 'The "Doctrine of Uniformity" in geology briefly refuted'. It argued that since the earth had begun as a molten ball, its age could be calculated by simply taking its temperature: the laws of thermodynamics – the physics of heat and cooling – were well understood, not least because these were the laws that governed that great Victorian workhorse, the steam engine (often then known as 'thermodynamic engines'). Thomson used the laws of thermodynamics, which he had helped formulate, to prove that the Earth was much younger than Lyell claimed, being at most 100 million years old (he later revised this figure down to just 20 million years; most modern estimates are around 4.5 billion years). Even 20 million years sounds a lot, but it was only a fraction of the time Lyell and Darwin had assumed they had at their disposal. If Thomson was right, there had simply not been enough time for natural selection to operate, at least, not at the stately pace Darwin had envisaged. Thomson's argument was seized upon by Darwin's critics; Fleeming Jenkin was one of many who had used it to undermine natural selection.

Trouble also came from Darwin's cousin, Francis Galton. His attempts to measure variability had also led him to the problem of blending, which he (characteristically) had expressed mathematically. If an intelligent man married a stupid woman, their children would, he argued, inevitably be closer to the average intelligence of the population than either parent was; hence his disparaging comment about the average person being 'an uninteresting being' and 'of no direct help towards evolution'.[8] In his book *Natural Inheritance* (1889) he called this principle 'the law of regression' and argued that it was inexorable: the superior qualities – mental and physical – of a person, animal or plant would always be diluted, generation by generation, until they disappeared.

Darwin's two problems, the overwhelming of new varieties by old ones and the relatively young age of the Earth, were closely connected. As we have seen, Darwin ruled out large, uncommon evolutionary leaps because their impact would simply be swamped. That meant evolution must proceed in small steps, steps so small that no one could ever observe them, any more than they could observe the rain reducing a jagged mountain to a gentle rolling hill. But vast changes could only be built by minute steps if the Earth was immensely ancient.

If evolution could be speeded up a bit, the odious spectre of Thomson would be banished, so some of Darwin's supporters (including Alfred Russel Wallace and the combative young zoologist Thomas Henry Huxley) simply assumed that it must move faster than Darwin thought. Galton also assumed that organisms must evolve in jumps; one aspect of his argument was that only certain combinations of features could produce a workable organism: a carnivore, for example, needs legs to run fast and catch its prey, claws and jaws adapted to kill and dismember it, and the right kind of digestive system to cope with meat. If evolution proceeded in immeasurably small steps, it would produce unworkable combinations, such as a carnivore's teeth leading to a herbivore's digestive tract. This kind of objection had been raised by others, who wondered, for example, how birds had evolved to fly; how could natural selection reshape scales into feathers, lighten bones, increase muscle and lung power, and produce working aerodynamic wings, all in a series of tiny steps? After all, what possible use to a bird could the minute early stages of a wing be, if they were far too small to lift it?

Galton addressed these problems by imagining an organism as an octagon lying on one of its sides. Each side metaphorically represented a stable combination of features, such as all the features that made-up a successful carnivore. The octagon could rest stably on any one of its faces, but there were no intermediate points; it could only move from one resting position to another in a single step. A force would be needed to tip the polygon abruptly from one position to another and for Galton that 'force' was a sport of nature, a plant or animal that somehow produced enough change in a single step to tilt the organism over, so that it engendered a new species.

However, Darwin rejected this approach. He uneasily accepted that evolution might occasionally proceed a *little* faster than he had at first envisaged, but in general he simply dismissed Thomson's claim; although he could not disprove it, he remained stubbornly convinced that he was right. Darwin felt that evolutionary change, like Lyell's geological forces, must proceed at a respectable pace; apart from anything else, more rapid change hinted at supernatural, perhaps even divine, intervention in the Earth's history – approaches that increasingly had no place in properly philosophical explanations.

In any case, explanations like Galton's were vulnerable to Jenkin's original swamping argument, and they raised the question why, if organisms could change so rapidly, had no one observed one change? As the century drew to a close, these questions perplexed and divided biologists. Some remained wedded to what they saw as true Darwinism: natural selection working slowly and gradually on the small, everyday variations each species threw up. Others were convinced that the objections to this approach were overwhelming, somehow nature must take leaps; however rare they might be, it was exceptional, large changes that must drive evolution, not the tiny ones.

Unexpectedly, both schools found justification for their views in Galton's work: some focused on Galton's octagon, convinced that it was not smooth, normal variability that drove evolution, but rare jumps, or saltations (from the Latin, *saltare*, to leap, as in *natura non facit saltum*). For this group, nature most definitely *did* make leaps; they became known as 'saltationists'. Meanwhile, their gradualist opponents drew on Galton's mathematical work, such as his statistical analysis of normal variability. They called themselves 'biometricians', practitioners of a science called 'biometry', the mathematics of life. The biometric school was founded and dominated by Karl Pearson and Raphael Weldon, who argued that mathematics could make the study of natural selection a rigorous, quantifiable science.

Carl Pearson had been a brilliant student at King's College, Cambridge, where his success in the mathematics tripos earned him a fellowship. He took advantage of the financial independence this brought him to spend some years in Germany, studying philosophy, law, biology and physics. His fascination with all things German was so strong that he began to use the German spelling of his first name, with a K instead of a C.

Back in Britain, Pearson was briefly an active socialist, lecturing to revolutionary clubs on German literature and contributing 'hymns' to the *Socialist Song Book*, but at heart he was too much of an intellectual snob to be whole-hearted about his politics. He became a firm believer in Galton's eugenics and argued that the way to 'regenerate' Britain was by means of a welfare state and controlled eugenic policies: ensuring that the British were fit enough to prevail in both military and economic contests with their imperial competitors. In 1884 Pearson became Professor of

Applied Mathematics and Mechanics at University College London (UCL). Not long after, he became interested in Galton's mathematical work, which reawakened his earlier interest in biology.

Walter Frank Raphael Weldon (always known simply as Raphael Weldon) was a couple of years younger than Pearson, but started his degree at UCL when he was only sixteen, taking classes in Greek, English, Latin, French and pure mathematics, though it was the botany and zoology lectures that most captured his attention. He briefly studied medicine, but soon dropped it to study zoology at Cambridge. Just like Galton before him, Weldon found the pressure of studying at Cambridge overwhelming and suffered a breakdown. Fortunately, his zoology professor, Francis Balfour, recognized young Raphael's talent and supported him in getting both a scholarship and a three-month holiday in Provence, which in turn helped him end up with a first class degree. He became a specialist in marine biology, especially of crustaceans, and eventually became a zoology lecturer at Cambridge.

Weldon was thrilled when he first read Galton's *Natural Inheritance*; the book suggested new ways of analysing the data he had been collecting about crabs and shrimps. He wrote enthusiastically to Galton and was soon applying his techniques. In 1891 Weldon also became a professor at UCL, which brought him into greater contact with Pearson. The two became close friends, sharing ideas, comparing solutions, answering each others' questions and asking new ones. They had both begun their mathematical work with the tools Galton had developed. When Weldon's analyses were plotted on graph paper, many produced the same shape – a single, symmetrical curve that looked like a bell and so became known as a bell-shaped curve. Such a curve described the way that – for many different measurements – most of the members of a population tend to cluster around the mean value, with departures from that average becoming more infrequent the larger they are.

Consider, for example, human height: if everyone in a city is measured, and the mean height calculated, most people's height falls within a few centimetres of that average – people who are extraordinarily tall or short are rare. This kind of distribution is so common that in 1877 Galton dubbed it a 'normal' distribution. The kind of small everyday variation that virtually every species possessed invariably produced a bell-shaped curve and soon became known as continuous variation, because there were no breaks or gaps in the range of variability. When, for example, the lengths of a sufficiently large sample of shrimp were plotted on a graph, there were always a few shrimps of every possible size; more average ones than exceptional ones, of course, but no gaps and thus no dips or steps in the smooth, bell-shaped curve. Nature made no leaps.

Part of Galton's argument against Darwin's version of evolution was that the bell-shaped curve represented the stable range of everyday variability within a species; for a new species to emerge, a completely new peak would need to appear, so that the single bell would become double-humped and eventually separate into two curves. Galton believed that natural selection acting on the small, everyday variations in a population could not produce such a drastic change. Weldon decided to test this prediction by carrying out a detailed study of common shrimps (*Crangon vulgaris*), which he collected in the waters off Plymouth. Long hours of collecting and counting shrimps were followed by months of careful statistical analysis. When it was finished, in 1890, Weldon had proved that Galton was right: selection might shift the average measurement for some feature of a shrimp, but the range of measurements remained normally distributed; the bell-shaped curve always reappeared. Such findings confirmed the biometricians' instinct that these statistical tools were essential to their work because they were the only effective way to grasp the patterns created by measuring so many thousands of individuals. Men like Weldon were working with wild populations, trying to study evolution in nature. There was no possibility of tracking individual shrimps and trying to work out which their parents had been or what their offspring would be like; the only practical approach was to catch large enough samples to allow for statistical analysis. Hence Weldon's claim, in a later paper that Pearson collaborated on, that it 'cannot be too strongly urged that the problem of animal evolution is essentially a statistical one'.[9] Statistics allowed Weldon to establish, for example, whether large shrimps were becoming more or less common in a real, natural environment. He concluded that 'the questions raised by the Darwinian hypothesis are purely statistical, and the statistical method is the only one at present obvious by which that hypothesis can be experimentally checked'.[10]

However, while the biometricians were finding mathematical evidence to support Darwin's vision of slow, gradual change, others remained unconvinced. The most prominent British sceptic was William Bateson, a close Cambridge friend of Weldon's, who had also studied with Balfour. Balfour introduced Bateson to the American biologist William Keith Brooks of Johns Hopkins University, and Bateson spent two summers there doing laboratory work with him. Brooks had just finished his book *The Laws of Heredity*, in which he had adopted a saltationist theory of evolution, in part because this allowed evolution to proceed more rapidly. As they worked together at the university's seashore laboratory in North Carolina, Bateson became deeply impressed with Brooks's ideas and absorbed his fascination with heredity: 'For me this whole province was new. Variation and heredity with us had stood as axioms. For Brooks they were problems. As he talked of them the insistence of these problems became imminent and oppressive'.[11] Bateson returned to Britain convinced that his friend Weldon was wrong – it was not selection on the small, everyday changes that drove evolution, but large jumps. Such variability would not produce a smooth, bell-shaped curve, but one with steps or dips in it; instead of the largest or fastest member of a population being linked to all the others via intermediate steps, there would be a gap, and so this kind of variation became known as *dis*continuous.

Bateson decided to look for evidence of evolution in wild populations; he set out for Central Asia, to see how a particular species of shellfish adapted to the variable levels of salt in the Aral Sea and nearby lakes. He spent eighteen months gathering and measuring shells and testing the salinity of each lake. Back in Britain, he published his results in the Royal Society's *Philosophical Transactions* and sent a copy to Galton, who complained that Bateson did not understand statistics properly. Stung by his distinguished older colleague's criticism, Bateson acquired a copy of *Natural Inheritance*, and he too became an enthusiast for Galton's ideas. However, while Weldon and Pearson had become fascinated with the smooth curves that represented gradual variability in a species, Bateson seized on Galton's unstable octagon. Primed by Brooks to perceive the importance of *dis*continuous variation, Bateson started looking for it; his first major paper on the subject concerned variation in flower symmetry. Examining a range of flowers within a group of closely related species did not reveal a gradual blending from symmetrical to asymmetrical – each flower was either one or the other – and Bateson argued that this kind of variation was both more common and more important than Darwin had appreciated. He admitted that some traits did vary slowly and continuously, such as human height or the size of Weldon's crustaceans, but nevertheless became convinced that natural selection could never produce new species by acting on such small variations. There were, Bateson decided, two different kinds of variations – small and large – and biologists needed to decide what role, if any, each played in the formation of new species.

Convinced that evolution proceeded by leaps and bounds, not small steps, Bateson produced a huge book, *Materials for the Study of Variation* (1894), in which he argued the saltationist case. He objected to the standard style in which orthodox Darwinians argued for natural selection, which he felt relied on too many assumptions and not enough evidence; once all the 'ifs' and 'suppositions' had been discounted, the arguments seemed entirely circular. Galton was delighted and published an enthusiastic article supporting Bateson. But not everyone was convinced. Though by now in his seventies, Wallace saw it as his duty to defend his late friend Darwin from any who dared criticize their theory. The grand old man of Darwinism was so alarmed by the implications of Bateson's book – and by Galton's enthusiastic reception of it – that he wrote a long article in the widely read *Fortnightly Review*, savaging the book and dismissing Bateson's claim that discontinuity was 'a final proof that the accepted hypothesis [i.e. natural selection] is inadequate'. Wallace regarded most of Bateson's examples as nothing more than 'malformations or monstrosities' without any bearing on the issue of the origin of species. It was not monstrosities, but 'slight differences' that accounted for most variation. He was equally dismissive of Galton's claims about regression back to the average, arguing that although Galton had 'admitted that there is such a thing as natural selection', he wrote as if it had no effect – forgetting that natural selection 'preserves and thus increases favourable variations by destroying the unfavourable'. To grasp selection's power, Wallace argued, one had to remember 'that it destroys about ninety-nine per cent. of the bad and less beneficial variations, and preserves about the one per cent. of those which are extremely favourable'.[12] The intensity of the struggle for existence ensured that even the smallest advantage would rapidly result in a new bell-shaped curve, whose average value would be quite different from the ancestral one. Although the curve would persist, it would also shift, gradually changing a species into one with entirely different characteristics from its ancestors.

Wallace was not alone in criticizing Bateson's book; Weldon's review was so unsympathetic that he and Bateson, once close friends, became increasingly bitter enemies. Their disagreements about evolutionary theory were exacerbated when they became rivals for academic appointments, and relations worsened further when Bateson became a member of the Royal Society's evolution committee – he used his growing influence to increase funding for his own research at the expense of Weldon's. Scientific disagreements and arguments about money were made worse by the sarcastic and aggressive tone Bateson used to criticize those he did not agree with. One of his contemporaries summarized their falling-out with a limerick:

> Karl Pearson is a biometrician,
> And this, I think, is his position,
> Bateson and co.,
> Hope they may go
> To monosyllabic perdition.[13]

Wallace's fury over Bateson's book must have been heightened by his sense that Darwinism was under attack on every side. By the end of the nineteenth century, the saltationists were not the only group convinced that orthodox Darwinism was wrong, or at best, incomplete. Although virtually all biologists accepted the fact of evolution, it was Darwin's mechanism of natural selection – seen as synonymous with the accumulation of small variations – that seemed increasingly inadequate. Some attempted to revive Lamarckism, others posited that some mysterious force drove evolution towards ever more perfect organisms, thus accounting for the apparent progress in the fossil record, which natural selection could not account for (the kind of view Nägeli had espoused). By 1903 criticism had reached such a point that one German botanist wrote: 'We are now standing at the death bed of Darwinism, and making ready to send the friends of the patient a little money to insure a decent burial of the remains.'[14] A few years later, the Stanford entomologist Vernon Lyman Kellogg – despite being a committed Darwinist himself – had to acknowledge that most current biology was 'distinctly anti-Darwinian in character ... the fair truth is that the Darwinian selection theory, considered with regard to its claimed capacity to be an independently sufficient mechanical explanation of descent, stands today seriously discredited in the biological world'.[15]

The evolving laboratory

These bitter disputes over Darwin's legacy were made worse by the problem that biologists were also divided over the nature of their discipline. At the beginning of the century, people with a scientific interest in plants and animals had called what they did 'natural history' and themselves 'naturalists', just as they had for many hundreds of years. Natural history was a huge, baggy subject, characterized by an unquenchable curiosity about nature's products, whether they were animals or plants, minerals or fossils, living or extinct. Naturalists were usually travellers, collectors, observers and classifiers. Anywhere in the world, they were instantly recognizable: their pockets full of bugs and beetles; a knapsack loaded with rocks and butterflies; some recently shot birds slung over one shoulder; and a portfolio bulging with drying plants and sketches over the other. The desire to collect, the passion to possess a sample of everything the world contained, drove naturalists on, into swamps and up mountains, across oceans and deserts, risking their health and often their lives to accumulate specimens – sometimes for profit, sometimes for fame and glory, but most often simply for the pleasure of collecting.

Gradually the menageries, botanic gardens, zoos, and private and public museums of Europe began to fill up with a vast accumulation of natural objects. Throughout the seventeenth and eighteenth centuries there had been a near universal assumption that however strange and wonderful the world's products were, they had one thing in common – God had made them all. For many Europeans, the study of nature was a pious activity, since it was the study of God's handiwork. The beauty and complexity of God's creations confirmed his existence and revealed his nature, since only a benevolent, loving God would fill the world with beauty for our enjoyment, food for our sustenance and cures for our diseases. This was a view known as natural theology, which ensured that there need be no conflict between natural history and religious faith. On the contrary, many saw the two as complementary, and hundreds of English vicars spent their Sundays in the pulpit and their weekdays in the hedgerows, collecting plants, birds and butterflies.

However, by the end of the century natural history was changing rapidly. Naturalists were becoming biologists; divided not merely into zoologists and botanists, but into physiologists and anatomists, and a dozen other specializations. Meanwhile, those who studied fossils and rocks were emerging as palaeontologists or mineralogists. Another difference between natural history and the new biological sciences was *where* they were done. Naturalists worked pretty well anywhere, often in their own homes. By contrast, biology was mainly being done in public institutions, especially in laboratories. As more countries industrialized, the competitive edge Britain had enjoyed merely because it had industrialized first began to fade. As the century wore on, science and technology became crucial tools in increasingly bitter imperial struggles for colonies and markets, and so government money increasingly went into science, funding new laboratories, employing more people and providing them with ever more complex and sophisticated equipment. As a result, by the end of the nineteenth century, the study of nature was dominated by a distinct group: they were almost all men; they had formal qualifications in science; they usually published in specialized journals, rather than in popular books and magazines; their work was often incomprehensible to the public, so that only their professional colleagues could judge its merits; their research was often funded by government or industry; they were paid to do science, so that 'amateur' had switched from being a term of praise to one of condescension; and, most significant of all, this new breed called themselves 'scientists', a word that had not yet been coined when Darwin set sail on the *Beagle*.

The biggest change was in *how* nature was studied. Natural *history* was mainly concerned with collecting and naming, while its more prestigious cousin, natural *philosophy* – the precursor to the studies we would now call physics – dealt with the underlying causes of things. During the nineteenth century, some began to ask what *caused* the patterns they had once been content to merely describe (this had been Lyell's main aim in devising his geological theories). For naturalists, observation had been the key to understanding – a naturalist went out into nature and sat, he waited and listened, writing notes and drawing sketches, but above all he watched. This all seemed rather too hit-and-miss to the new biologists; instead of just watching nature, biology set out to bring nature to a place where it could be controlled and managed – a laboratory. Once plants or animals were there, it was no longer necessary to wait for something interesting to occur; experiments could be designed that created interesting phenomena, and then re-created them repeatedly if necessary. While naturalists like Darwin has mostly been content to wait for nature, Mendel and Galton had wanted to move it towards the laboratory, enabling them to control the conditions under which incidents occurred and – in theory at least – that made it much easier to establish their causes.

Of course, these changes did not take place suddenly; the speed with which they took hold varied in different countries and in different sciences. But gradually, the claims of laboratory biology became increasingly hard to resist; Galton's decision to found an Anthropometric *laboratory*, instead of relying on purely anecdotal evidence, was partly prompted by his awareness that scientific standards were shifting and he needed firmer evidence for his theories. Mathias Jacob Schleiden and Theodor Schwann had formulated the cell theory fifty years before, based on work done in laboratories. The laboratory revolution took off in the German-speaking world long before it affected Britain, where the English tradition of resisting pernicious European innovations ensured that the gentlemanly amateur clung on much longer than on the Continent. Whatever else it might be, an Englishman's home was no laboratory and German biologists in particular began to question the reliability of the kinds of experiments men like Darwin were still doing in their back gardens.

Lab workers referred disparagingly to field naturalists as 'bug-hunters'; and the bug-hunters responded by calling the laboratory scientists 'worm-slicers', who sometimes did not know if they were dissecting a platypus or a pearly nautilus. The worm-slicers responded that the bug-hunters were on occasion unsure as to whether 'the hen came from the egg or the egg from the hen, and by what kind of process'.[16] The worm-slicers felt that only the results of properly conducted experiments could resolve the way in which chickens produced eggs, or anything else, by supplying reliable scientific data; without the control that a lab offered, the causes of variation in a wild population must remain forever obscure, regardless of how much data was amassed. To many of the lab workers, evolutionary theory seemed to have run aground, mired in pointless controversy over the meaning of dubious anecdotes. By contrast, the bug-hunters, who studied real populations in the wild, often dismissed lab work as artificial since its results were shaped by its unnatural conditions. Such abnormal surroundings would never shed light on the evolution of real wild organisms because evolution was too slow to be studied in the lab. These differences made the disputes over evolution almost impossible to resolve, since there was so little agreement about what kinds of evidence were relevant.

From pangenesis to mutation

With the help of Oenothera, Hugo de Vries was determined to end the sterile battle between the different evolutionary camps. He had loved plants since childhood; when he was only twelve years old, he had a herbarium of over 100 specimens. Like most botanists of the time, de Vries began his studies with taxonomy – the classification of plants – which had preoccupied naturalists for centuries. But as a student he read a new book by the German botanist Julius Sachs which transformed his view of plants, turning him into a dedicated laboratory experimentalist. He more or less abandoned classification and began to conduct his own experiments, winning a prize for his investigations of the effects of temperature on plant roots. His final doctoral dissertation on temperature and plants earned him an additional year of study in Germany, working with Sachs himself. There, de Vries became even more strongly steeped in the German tradition of laboratory botany founded by Schleiden. Back in the 1840s, before de Vries was even born, Schleiden had been urging his students not to waste their time collecting herbarium specimens (which he dismissed as 'hay'), but to buy the best microscopes they could afford. As he put it, anyone 'who expects to become a botanist or a zoologist without using the microscope is . . . as great a fool as he who wishes to study the heavens without the telescope'.[17]

In 1877 de Vries came back to Holland, as Professor of Botany at the University of Amsterdam, determined to apply the cell theory, the microscope and the other modern German methods to understanding how plants worked. Why, for example, did plants wilt when they did not get enough water? He performed careful experiments that helped unravel the way that water pressure within cells helps maintain the rigidity of stems and leaves.

Yet, despite the success of his physiological experiments, de Vries gradually abandoned them to work on variability and how it was inherited. These problems struck him (and many others) as the really tough ones; whoever could solve them would have completed Darwin's theory, a prospect that fascinated the ambitious de Vries. In 1881 he wrote to Darwin, telling him that 'For some time I have been studying the causes of the variations of plants, as described in your Treatise on the variations of animals and plants under domestication, and have endeavoured to collect some more facts on this theme.' In contrast to most of those working on variation, de Vries told Darwin that he was 'especially interested in your hypothesis of Pangenesis, and have collected a series of facts in favour of it'.[18]

However, de Vries modified Darwin's theory by dividing it in two. He was convinced by the idea that inheritance was controlled by separate gemmules, each of which was inherited independently of the others. But Darwin had also argued that the gemmules circulated within the body, accumulating in the reproductive organs, ready to be passed on to the next generation. De Vries argued that this second theory, which he called the 'transportation hypothesis', was not essential. Transportation had only been included to explain how acquired characteristics were inherited (modified organs produced modified gemmules), but this was an idea that was rapidly falling into disfavour among biologists of de Vries's generation. August Weismann's concept of the 'continuity of the germ-plasm' – the separation of the reproductive cells from the rest of the body – made the inheritance of acquired characteristics impossible, and Galton's rabbit experiments seemed to have proved that there were no gemmules in the bloodstream. And, as the cell theory became increasingly widely accepted, biologists became convinced that cells were always formed by the division or fusion of other cells; which made it hard to imagine how a swarm of gemmules was supposed to coalesce into a new egg or sperm.

So de Vries simply abandoned the transportation hypothesis, while retaining the idea that some kind of minute particle was responsible for the inherited characteristics of each separate aspect of an organism. He published the modified theory as *Intracellular Pangenesis* (1889), a title that paid tribute to Darwin while emphasizing the differences in the two theories; de Vries rechristened the hereditary particles 'pangenes' for the same reason. Pangenes were, he hypothesized, found in every cell, probably in the enigmatic nucleus that Robert Brown had first identified. Each nucleus stored every type of pangene, but only some left the nucleus to become active. De Vries hypothesized that the active pangenes depended upon the kind of cell they were in: pangenes for eye colour were active in the cells that made up the iris, but not in the skin of the eyelid, even though they were present in both.

Since Brown's time, microscopes and the techniques for using them had improved, and laboratory biologists had observed strange, threadlike objects within the nucleus. They were eventually dubbed 'chromosomes' (derived from the Greek *chromo* – colour and *soma* – body), a name that reflected the bafflement the microscopists felt. No one knew what these 'coloured bodies' were or what they did, but during the 1880s a series of researchers gradually unravelled the way chromosomes worked. They were aided by new tools, such as improved microtomes (which cut very thin, even sections through specimens, producing better microscope slides), and a worm parasitic on horses, Ascaris, which briefly became a star organism for the study of chromosomes. The worm revealed enough about the role of chromosomes for the German embryologist Wilhelm Roux to argue in 1883 that they must carry the hereditary material.

De Vries had no direct evidence as to what his hypothetical pangenes were, but he guessed they might be connected to the chromosomes – further research would show. But whatever they were, pangenes were nothing like the little free-swimming animalcules that Darwin seems to have imagined his gemmules to be. This was a key point for de Vries: pangenes were more than just gemmules under a new name; they were the hereditary elements, the true building blocks of species. He assumed that, just as the same chemical elements can be combined to form many different molecules, each with unique properties, so the same elementary pangenes must occur in many different species, but each species had a unique mixture of them, hence its unique identity. Otherwise, he reasoned, it would be impossible to hybridize different species to produce new types.

De Vries supposed that pangenes were inherited entirely independently of each other – which explained how you could inherit your mother's eye-colour but your father's hair. He argued that they must exist in both active and latent forms and could switch between these states; which explained how characters could disappear but later reappear, allowing your grandfather's nose to re-emerge as one of your inherited characteristics. Even more interestingly, the pangenes might be changed in some way during cell division – mistakes or irregularities could arise and these might potentially be radical enough to create new species at a stroke, in a single generation. Such rapid change could overcome many of the problems posed by natural selection's reliance on small variations.

De Vries published *Intracellular Pangenesis* with great excitement, convinced he had completed Darwin's unfinished revolution, but the book was a failure. The few reviews that appeared were very critical, attacking almost every aspect of the theory, but focusing particularly on how little solid evidence de Vries had offered for most of his claims. He was already a well-known plant physiologist, yet had failed even to hint at a physiological basis for his theory. It all seemed too speculative. The book sold poorly and won him few converts.

De Vries was disappointed, even angry, but not ready to give up. He realized that in the new world of laboratory biology he needed more convincing evidence for his theories, so he began to collect it. He had been carrying out plant-breeding experiments for some time, but it was only in the mid-1880s that he began to use statistics to analyse them. In 1893, he read one of Weldon's papers, which acknowledged Galton's influence and assistance. De Vries immediately sought out Galton's books, including *Natural Inheritance*; he was fascinated by what he read and started referring to 'Galton curves' in his notes.[19]

De Vries soon became convinced that Galton's statistics were useful for analysing what he called 'fluctuating variability' – the small everyday variations in a plant's normal features that others referred to as continuous variation. These fluctuations, he decided, were governed by the numbers of pangenes: a few more or less 'height' pangenes produced a larger or smaller plant, but they could only affect a character's intensity, they could not generate new features, much less new species. That needed larger, more dramatic variations.

Large changes must be rare, so rare that no one had yet observed them, but de Vries was confident that he could find them. He wrote that 'what was known at this time was sufficient to convince me that the formation of species should lend itself to experimental investigation', but realized that his view 'was certainly directly opposed to the *reigning* opinion . . . of a slow and gradual origin'. (As we have seen, this view – far from being the reigning one – was under threat in the late nineteenth century, but de Vries liked to present himself as battling against the weight of received opinion.) He began to search systematically for a plant that would prove him right. Seeds of hundreds of wild flowers were collected and grown in Amsterdam's beautiful old botanical garden, De Hortus Botanicus. The odds were, he recognized, heavily against him, 'Yet I was lucky enough to find the very thing wanted': *Oenothera lamarckiana* – the only one among over 100 species tested that showed the sudden leaps he had been looking for. In an abandoned potato field not far from Amsterdam, he found new types of Oenothera growing side-by-side with the familiar form. The differences were subtle but, to his practised eye, unmistakable. The most significant point was that the new types grew happily alongside the parental form – they were not swamped by cross-pollination with the original type. Certain that the end of his quest was near, de Vries decided to 'abandon nearly all other experiments and to study this one plant as thoroughly as possible'.[20]

The Mutation Theory

De Vries later described how in 1886 'I collected a quantity of seed from wild plants of *Oenothera Lamarckiana*', which, together with growing plants, was brought to the Amsterdam botanical garden. The seed was sown 'and yielded at once what we desired' – proof that he had been right all along and that those who had dismissed Intracellular Pangenesis had been mistaken. 'Among the plants obtained from [the seed], there were three which, though agreeing among each other, possessed characters entirely deviating from those of the rest.' The new plants were initially nicknamed 'roundheads' because of their tightly bunched leaves. They had other peculiarities, the most important of which was that they did not produce fertile pollen. De Vries named these plants *Oenothera lata*, certain that he had found a new species.[21]

Over the next few years, de Vries continually visited the field where he had first collected his seed, looking for more odd varieties and collecting anything that looked promising. *O. lata*, which initially had only been found in the garden, was discovered in the wild as well. Other types soon followed, but some were 'too weak to live a sufficiently long time in the field'. That was worrying, because it might mean they were merely diseased or degenerate forms, rather than new species.[22] *O. lata*'s apparent sterility also suggested that it could hardly be the progenitor of a new species. Fortunately, de Vries soon found a more promising sport which he called *Oenothera gigas*, describing it as 'a splendid, exceedingly robust plant, which, with a rich crown of very large flowers, easily excels the mother species'.[23] But the crucial point for these species was that they all bred true – their offspring resembled them and showed no signs of reverting to the original type. They were 'new forms, which are sharply contrasted with the parent, and which are from the very beginning as perfect and as constant, as narrowly defined, and as pure of type as might be expected of any new species'.[24]

These new types were rare – 15,000 seedlings produced only ten plants with unusual features – but that was only to be expected; it explained why this crucial phenomenon had been overlooked in the past. Crucially, there were no intermediates between the new and the old, just a clean jump. As de Vries wrote:

They came into existence at once, fully equipped, without preparation or intermediate steps. No series of generations, no selection, no struggle for existence was needed. It was a sudden leap into another type, a sport in the best acceptation of the word. It fulfilled my hopes, and at once gave proof of the possibility of the direct observation of the origin of species, and of the experimental control thereof.[25]

For de Vries, these were the crucial points: first, 'direct observation of the origin of species' – evolution was no longer a matter of speculation, but something that could be done under controlled conditions. As he announced on the title page of the first English-language book that explained his theory:

The origin of species is a natural phenomenon.

LAMARCK.

The origin of species is an object of enquiry.

DARWIN.

The origin of species is an object of experimental investigation.

DE VRIES.[26]

The second exciting implication of the evening primrose work was the possibility of the 'experimental control' of speciation; once the mechanism that created new types was understood, it might be possible to produce new species to order. Instead of waiting for nature to throw up a useful variation, human plant-breeders would be able to create them. Nature itself would come under direct human control.

After more than a dozen years of experiments, de Vries produced two sturdy volumes, originally published in German under the title *Die Mutationstheorie* ('The Mutation Theory', 1901–3). In them, he boldly announced that the long-held view of evolution as slow and gradual was incorrect and that 'The object of the present book is to show that species arise by saltations and that the individual saltations are occurrences that can be observed like any other physiological process.'[27] He gave these new types the name 'mutations', giving a new meaning to an old word and arguing that they were the key to evolution. Because *Intracellular Pangenesis* had been received so badly, de Vries did not refer to it in his new book, but it was still the basic theory underlying his work.

Inevitably, some wondered why, if mutation was such an important phenomenon – and so common in Oenothera – it had not been noticed before. De Vries's answer was to assume that every species enters into periods of mutation, but these are very rare. Perhaps in response to dramatic new environmental conditions (de Vries was vague on this point), they began to generate new forms that were filtered by natural selection, allowing the species to adapt to its new conditions. As the environment stabilized, so did the species, which explained why rapidly mutating species like Oenothera were so rare. It also explained why the fossil record revealed so few intermediate forms – it would be very surprising if a fossil preserved a species in the middle of its intense, short-lived burst of mutation. And the idea of mutation periods also overcame Fleeming/Jenkins's objection that new, improved forms must be swamped by the old, unimproved forms. As Oenothera demonstrated, 'these new species are not produced once or in single numbers, but yearly and in large numbers', which meant that – during a mutation period – there were enough specimens of the new type around to allow them to breed with each other, which was why they were able to breed true without blending.[28] This explanation satisfied de Vries's supporters, but critics remained dubious, wondering what caused these hypothetical mutation periods given that *Oenothera lamarckiana* did not appear to be experiencing any unprecedented environmental conditions.

Creating a community

The Mutation Theory hardly seemed designed to generate excitement: two thick volumes of detailed experiments laboriously described in de Vries's heavy German prose. Yet it did excite. In 1905, the American Society of Naturalists – the USA's leading biological society – devoted a special meeting in Philadelphia to the Mutation Theory and the evening primrose. Some of those who attended were probably attracted by de Vries's assertion that his theory offered a clear, unambiguous definition of species, one that would end the endless debates of the classifiers. He argued that once biologists learned to identify the large, progressive mutations that created new species, they would be able to identify what he called the 'elementary species', nature's true units. Just as pangenes were the building blocks of organisms, elementary species were the key to its diversity. The evening primrose promised to end centuries of futile debate over naming and classi-fication because, de Vries claimed, most of the species named by classifiers were really false species, an untidy amalgam of several elementary species. Identifying these would also allow breeders to purify and thus improve the crops on which they relied.

Although many of the details of how this promised trans-formation was to be achieved were still hazy, the potential was obvious. Charles Benedict Davenport, the influential director of the Cold Spring Harbor Laboratory on Long Island hailed the book as marking 'an epoch in biology as truly as did Darwin's *Ori-gin of Species*', adding that 'there was need of a revolution in our method of attacking the problems of evolution'. However, the two books were based on different kinds of evidence: Darwin's relied on observation, anecdote and speculation; whereas de Vries's book was 'founded on experimentation', which – in Davenport's view –

meant that it was only by repeating and confirming the experiments that biologists could 'judge of its merits'. Whether the theory itself was confirmed or not, the book would prove invaluable if, as Davenport expected, 'it creates a widespread stimulus to the experimental investigation of evolution'.[29]

De Vries's book carried the laboratory revolution with it, stimulating new experimentation faster than anyone could have anticipated: from meetings like the one in Philadelphia, biologists returned to their labs to set up experiments, read de Vries, apply for funding and test hypotheses, but most of all, they planted evening primroses. The crucial question everyone hoped to answer was what exactly was heredity? How were the characteristics of parents transmitted to their offspring? As Karl Pearson had written, 'If Darwin's evolution be natural selection combined with *heredity*, then the single statement which embraces the whole field of heredity must prove almost as epoch-making to the biologists as the law of gravitation to the astronomer.'[30] Whoever solved the problem would be biology's Newton, and the evening primrose might prove to be a real-life counterpart to Newton's apocryphal apple. Ten years after the Philadelphia meeting, the journal *Nature* observed that 'Since the publication of de Vries's classic work the Oenotheras have attracted more scientific attention than almost any other plant or animal.'[31]

All kinds of attempts were made to identify mutations in other plants, as well as in animals, but Oenothera remained the star. A community of researchers – in both Europe and America – rapidly grew up around the plant; they were often divided by language, came from different backgrounds, had varied specializations and diverse interests, but the evening primrose provided a common language that allowed them to communicate and collaborate. This was the first time in the history of the life sciences that an organism had played this role; Oenothera promised a fresh approach to biology. That was perhaps what was most fascinating about the Philadelphia meeting: the range of biologists who attended. Laboratory workers left their microscopes to rub shoulders with plant- and animal-breeders, and with field naturalists and embryologists. These different communities were interested in different aspects of the theory, but all shared de Vries's belief that 'we may hope to realise the possibility of elucidating, by experiment, the laws to which the origin of new species conform'.[32] The nineteenth-century natural history tradition had been marked by tireless eclecticism; Darwin had worked not just on dozens of different plants and as many animals, but was also a well-known geologist. By contrast, twentieth-century biology was becoming rapidly fragmented into new specializations, but Oenothera offered a chance to bring some of these specialists together, to build an innovative kind of biology, centred around an organism instead of a single discipline.

The attention paid to Oenothera in the United States in part reflected the country's economic needs; in the last decade of the nineteenth century, the USA had suffered a severe economic depression, which provoked fears of social unrest that had followed similar depressions in Europe. Lacking any substantial empire, the United States increasingly turned to science to restore its economic prosperity; in 1900, leading companies like General Electric and Du Pont founded research laboratories on the German model. But as the USA economy recovered, its rapid industrial growth was accompanied by a rapidly growing population who needed feeding. Much of the interest in understanding inheritance was driven by this practical consideration. In 1910 the first issue of *American Breeder's Magazine* announced that 'heredity is a force more subtle and more marvelous than electricity. Once generated it needs no additional force to sustain it. Once new breeding values are created they continue as permanent economic forces.'[33] The potential rewards for understanding this mysterious, profitable force went far beyond the satisfying of scientific curiosity. In addition to Davenport, De Vries's American supporters included Jacques Loeb, another disciple of Sachs with an interest in the control of life's fundamental processes, and Thomas Hunt Morgan of Columbia University. John H. Schaffner, at Ohio State University, claimed to have identified a mutant Verbena, while the beetle collector Colonel Thomas Lincoln Casey used the Mutation

Theory to explain palaeontological evidence of the rapid appearance of fossil molluscs and even argued that the Mutation Theory was more theologically acceptable than conventional Darwinism. But in the US, Daniel Trembly MacDougal was the leading Oenothera biologist.

MacDougal had trained as a plant physiologist at Purdue University, Indiana, and had subsequently spent a year working in Germany, where he got his PhD and – even more importantly – where he met both Sachs and de Vries, who rapidly became MacDougal's mentor. The young American returned home determined to adopt the European laboratory approach and apply it to the evolutionary topics he was most interested in, but which had previously been exclusively the focus of field work. MacDougal ended up working for the New York Botanic Gardens as Assistant Director of its labs. He also became an adviser to the Carnegie Institution of Washington, a philanthropic body established by the wealthy industrialist Andrew Carnegie to promote science. One of the first funding proposals it received came from Frederick V. Coville (the man who had collected the vomiting coyote story) who wanted to found a laboratory to investigate how plants coped with the harsh conditions of deserts. The Carnegie Institution agreed and it was established in 1903 at a site selected by Coville and MacDougal, near Tucson, Arizona. Three years later, MacDougal left New York to became the new lab's director.

Oenothera seemed to offer MacDougal just what he too had been looking for, a way to investigate evolution in the laboratory. He began a series of breeding experiments and pedigree studies, originally in New York, which were intended to reproduce de Vries's primrose experiments, partly in order to establish that it was the plants themselves, not some peculiarity of the growing conditions in Amsterdam, that were responsible for the mutations. He also wanted to confirm that the mutants really were new species which bred true, rather than hybrids that would revert to the parental form. MacDougal got seeds from de Vries and began to turn the botanic gardens' flowerbeds into an extension of its laboratory; he even had the soil sterilized in a medical autoclave before the seeds were planted, in an effort to control every factor that might complicate his results. When the seeds germinated, he was delighted to report that all de Vries's mutants were observed and their appearance and other characters conformed to de Vries's descriptions.

MacDougal became a crusader for the lab approach: he began to campaign before he had even started his own Oenothera experiments. He was fired by a vision of a revolutionary new biology – based on the Mutation Theory, the European laboratory approach and most of all, on Oenothera. He believed that the benefits of his proposed revolution would extend well beyond pure science: the tantalizing prospect of the eventual control of mutation promised economic rewards that would dwarf those offered by traditional plant and animal breeding. MacDougal toured the United States, giving lectures, visiting research institutions and promoting the evening primrose. In 1904 he arranged for de Vries to give a series of lectures on mutation at the University of California, which MacDougal then edited into a successful book, *Species and Varieties* (1905). This was the first English-language version of De Vries's theory and – as some reviewers noted with relief – shorter and rather more accessible than the original, which undoubtedly helped promote Oenothera and the Mutation Theory. In sharp contrast to *Intracellular Pangenesis, Species and Varieties* went into a second edition within a few months, featuring a frontispiece showing de Vries visiting MacDougal's Desert Botanical Lab, symbolically giving his blessing to innovative American biology.

For MacDougal, the support of the distinguished European botanist helped attract patronage and financial support, and de Vries was thrilled by the American enthusiasm for his ideas. He praised America's scientific progress, reminding his hosts that the evening primrose 'which yielded these important results is an American plant. It is a native of the United States', yet one which 'by a strange but fortunate coincidence bears the name of the great French founder of the theory of evolution', Lamarck.[34]

For MacDougal, as for his colleague Jacques Loeb, biology in the twentieth century was going to be about much more than simply understanding life; their goal was controlling life, engineering new organisms for the benefit of humanity. The Carnegie Institution, which funded both MacDougal's work and the Cold Spring Harbor Lab, was crucial in promoting this new approach. Robert S. Woodward, the Institution's president, told the institute's trustees that MacDougal meant 'in the future to apply invention to living things' – nothing they were funding was more exciting. The practical applications of the new theory were especially attractive to Americans, with a rapidly growing population to feed, and the promised speed of controlled evolution – based on generating new mutations – appealed greatly to a nation that prided itself on progress and seemed permanently in a hurry to pursue it.

No sooner had MacDougal started on his Oenothera experiments than he had exciting results to announce: the first artificially created mutations. He and a colleague, George Harrison Shull, had used the mysterious new force of radiation to produce mutations in a closely related genus of evening primroses. Their work had been inspired by experiments performed by one of MacDougal's colleagues, Charles Stuart Gager, who had used radium to produce what appeared to be heritable changes. Gager was cautious about his results, unsure whether these changes were the same as de Vries's mutations, but MacDougal had no such doubts. Radium clearly affected the chromosomes, which were increasingly accepted as the carriers of the inheritable characteristics. MacDougal leapt to the conclusion that radiation would provide a 'ready means of suppression or substitution of characters'. Meanwhile, field studies at the Desert Botanical Lab suggested that beetles produced heritable changes in response to intense environmental stimuli. This work was also incomplete and its results inconclusive, but MacDougal seized on it as confirming the radium work in the lab. The engineering of life seemed just around the corner.[35]

In the first decade or so of the twentieth century, the Mutation

Theory seemed to go from strength to strength, but not everyone was convinced by this new, quickfire version of evolution. There were plenty of defenders of traditional Darwinism still around. Among the Mutation Theory's opponents were William Bateson and August Weismann, and a number in America, where the most vociferous opponent was probably Clinton Hart Merriam. Merriam was a highly respected member of the US Biological Survey; he took the opportunity of his vice-presidential address to the 1905 meeting of the American Association for the Advancement of Science to attack the mutationists, claiming that their retreat to the laboratory had left them ignorant of real, wild species. He observed that until the arrival of the Mutation Theory, 'it was the practically unanimous belief of zoologists and botanists the world over' that species arose in just the way Darwin described. 'Has any reason been brought forward to justify – much less necessitate – a change in this belief?' Merriam demanded. 'Are we, because of the discovery of a case in which a species appears to have arisen in a slightly different way – for after all the difference is only one of degree – to lose faith in the stability of knowledge and rush panic stricken into the sea of unbelief, unmindful of the cumulative observations and conclusions of zoologists and botanists?'[36]

Merriam described the mutationists as suffering from 'enthusiasm run wild'. Even if species do sometimes arise as a product of extremely rare mutations, that in no way proved that they *could not* arise through the gradual selection of the much more common continuous variations. As Merriam asked, 'Why can not species originate in both ways?'[37] However, MacDougal – who, unlike many laboratory biologists, *did* have extensive field experience – fought back in a lecture the following year. As the laboratory approach spread, US naturalists were increasingly on the defensive in the first decade of the twentieth century; at the 1908 meeting of the Botanical Society of America, MacDougal led an attack on the failure of traditional naturalists to define species clearly, arguing that only laboratory evidence could uncover the real relationships between species.

However, the Mutation Theory was not merely opposed by field naturalists and traditional Darwinists. A new and increasingly influential group of biologists were emerging who believed their theory would supplant both Darwin and de Vries. They were the Mendelians.

Mendel's rebirth

Although Mendel's work was both known of and occasionally cited by botanists during the decades after his death, it was generally overlooked. The supposed 'rediscoverers' were Carl Correns, a German, Erich von Tschermak, an Austrian and Hugo de Vries. All three were supposedly working independently on plant breeding and hybridization when they hit on the same ratios Mendel had found, which led them to go back and look again at Mendel's 1865 paper. The rediscoverers maintained that they had each arrived at Mendel's laws independently, but historians have increasingly shed doubt on some of the details of their claims. What is clear that while all three reread Mendel's pea paper, they did so for rather different reasons and followed very different and circuitous routes back to Mendel. None of the three seem to have experienced a 'Eureka!' moment when they read his work.

Although these three botanists decided to agree that Mendel was the true founder of this new science, which rapidly became known as Mendelianism or Mendelism, they soon began developing Mendel's ideas in new directions. For example, Correns immediately took Mendel's *Anlage* and recast them into a more precise and familiar form. The characteristics of the plant , such as pea colour or height, were defined as its characters, each of which was controlled by a pair of what Correns called 'factors'. Mendel himself had never made this assumption; it is clear from the way he recorded his experiments that he never imagined that each character was controlled by dual factors. As a result, what became known as *Mendel's* Law of Segregation – the two factors which control a particular trait are separated in the formation of eggs or sperm – was clearly not Mendel's law at all. So while the

Mendelians claimed to be the true inheritors of Mendel's tradition, they found it necessary to recast his work in a form he probably would not have recognized or understood; just as the biometricians reworked Darwinism into a statistical science, even as they claimed to be his truest defenders.

The biometricians were, unsurprisingly, immediately hostile to the new Mendelian approach, but so too were some supporters of Weismann's work, which assumed that *multiple* factors controlled each character, not a simple pair as the Mendelians initially believed. These conflicts help to explain why Correns, Tschermak and de Vries found it useful to give the real credit for the new science to Mendel: a heroic, neglected and safely deceased founding father was much more useful to the promoters of the new approach than a bitter priority dispute between them; a squabble among the Mendelians could only have served their opponents. In de Vries's case, priority over the rediscovery mattered even less because he did not believe Mendel's ratios were all that important. He repeatedly stressed that Mendel's laws only applied to hybridization between species, just as the biometrician's curves only mapped insignificant everyday variation in a population. To de Vries, neither was crucial, since neither could lead to the creation of species. He did not deny that natural selection existed, but insisted that natural selection is 'a sieve which decides which is to live, and what is to die', but cannot create the variations it selects between. As one wag observed, in de Vries's view, natural selection explained the *survival* of the fittest, but not the *arrival* of the fittest.[38]

However, if de Vries was happy to relegate Mendel's contribution to a subordinate position, others were more excited. Bateson became Britain's foremost exponent of the new theory, eagerly seizing on the idea of independent factors, or unit characters as they became known, as proof that he had been right all along – *dis*continuous variation was the key to evolution. Mendel's peas did not exhibit the blending that seemed so fatal to Darwin's original formulation of natural selection; they were clearly yellow or green, wrinkled or smooth. In Bateson's hands,

Mendelism became another stick with which to beat the biometricians and his eager promotion of the new theory further exacerbated his conflict with them. De Vries wrote to Bateson to urge him not to get carried away with Mendelism: 'it becomes more and more clear to me that Mendelism is an exception to the general rule of crossing. It is in no way *the* rule!'[39]

In Bateson's case, this plea fell on deaf ears, but supporters of the Mutation Theory tended to be suspicious of what they regarded as the inflated claims of the Mendelians. British Oenothera research was dominated by the Canadian-born Reginald Ruggles Gates, who had completed his PhD on evening primroses at the University of Chicago. In 1910 he had met the pioneering birth-control advocate Marie Stopes at a scientific meeting. They were married the following year and settled in London, but – despite his rapidly growing reputation – Gates had trouble getting a permanent academic position. He served in the British air force during the First World War and somehow found time to write a major book, *The Mutation Factor in Evolution* (1915), focused on Oenothera. However, according to his wife, Gates never found time to consummate their marriage and it was annulled in 1916. Gates remained bitter about Stopes's claims and continued to deny them until the end of his life (in private, he always insisted that she was not the innocent virgin she had made herself out to be in court). But although Gates married twice more he never produced any children, presumably a disappointment to an ardent eugenicist. After the war, Gates settled at King's College, London, where he worked for many years on a wide variety of plants and animals, in addition to Oenothera.

Gates's *Mutation Factor* was widely reviewed, with one commentator observing that 'the book may be described as an entrenchment of the Mutationists, from which they hurl explosive and very damaging missiles at their opponents, the Mendelians'.[40] Gates emphasized that the Oenothera mutants did not in any way conform to the Mendelian ratio, which suggested that mutations were something quite distinct from an organism's normal variability. Other reviewers were less entertained by Gates's missiles:

the *Botanical Journal* observed that 'The book is by no means clearly expressed', but the reviewer recognized Gates as 'a close disciple of De Vries and strong opponent of the Mendelian School, at any rate, so far as the Oenotheras are concerned'.[41] And there, indeed, was the rub: 'so far as the Oenotheras are concerned'. Despite strenuous efforts, even vaguely comparable examples of mutations in other plants or animals proved hard to find. To the increasing embarrassment of the theory's supporters, mutation appeared to be restricted to Oenothera, which rather weakened its claims to be a universal theory of evolution.

As we have seen, de Vries argued that the rarity of mutation periods explained the failure to find more examples, but to critics this seemed a rather ad hoc explanation. The other problem that continued to confront the Mutation Theory was de Vries's failure to demonstrate the material basis of mutation; *what* exactly was mutating and where? In reviewing the first English translation of the *Mutation Theory*, Gates commented that 'while the supreme importance of De Vries's investigations on mutation in Oenothera is fully recognized', there were aspects of his approach that had 'always seemed to the reviewer unsatisfactory'. Although Gates was a taxonomist and not primarily a laboratory worker, he and his collaborators began to investigate the material basis of mutation. Once their role in heredity had been confirmed, chromosomes understandably became an object of fascination for many researchers, including Gates and his team. Gates observed that 'the whole question of the relation of the mutants to their parent will be found to be much more complex than at present supposed', which proved to be a classic example of English understatement, as we shall see.[42]

Too big, too slow?

By about 1915, the enthusiasm for Oenothera was losing its momentum. It was becoming obvious that, whatever it was that caused the new forms to arise, it was something unique to the evening primrose. As an experimental organism, *Oenothera lamarckiana* suffered an eclipse almost as complete as its most famous exponents. For the first decade or so of the twentieth century it was a botanical superstar, one of the most widely studied plants in the world, but interest in the plant died away rapidly after the First World War. And as the Oenothera community declined, so did the Mutation Theory: despite dedicated hunting, comparable 'mutations' failed to turn up in other organisms and gradually the biologists' fields of Oenothera were abandoned as researchers turned their attention elsewhere.

Oenothera had briefly promised a new botanical approach, based on a variety of specialists all working with the same plant, which promised to provide a common language that would allow ecologists, embryologists, laboratory and field workers to communicate with each other. However, this was a promise that – for the botanists – would not finally be fulfilled until the last decades of the twentieth century, and by a very different plant, as we shall see. Perhaps Oenothera was unlucky in the researchers it attracted. Hugo de Vries was a difficult man, stubborn, sarcastic and often guilty of blind favouritism. His sharp tongue often alienated both colleagues and students, and once he had taken against a fellow biologist he seldom changed his mind. He never admitted that he had been wrong about the Mutation Theory and, after he retired, he continued to grow Oenotheras in his own garden, remaining convinced that he would eventually persuade the doubters, right up until his death in 1935. Daniel MacDougal was an equally difficult character, always ready to attack naturalists, palaeontologists, taxonomists and anyone else who disagreed with him. He seemed determined to alienate those he considered mistaken, by insisting that there was only one way to do biology: he even managed to annoy many of his fellow experimentalists, by maintaining that the Mutation Theory was the only theory worth considering. His patrons at the Carnegie Institution became anxious not to be seen to be promoting faction within biology and quietly began to support researchers with a more conciliatory attitude, in an effort to build bridges between different biological approaches. Reginald Ruggles Gates was also a complex character, not an obvious leader for a research group.

When *The Times* published his obituary in 1962, the obituarist felt obliged to acknowledge that 'as a colleague he was difficult to get to know and to be friendly with', admitting that 'when occasionally he became involved in unworthy controversies, sometimes on the wrong side, he rarely admitted to error or retracted any statement previously made'. Given that obituary writers tend to play down the faults of the deceased even today, one wonders how his opponents in those 'unworthy controversies' would have described him in private. And perhaps some whiff of scandal remained from the humiliation of his first divorce; it is certainly striking that the obituarist described him as 'almost effeminate in his gentleness'. At a time when homosexuality was still illegal in Britain, this may have been a coded message about the real reason for his failed marriages.[43] Had Oenothera attracted a less contrary and more modest group of workers, its research community might have grown more vigorously; as it was, many biologists were eager to see the plant's downfall.

However, even if the Mutation Theory had proved correct, Oenothera would probably have been doomed in any case. The evening primrose can grow to six feet (183cm) and most forms are biennial, meaning they take two years to produce seeds; if similar mutations had been found in a smaller, faster-growing organism it would quickly have become the main focus of research. Why would researchers fill fields with slow-growing plants if they could generate answers (and publications) more quickly by using some other organism?

Partly as a result of Oenothera's unique qualities, Hugo de Vries has the rather sad distinction of being famous for two failed scientific theories: Intracellular Pangenesis and the Mutation Theory, which were – as we have seen – really the same theory presented in two different ways. He is also remembered for something that he did not really do and which he certainly did not consider very important – the rediscovery of Mendel. As late as 1922, when the Mutation Theory was largely dead and Mendelianism was enjoying increasing success, de Vries was still insisting that 'the glorification of Mendel is a fashionable article in which everyone, even those without much understanding, can join; this fashion will surely pass'.[44]

What de Vries perhaps ought to be remembered for is accelerating the laboratory revolution in the study of evolution and emphasizing the immense possibilities science offered for improving the plants and animals we depend on for food. Although the Mutation Theory faltered, laboratory biology proved unstoppable: whether in universities or industry, funded by governments, individuals or corporations, laboratories continued to expand and the tools and techniques lab workers had at their disposal became more and more powerful. Within the labs, mutationism rapidly gave way to Mendelism, but biology was still divided. Outside the lab, there remained field workers, naturalists, taxonomists and biometricians – many of them scorning the 'artificial' conditions of the lab – who remained convinced that the Mendelians could not cope with the complexity of wild populations.

The American biologist Thomas Hunt Morgan, initially one of de Vries's supporters, was typical of the anti-Mendelians. He argued that Mendel's laws might apply to peas, but had not been demonstrated in other plants, much less in animals, and that in wild populations, dominant and recessive traits were seldom as clear-cut as in Mendel's simple contrasts between green and yellow peas. In the complex living world outside the artificial constraints of the experimental garden, many crossings produced intermediate forms, which left the whole problem of blending inheritance still unresolved. And many biologists agreed with Morgan that there was no more proof of the physical existence of Mendelian factors than there was of de Vries's unfashionable pangenes – they were all hypothetical constructs. His argument was bolstered by the tendency of enthusiastic Mendelians to simply postulate the existence of any factors they needed to make sure their sums came out right. As Morgan sarcastically commented, 'facts are being transformed into factors at a rapid rate. If one factor will not explain the facts, then two are invoked; if two prove insufficient, three will sometimes work out . . . I cannot but fear that we are rapidly developing a sort of Mendelian ritual by which to explain the extraordinary facts of alternative inheritance.'[45]

Concerned by the limitations of the 'Mendelian ritual', Morgan became an enthusiastic supporter of the Mutation Theory, but he realized that it could never really prevail until mutations could be demonstrated in other organisms. Based in an overcrowded lab in uptown Manhattan, Morgan had no room for fields of 'a beautiful, freely branching plant, often attaining a height of five feet or more', so he decided to look for mutations in a small, fast-breeding animal and selected *Drosophila melanogaster*, a tiny fly that was destined both to change Morgan's mind about Mendel, and to finally solve the Oenothera mystery.

Drosophila melanogaster

FRUIT FLY

Chapter 6

Drosophila melanogaster: Bananas, bottles and Bolsheviks

Like most children, my son's favourite fruit is the banana, which gives him something in common with Darwin, who also loved them. In 1876, when Joseph Hooker (by then director of Kew) sent his sweet-toothed friend bananas from Kew's hothouses, a delighted Darwin responded that 'You have not only rejoiced my soul, but my stomach, for the bananas are simply delicious. I never saw any like them.'[1] Bananas were then a rare treat, even for the comparatively wealthy Darwin – most Victorians would never have even seen one. Many Americans tasted their first banana in the same year, at the Philadelphia Centennial Exposition, where they could have bought a single banana, carefully wrapped in tinfoil, for 10¢. Despite the cost (over $10 in today's money), the bananas were so popular with visitors that guards had to be posted to prevent visitors pulling the trees apart for souvenirs.

Inspired by their popularity, an American ship's captain, Lorenzo Dow Baker, bought 160 bunches of bananas in Jamaica and freighted them to Jersey City in the US, where he sold them for a 700 per cent profit. Inspired by this success, he teamed up with a Bostonian merchant, Andrew Preston, determined to develop the banana market; the long-term result of their efforts was to be the United Fruit Company (now Chiquita).

It is unlikely to be a coincidence that in 1875, when bananas were starting to become common on the east coast of the United States, another exotic foreigner was first spotted in New York, a species of fruit fly which bug-hunting naturalists identified as *Drosophila ampelophila*. The name *ampelophila* means 'lover of grapes', reflecting its close association with fruit, especially with wine, but the species is now known as *melanogaster*, which – rather less glamorously – simply means 'black-bellied'. No one knows how this species arrived in New York, but the timing suggests they came in with shipments of fruit, probably from central America and the Caribbean, where most of the USA's tropical fruit still comes from.

Drosophila and bananas go together. As Groucho Marx once observed, 'Time flies like an arrow; fruit flies like a banana.' Together, flies and bananas created modern genetics by developing Mendel's vague *Anlagen* into genes, but they could not have done so (nor would bananas have become part of either my son's diet or Darwin's) without slavery, the impact of colonialism and a rather curious genetic behaviour called polyploidy.

As humans have moved across the globe, we have carried seeds with us, spreading our crop plants all over the world, so it is now difficult to know exactly where and when the wild ancestors of many of our domesticated plants evolved. Bananas are no exception – we have been eating them since before we could write, so there are no records of when and where we first found them. However, we can get a pretty clear idea by looking at where their wild relatives grow today. Modern commercially grown bananas are derived from two species of the genus *Musa* that are widespread across south-east Asia as far as India. By studying the overlap of the wild species, botanists have concluded that they almost certainly originated in what is now Malaysia. A similar exercise for Drosophila reveals a similar pattern: they too almost certainly evolved in the jungles of south-east Asia, so it is more than likely that as bananas and other tropical fruits were spread around the world by humans, the flies came along with them.

Most scholars credit the Arabs with bringing bananas to the Middle East and thence to Europe. They are considered a holy plant in the Koran – indeed both Islamic and some early Christian scholars considered them to be the first fruit that humans ate; there is even a Christian tradition that identifies bananas as the 'forbidden fruit' in the garden of Eden, a story which inspired Linnaeus to name one species *Musa paradisiaca*. Bananas were spread throughout sub-Saharan Africa by Arabic traders; western explorers reported them growing there in the late fifteenth century. Evidence of the Arabic influence comes in the name itself: 'banana' derives from West African names for the fruit, which were in turn derived from the Arabic *banan*, or 'finger'.

Bananas reached the Caribbean in 1516, thanks to Friar Tomás de Berlanga, a Catholic missionary of the Dominican order of Predicadores (preachers), who is best known for discovering the Galápagos islands. He brought African bananas to the island of Hispaniola (the present-day Dominican Republic and Haiti), where they were intended as a cheap food for the island's growing population of African slaves. When Berlanga became bishop of Panama he probably took bananas with him to the mainland, where they spread so rapidly that many later travellers assumed they were indigenous to central America.

Drosophila probably reached the Americas at about the same time as the banana, but the flies had been known to Europeans much earlier. They are first mentioned in Aristotle's *Historia animalium* ('History of Animals'), where he describes an insect whose grub 'is engendered in the slime of vinegar'.[2] Aristotle referred to this insect as 'conops', which translates as gnat or mosquito. ('Conops' is the source of the English word 'canopy', from *conopeum*, which originally meant a bed with a mosquito net.) The various references to conops in Aristotle seem to conflate several different insects, since he says some draw blood while others will only eat sour things, like vinegar, not sweet things. So Aristotle perhaps confused mosquitoes with vinegar flies (one of the common names for Drosophila).[3]

It might seem unlikely, given how different bananas and vinegar are, that they are eaten by the same insect, but Drosophila are in fact not fruit flies at all; they are not even closely related to the real fruit flies, which devastate fruit crops. Drosophila live on yeasts, the products of fermentation and decay. It is not fresh bananas but over-ripe and rotting ones that attract the flies; as they go bad, they ferment – producing yeasts for the flies to feed on. Anywhere that fruit is going bad, there are Drosophila.

Of course, humans do not only ferment fruit by accident. As *Drosophila melanogaster*'s original name, *ampelophila*, records, humans like to ferment fruit to produce alcoholic drinks. (Another name for them, common in the early twentieth century, was pomace fly – pomace being the mashed apple that is left after cider-making.) Shipments of rum probably had as much to do with Drosophila's arrival in the United States from the Caribbean as did shipments of fresh fruit. In the early nineteenth century, two English naturalists, William Kirby and William Spence, named these insects Oinopota, which means 'wine-drinker', christening one species *Oinopota cellaris*, because they believed it was only found in cellars where wine and beer were stored.[4] *Oinopota cellaris* found a place in one of the most remarkable books of the nineteenth century, the best-selling *Vestiges of the Natural History of Creation* (1844). Its anonymous author used the existence of this fly which 'lives nowhere but in wine and beer, all of these being articles manufactured by man' as evidence of the spontaneous generation of life.[5] The author, now known to have been the Edinburgh publisher Robert Chambers, reasoned that the flies could not have existed before humans were helpfully stocking cellars for them. And since they could not survive outside the cellars, they must have arisen, *in situ*, over and over again; whenever and wherever the right conditions existed, the flies were spontaneously generated.[6] (Obviously the flies were not spontaneously generated in cellars, any more than they are in your fruit bowl; their eggs are simply too small to be visible to the naked eye.)

Despite their cellar-dwelling and obvious fondness for alcohol, the name Oinopota did not stick. Although they evolved in the tropics, these insects cannot cope with too much heat – as the temperature rises, they become sterile and eventually die. So, unlike mad dogs and Englishmen, the flies avoid the midday sun; they are usually active around dawn and dusk, which is why in 1823 the Swedish naturalist Carl Frederik Fallén renamed them Drosophila, which literally means 'dew lover'.

Thanks to the enterprising efforts of Catholic missionaries, slave traders and the United Fruit Company, bananas – and fruit flies – became increasingly common in American cities towards the end of the nineteenth century. By the first decade of the twentieth century, United Fruit effectively ruled the Caribbean 'banana republics' it dealt with, owning everything from the plantations, the ships and railroad cars, to the markets where the fruit was sold. They had the growing and shipping of bananas so well organized that they were able to supply them all year round, right across the United States. By 1905, the company imported forty bananas a year for every man, woman and child in the USA; by 1920, it had become one of America's largest corporations. A combination of steamships, railroads, refrigeration and ruthless business tactics turned the banana from a luxury item into America's most popular fruit.

Race, class and fruit flies
The speed with which bananas – and flies – could travel from Jamaican plantations to New York kitchens was just one symptom of the pace of the twentieth century. New technologies such as refrigerated transport made it possible for people to eat new things, but the growing prosperity of the United States by the early twentieth century was also crucial in creating the demand for exotic technologies and foods. In the late 1890s, appalling harvests in Europe had coincided with bumper ones in America; as exports soared, American factories and farmers found they could not keep pace with demand across the Atlantic. Unemployment fell so fast that a labour shortage seemed possible and American producers worried that their domestic market was becoming saturated. Immigration was the solution to both problems, and millions of Europeans poured across the Atlantic, escaping pogroms and hunger, hoping for freedom and

prosperity. A second great migration brought hundreds of thousands of African-Americans from the south to the north, leaving behind lynch mobs and plantations for work in the seemingly ever-growing factories. The USA was transformed.

The millionaire Andrew Carnegie marked the new century with a new edition of his book *The Gospel of Wealth*. The title brashly expressed the self-confidence of US industry that having put millions to work – and put bananas on every table – American business savvy held the key to solving the world's problems. If everywhere could be like the USA, everyone could be clothed, housed and fed. Carnegie backed his judgement by investing millions in ambitious philanthropic schemes, such as the Carnegie Institution of Washington, which – as we have seen – invested heavily in biological research, intent upon vanquishing disease and hunger. Like most of his contemporaries, Carnegie assumed that competition was the key to progress: 'It is to [the law of competition] that we owe our wonderful material development, which brings improved conditions in its train.' He argued that competition was 'not only beneficial, but essential to the future progress of the race'.[7] Carnegie's contemporaries also saw competition as a law of nature, as Darwinism in action; as one of them wrote, 'millionaires are a product of natural selection'.[8]

Darwin had borrowed his metaphor of natural selection from industrial capitalism; believing, like most Victorian gentlemen, that competition between rival businesses led to the best products dominating the market, he reasoned that similar competition between organisms would lead to the best-adapted dominating the reproductive market – by producing the most offspring. So, using Darwinism to prove that capitalism was natural was perhaps rather a circular argument, but men like Carnegie were not daunted by such considerations. They rather tended to worry that there might not be enough competition to ensure continued progress. As prosperous nations like the United States got even wealthier and technology made life easier (at least for some), Carnegie and others were concerned that the sharp edge of natural selection was being blunted and that instead of progressing, Americans might start to degenerate. Similar fears were also widespread in Europe. In 1900, the London *Times* told its readers that 'An Empire such as ours requires as its first condition an Imperial Race – a race vigorous and industrious and intrepid. Health of mind and body exalt a nation in the competition of the universe. The survival of the fittest is an absolute truth in the modern world.'[9] The question was, were the British still a fit race? During the Boer War a high proportion of the recruits had turned out to be unfit for service; the squalor and disease of the urban slums in which they had been raised were widely blamed. Other critics were more concerned with the officer class, claiming not only that society's elite was going soft, but also deploring the fact that the birth rate among the elites was so low (why this was so remained a subject not fit for polite conversation, though knowledge of effective contraception was widespread). The result was that the less fit were outbreeding their betters, reducing the quality of the race as a whole.

While the British, characteristically, worried about class, the Americans were more concerned with race. The Statue of Liberty extends its welcome to the Old World's tired, its poor, its 'huddled masses yearning to breathe free', but also offers a home to the 'wretched refuse of your teeming shore'. But some were apprehensive that the new arrivals pouring in through Ellis Island were indeed Europe's rejects – lazy, stupid and immoral. Was the hardy pioneer stock that had tamed the wilderness becoming fatally diluted by mass immigration?

Not surprisingly, these anxieties led to a revival of interest in Galton's theory of eugenics. When he had first proposed it in the 1860s he had been widely mocked, but in 1904, Galton – by now a rather grand old man of seventy-eight – gave a public lecture on eugenics to the Sociological Society of London, drawing a large, influential audience. One result was the founding of the Eugenics Education Society, which publicized his ideas and promulgated the benefits of controlled breeding. The revival of Galton's ideas led to new answers to his old question: was it nature or nurture that defined character and behaviour? Britain's infant Labour Party and its allies in the Liberal Party argued for nurture: the key to progress was eliminating poverty and improving the living conditions of the poor. But the eugenicists rejected this approach: Karl Pearson, the first occupant of the Galton-funded chair of eugenics at University College London, argued in 1909 that if the government were to 'Give educational facilities to all, limit the hours of labour to eight-a-day – providing leisure to watch two football matches a week – give a minimum wage with free medical advice', they would 'find that the unemployable, the degenerates and the physical and mental weaklings increase rather than decrease'.[10]

For many, the only way to resolve these arguments was through biology, by understanding the nature of heredity. The new science of Mendelism seemed to offer a solution: could controlled laboratory experiments finally prove, once and for all, whether you could choose to create a Prospero or a Caliban?

At Harvard, Professor William E. Castle, a modest Mid-westerner who was to become one of the century's most influential biologists, was one of many who took up this challenge. Castle was especially interested in the question of inbreeding: although repeated inbreeding was deliberately used by breeders as a way of 'fixing' a desirable trait, would this eventually prove harmful to the breed? Castle certainly found so when he carried out repeated brother-sister matings in rabbits – undesirable traits seemed to build up rapidly, eventually producing animals that were too sickly to survive. He was mainly interested in inheritance in mammals: mice, horses, rabbits and – as we will see – guinea pigs. However, in his efforts to find an organism that could withstand repeated inbreeding, he looked further afield, eventually settling on a creature that could survive twenty generations of close inbreeding with no loss of vigour – *Drosophila melanogaster*.

Castle's central focus remained on mammals, but several of his students took up the fly in the early 1900s and they were soon to be found in biology labs at Harvard, at Indiana University and at the Cold Spring Harbor Laboratory, where Frank Lutz, another of Castle's students, was using flies for work that was funded by the Carnegie Institution. It was from Lutz and Castle's other students that Thomas Hunt Morgan, at this time still a supporter of de Vries's Mutation Theory, first learned how useful the flies could be. While some researchers latch on to a particular creature and work with it their whole lives, Morgan loved finding new organisms to work on; at various times he studied pigeons, Hawaiian land snails and Western song sparrows. As one of his colleagues joked, Morgan 'has more irons in the fire than an ordinary man has coals'.[11]

In 1907, when a new doctoral student in Morgan's lab at New York's Columbia University was interested in experimenting on the inheritance of acquired characteristics, Morgan suggested he tried using Drosophila probably, at least in part, because his lab was small and already crowded – and the flies were very small. Castle may well have had similar motivations: in 1910, his lab had 400 rabbits, 700 guinea pigs, 500 mice, 1,000 rats, 400 pigeons, eight dogs and more frogs than he could count. Flies were also cheap to acquire and keep, and Morgan was notoriously mean with institutional money (despite being very generous with his own). The student later recalled how he had obtained his first stock of flies: 'I used the simple procedure of laying some ripe bananas on the window sills; the flies thus caught were the start of my experimental work.'[12] The long-established relationship between bananas and Drosophila was about to acquire a new partner, biologists. Together, they would do astonishing things.

Mutation hunting

The relationship between flies and biologists is an example of what is sometimes referred to as pre-adaptation: a clumsy term, but one that describes an important phenomenon. Sometimes a species that evolution has shaped for one environment proves – purely by accident – to be ideally suited to an entirely new one. As the historian Robert Kohler has shown, Drosophila proved to be pre-adapted to academic biology labs: the flies are plentiful in the autumn – the beginning of the academic year – thanks to rotting fallen fruit. As long as they were kept indoors and warm, the flies continued to breed throughout the winter – producing a new generation every couple of weeks – and bananas provided a cheap, convenient source of food. Best of all, especially for those trying to teach biology, if a careless student killed them off, Drosophila were inexpensive to replace; colonies of larger animals, by contrast, were far too valuable to be entrusted to inexperienced students.[13] Morgan found that the flies made especially good New Yorkers since they were happy to live in tiny, cramped apartments: dozens of flies would live and breed happily in an ordinary half-pint milk bottle, which gave them another advantage – Morgan and his students could 'liberate' a few milk bottles from their neighbours' doorsteps on their way into the lab in the mornings, so the cost of keeping flies was little more than the price of a few bananas.

At the time Morgan entered the fruit flies' story he was, as we have seen, an advocate of both the Mutation Theory and the European style of laboratory biology. In 1904 he married one of his former graduate students, Lilian Vaughn Sampson, daughter of a wealthy Philadelphia family. They spent their honeymoon at marine research labs in California, where Morgan spent the summer studying the shift from asexual to sexual reproduction in aphids. At the end of the summer they moved to New York, so that Morgan could take up his new post at Columbia's zoology department, working alongside his friend Edmund Beecher Wilson, who was chair of the department.

Columbia proved the perfect place for Morgan, because a key step towards understanding chromosomes, the still rather mysterious 'coloured bodies' of the cell nucleus, had been made in Wilson's lab at Columbia by one of his graduate students, Walter Sutton. Sutton was a farm boy from Kansas, where he and his teacher Clarence McClung had made a small contribution to reducing the state's abundance of grasshoppers by using them as a standard organism for work on cells and chromosomes; they had discovered that the large lubber grasshoppers (*Brachystola magna*) had very large testicles (by insect standards), making them easy to study. Sutton took his grasshoppers to Columbia with him, where he hoped to use them in investigating the connection between chromosomes and inheritance that had been suggested by several distinguished European biologists.

Sutton showed that the behaviour of chromosomes during cell division clearly mirrored the evidence from Mendelian breeding experiments. His argument centred on the suggestive fact that when viewed under a microscope, the normal body cells of an organism all have their chromosomes arranged in pairs that are similar but, as we shall see, not quite identical. Normal body cells are referred to as diploid (from two Greek words meaning 'double form'); in *Drosophila melanogaster* each cell has four pairs of chromosomes, giving eight in total. As a plant or animal grows, each of its cells divides to create new cells, and each pair of chromosomes is duplicated so that each of the new cells gets a full set (and so every cell in the organism is diploid). However, the sex cells of an organism – eggs and sperm, or ova and pollen in plants – have only one copy of each chromosome (and are known as haploid, from 'single form'). When an organism's sex cells, or gametes, are being created a different kind of cell division occurs, in which the number of chromosomes is halved; in Drosophila, the eight chromosomes are reduced to four, one from each pair. Without this 'reduction division' (now known as meiosis) the number of chromosomes would increase, generation after generation; if each cell already contained a full set, the fused cell would contain two sets, in the next generation there would be four, and so on. Instead, the chromosome number in the gametes is halved, so that when the two haploid sex cells join during fertilization, the two half sets match up again to create a new full set. Obviously, half of the chromosomes in the newly fertilized egg have come from the ovum (the maternal chromosomes) and half from the sperm (the paternal chromosomes), and the fertilized egg is now diploid again, ready to begin growing by dividing. One final point about chromosomes is that they come in matching pairs; when gametes are formed, there is only one member of each pair of chromosomes in each gamete. For convenience, biologists normally number them, so Drosophila's chromosomes are known as chromosome 1, chromosome 2, and so on; when two gametes fuse, chromosome 1 from the father pairs up with chromosome 1 from the mother, and the same applies to each of the other chromosomes.

While Sutton was in the middle of working out what was going on with his grasshoppers' chromosomes, William Bateson visited New York to promote the newly 'rediscovered' Mendelian theory. Listening to Bateson speak, Sutton suddenly saw how Mendel's principles related to his work; the behaviour of the chromosomes mirrored the behaviour of the still-hypothetical Mendelian factors. When he published his experiments, he noted 'the probability that the association of paternal and maternal chromosomes in pairs and their subsequent separation during the reducing division as indicated above may constitute the physical basis of the Mendelian law of heredity'.[14] Sutton took his results to Wilson, who was initially baffled, then stunned as he realized their implications: if Sutton was right, chromosomes must indeed be the physical location of the elusive particles of inheritance.

When Morgan arrived at Columbia, he found that Wilson and one of Morgan's own former graduate students, Nettie Stevens, were working independently on following up Sutton's work, by investigating whether chromosomes determined the sex of the offspring, a topic that intrigued Morgan. The key seemed to lie in a mysterious chromosome, which McClung had first identified, which appeared not to have a pair and so was referred to as the 'accessory chromosome'. McClung marked these mysterious chromosomes with an X in his drawings of them, which is how they got their modern name, the X-chromosome. McClung thought the accessory chromosome might determine the sex of the offspring and when Sutton found X-chromosomes in only one half of the grasshopper sperm he examined, he believed he had confirmed his teacher's theory. Since the males had an X-chromosome and the females did not, it seemed that this chromosome did indeed determine their sex. In fact, this turned out to be wrong, but it was a perfectly understandable mistake: grasshoppers are unusual in that the males have one chromosome fewer than the females (they have no Y-chromosome). A couple of years later, Stevens identified the far more common pattern, which is that the X-chromosome does have a partner, the much smaller Y-chromosome. In most animals and plants, two copies of the X chromosome make the organism female, while one X and one Y make it male.

This all took a few years to unravel; in the meantime, although Morgan was intrigued by Sutton's work, he was still convinced that sex determination must be more complicated than this. Like so many of his contemporaries, Morgan was both fascinated and baffled by the problems of inheritance. He did not believe that ordinary continuous variations were inherited – such minor fluctuations were, he thought, most likely to be due to minor variations in the organism's environment. But even if continuous variations were inherited, they would soon be diluted until they disappeared. However, he realized that there were problems with saltationism too, since an organism that possessed one of the larger discontinuous variations (whether one called it a mutation or a saltation) would still need to find another organism to breed with; the chances of it finding a partner which possessed the same rare mutation must necessarily be small. As a result, Morgan concluded that with continuous *and* discontinuous variation, 'the swamping effect of intercrossing would in both cases soon obliterate new forms'.[15] Only a significant number of fairly large variations, all occurring together and all of the same kind, could overcome these problems. Finally, but perhaps most importantly, Morgan made the standard worm-slicer's objection that Darwinism could not be verified experimentally in a laboratory.

These were the objections that had persuaded Morgan of the value of de Vries's Mutation Theory. De Vries's mutations were qualitatively different from normal continuous variations (they were large enough to create some degree of infertility with the original, unmutated form, which protected them from swamping). Also, de Vries's hypothetical 'mutation periods' ensured that many mutations occurred simultaneously; there was thus a greater chance of two organisms with the same mutation mating – and passing that mutation on. Morgan seems to have first taken an interest in Drosophila in the hope that the flies would finally provide clear evidence of de Vriesian mutations in a species other than Oenothera. When de Vries visited the United States in 1904, Morgan heard him speak about the strange new 'Röntgen' or X-rays produced by the decay of that equally mysterious new substance, radium.[16] Perhaps they might produce artificial mutations in plants and animals? This suggested a new approach to identifying mutations and Morgan did some experiments, including some on insects, but did not get any results he thought worth publishing and dropped the idea.

A few years later, Morgan once again tried inducing de Vriesian mutations artificially. He now subjected Drosophila to acids, alkalis and other chemicals, varied their diets and conducted more radium experiments. The results were still disappointing, so Morgan tried another approach. One of de Vries's hypotheses was that intense selection – such as would be caused by a dramatic change in living conditions – could induce a mutation period. It may have been this idea that prompted Morgan to try large-scale breeding experiments with his flies.

The flies bred enthusiastically, so much so that Morgan soon complained he was 'head over ears' in flies and recruited some undergraduates to help out. In the fall of 1909 he taught the introductory biology course at Columbia – for the one and only time in his career. Among his students were Alfred Henry Sturtevant and Calvin Blackman Bridges. Despite not being a particularly effective lecturer, Morgan managed to convey – to these two at least – the excitement of biological research, and communicate a sense of the vital problems still to be solved. It was his personality, rather than what he said, that persuaded them both to approach him and ask if they could help in his lab. Morgan accepted their offers gratefully. Sturtevant rapidly became Morgan's favourite student; his father bred horses on a farm in Alabama and he had written a paper on the inheritance of coat colour in horses, which had impressed Morgan so much that he gave Sturtevant his own desk in the lab. Knowing Bridges needed money, Morgan hired him as a bottle-washer, cleaning the rotting banana and dead flies out of the purloined milk bottles, ready for the next experiment.

However, Morgan and his 'boys' – as Sturtevant and Bridges soon became known – faced an unusual problem when they started their breeding experiments: not being a domesticated species, Drosophila had no well-defined, clearly visible characteristics to select for – they all looked pretty much alike. By contrast, human 'fanciers' had lovingly cherished and bred animals such as dogs and pigeons to create well-marked and unusual characteristics, such as extravagant tail feathers. So Morgan's team became the world's first fly-fanciers: they selected a visible characteristic, a pattern on the fly's thorax (its chest region) that looked like a trident, and began selective breeding, crossing large tridents with large tridents, and small with small.

Morgan's early experiments with Drosophila seem to have been intended to push the flies into a mutation period like the one de Vries believed he had observed in Oenothera. But the hoped-for mutations failed to appear and after a couple of years Morgan was on the brink of losing interest when, in January 1910, a much darker trident, pattern appeared. Finally, a mutation, which Morgan dubbed '*with*'. The mutant fly appeared at about the same time as Morgan and Lilian's third child (a daughter who was also called Lilian). When Morgan went to meet the new arrival in hospital, his wife's first question was, 'Well, how is the fly?' He launched into an excited description of his research and it was several minutes before he remembered to ask 'And how is the baby?'[17] Thirty generations later, in November of the same year, an even more pronounced mutant – *superwith* – showed up; various others had been spotted in the intervening months: in March a mutant with a dark blemish at the junction of wing and thorax – *speck* – had appeared and the body-colour mutant – *olive* – emerged in the same month. May saw the arrival of *beaded* wing, and a different *olive* body-colour mutant. Morgan quickly produced a paper claiming that – for the first time – a de Vriesian mutation period was under way in a species other than the evening primrose.

However, just as the flies seemed to be confirming de Vries's approach, doubts set in. These new 'mutants' were not in fact mutants, at least not in the sense de Vries had used the term. A genuine de Vriesian mutation was supposed to be a major jump, possibly large enough to produce a new species in a single leap (de Vries was always a little vague on this point). A large enough leap would produce a new organism that could no longer interbreed with the old version, thus allowing it to be defined as a new species: it was the survival of the Oenothera mutants alongside the parental forms, resisting blending and swamping, that had first excited de Vries. Yet Morgan's mutants were different. They were definitely leaps – not examples of the smooth, continuous variations of classical Darwinism – but the jumps seemed too small. Also, the new mutants could be successfully crossed with each other, so they could not be considered new species.

The new, small mutations might not be what de Vries had in mind, but they provided Morgan's fly workers with a glimpse of the still mysterious Mendelian 'factors' at work, most of which were normally hidden. Take eye colour, for example: human eyes come in many colours, whose patterns of inheritance allow us to deduce something about the factors involved, such as the fact that the factor for blue eyes is recessive to that for brown. But all Drosophila have red eyes, so Columbia's fly fanciers could not say anything about the inheritance of eye colour in flies until a new mutant showed up – which indeed it did: one that had white eyes. When a male *white* mutant was crossed with a normal red-eyed female, Morgan got a whole generation with red eyes, but when he crossed these, he got three times as many reds as whites in the next generation. Despite his scepticism about Mendelism, Morgan had done some mouse-breeding experiments a few years earlier to test the theory, so he knew a Mendelian ratio when he saw one: not only was the *white* mutation not a new species, it was clearly behaving like a standard Mendelian recessive factor. Although other labs were working on Drosophila, no one had seen the Mendelian ratio before, partly because the flies lacked clear characters, comparable to the yellow and green colour in peas, which could have been easily contrasted. Just as Mendel had carefully selected his peas to produce clear-cut characters for his experiments, Morgan and his boys had begun remaking the fly into something that could be used for experiments, and the mutations were what made this possible: *white* revealed patterns of inheritance that could not have been observed in wild flies. As the fly workers selected and bred flies with visible mutations for their work, a wild organism was being domesticated – turned into a tool.

Morgan recognized that he was on to something interesting and was intelligent enough to admit that he had been wrong about the Mutation Theory, which he quickly dropped, starting to follow up his new Mendelian leads instead. Back in Europe, de Vries had reason to be concerned. The Mutation Theory seemed to be crumbling; when Morgan called his tiny *white* fly a 'mutation', he knew that however enthralling it was, it clearly was not what de Vries would call a mutation.

Mass production

As the flies bred and mutated, Morgan's lab, on Manhattan's Upper West Side, was transformed from a general-purpose biology laboratory – full of starfish, pigeons, mice and a host of other creatures – into a factory, a production line for churning out Drosophila. It became known as the fly room. Compared with Henry Ford's new 2,000-acre car factory at River Rouge, which was being built at about the same time, the fly room was tiny, roughly 16 by 23 feet (accounts differ). But the fly room's eight desks were as devoted to modern, standardized mass production as Ford's assembly lines.

In the early twentieth century, America was rapidly learning the benefits of standardization. The individual brilliance of the previous century's heroic inventors, men like Thomas Alva Edison, was admirable, but light-bulbs were useless without a system for generating and distributing electric power, and that involved defining and setting standards for everything from wires to voltages. The power grid allowed electric motors – which were smaller, cleaner and more flexible than steam engines – to be used in factories. It was that flexibility which made Ford's production lines possible: instead of arranging the machinery and the workers around massive, noisy steam engines, electric motors allowed the machines they powered to be arranged according the stages of a car's production. Each type of car component was identical, built from standardized parts, with one section of a standardized conveyor belt devoted to each assemblage; electricity made it possible for raw materials to flow into the factory at one end, while cars flowed out the other.

In 1911, as new mutants were appearing up in the fly room, Frederick Winslow Taylor published his *Principles of Scientific Management*, a book that exemplified the new spirit of business efficiency. It was an immediate success, quickly translated into half a dozen languages, and it carried Taylor's message of 'scientific management' around the world. He argued that every business needed to be reshaped like a machine, each individual worker becoming a small, standardized and easily replaceable part. As he put it, 'in the past, the man has been first; in the future the system must be first'.[18] It was a creed that horrified labour leaders and trades unionists, for whom 'Taylorism' encapsulated the soul-destroying tyranny of mechanized mass production: labour without skill, creativity or pause for breath. But for manufacturers, Taylorism offered the prospect of vastly increased productivity and profits.

Taylor, though coming from a wealthy Philadelphia family, had chosen to work in the Midvale Steel Company's machine shop, originally because his doctor had recommended manual labour for his health. He was deeply impressed by the company's president, William Sellers, a prolific inventor who had helped devise the screw threads used in all American factories. The Sellers thread was promoted as the perfect example of the benefits of standardization: instead of every machinery maker and factory-owner having his own screws precision cut by master craftsmen, agreed standards allowed millions of identical screws to be cheaply mass-produced.

Taylor's gospel of scientific management and standardization caught the imagination of Milton J. Greenman, director of Philadelphia's Wistar Institute for Medical Research. Greenman decided that Taylor's system of time management could be applied to research labs, just as easily as to factories. He took the Sellers thread as a model of the virtues of standardization; 'such standards', he wrote, might 'result in immense economies in science as well as in commerce'.[19] Greenman set-out to Taylorize the Wistar Institute's rat colony, to produce standard, experimental rats by standard methods. By 1912, Wistar was to rats what United Fruit was to bananas: it was turning out over 6,000 of its carefully bred 'standard' rats a year and shipping them all over the country.

The Wistar rats showed American scientists what was possible when an organism was mass-produced. With standard rats came standardized data: the rats were supplied with a copy of the Institute's book *The Rat: Data and Reference Tables* (1915), the operating manual for the Wistar rat. No comparable statistics existed for any other animal, even humans. This allowed the results from one lab to be readily compared with those from another. Mass-produced animals would push the laboratory revolution further and faster than anyone could have imagined.

Morgan's fly room, funded in part by the Carnegie Institution, certainly did not look anything like one of Taylor's scientifically managed modern factories: the desk drawers were full of cockroaches, living off the Drosophila food, the whole place was chaotic and noisy, filled with the buzz of flies and talk about flies. Morgan's team had had no intention of getting into the mass-production business, yet – despite the racket and the squalor – Columbia's fly production soon outstripped both Wistar and River Rouge. It became clear that what the fly room was witnessing was not a de Vriesian mutation period at all; the rapid discovery of new mutants was simply the result of mass production, a reflection of the sheer numbers of flies involved. No one had noticed these small but distinct mutations before because they were rare. If only one fly in 100 produced a mutation, mutations were unlikely to show up in an experiment with only 100 flies; in a colony of thousands, they become common and patterns of mutation become obvious. That was also why the early experiments with acids and radiation had apparently not produced results. More flies meant more mutants. More mutants meant more publications, more prestige, more graduate students to carry the fame of the fly across the world, and more honours, funding and publicity for Columbia's fly workers.

Morgan was Drosophila's Henry Ford, presiding over a mass-production system that turned out new research papers almost as quickly as it turned out flies, but Calvin Bridges was the fly room's Frederick Taylor. While Henry Ford always denied any debt to Taylor, Morgan and Bridges formed an ideal team, although Morgan did not always realize it. As we have seen, he made a virtue of making do with inexpensive, improvised equipment; Bridges loved to tinker, constantly looking to improve the fly room's efficiency, as if intent on increasing the fly-hours worked.

Bridges's ingenuity was applied to every aspect of the fly production line. In the early days of the fly room, when a researcher wanted to trap a single interesting fly, he would take advantage of the fact that Drosophila move instinctively towards light. A researcher would have to take the top off a bottle swarming with flies, hoping the target fly did not escape in the process, and slap a clean bottle upside down on top. Then he would hold the bottles up so that the light shone through the bottom of the empty bottle. Gradually, the tiny insects would crawl towards the light, into the new bottle. As soon as the target fly was in the new bottle, it was capped. The process was then repeated, over and over again, until the target fly was isolated (or had escaped in the process). This tedious process outraged the engineer in Bridges. He discovered that a carefully measured dose of ether was enough to knock the flies out for a few minutes, so that specific flies could simply be picked out and placed in a new bottle before they came round. Not content simply to pour ether over his flies (too much would kill them), Bridges designed a fly-etherizer, which gave the flies a carefully measured dose.

As in other factories, the workers in the fly room went on strike when the place got too hot or too cold; cold weather stops Drosophila breeding, high temperatures can kill them. So in 1913 Bridges turned some old bookcases, incandescent lights and thermostats into constant-temperature cabinets. A few years later he constructed improved versions, with ventilation and humidity controls. Never able to resist a chance to improve things, he introduced further refinements in 1930, and his home-made cabinets were still substantially cheaper than commercial incubators. He also built the first fly morgues (to dispose of dead flies). However, even with all Bridges's improvements, a state-of-the-art fly room in the mid-1920s was much cheaper to equip and run than a guinea pig colony or a plant-breeding station.

Although Bridges's ingenuity saved both time and money, Morgan was a little dismissive of what he called the 'folderol' that Bridges was constantly introducing into the work – such as the sophisticated binocular microscopes that were replacing the hand lenses Morgan preferred. Ignoring Bridges's elegant fly morgues, Morgan simply squashed flies that carried no interesting mutations on his desk or notebooks. He also rejected the etherizer, preferring to simply pour ether on the flies and risk killing them, but the other fly workers were more appreciative.

Standard flies and fly people

So, what exactly were Morgan and his students doing with all these flies? When Morgan first experimented with the white-eyed fly, he had noticed something interesting – the white-eyed flies were always male. Careful crossings confirmed that the white mutation was always linked to the sex of the fly. That was interesting because of its implications for Morgan's work on sex-determination; it suggested a connection with the work Wilson and others had done on chromosomes. One of Morgan's initial objections to the idea that chromosomes might carry the hereditary factors was that the flies have far more visible characteristics (such as eye colour, body colour and trident pattern) than they have chromosomes. If the hereditary factors were located on the chromosomes there would have to be several on each one, in which case all the factors on a single chromosome would always be inherited together. The connection between eye colour and sex suggested exactly the kind of physical linkage, or 'coupling' as it was initially known, that the chromosome theory would predict. Within a year, Morgan had found two more examples of what he called 'sex-limited' mutations: one for yellow body colour and one for miniature wings. One link might be a coincidence, but three seemed to prove that factors were indeed inherited together, on the same chromosome.

But no sooner had linkage been discovered than it began to break down. Factors that were normally linked were sometimes inherited separately. Intriguingly, for any given pair of factors the frequency with which linkage broke down was constant, but it varied between one pair of factors and another. For example, factors A and B – which were almost always inherited together – would separate in 1 per cent of crosses, while factors C and D, also usually linked, separated 2 per cent of the time. As Morgan's team struggled to understand this anomaly, they were able to draw on an increasingly comprehensive body of work on chromosomes, whose role in heredity was gradually becoming clear.

In 1909 Morgan read a paper by Frans Alfons Janssens, a Belgian Jesuit priest and gifted microscopist, who taught biology at the Flemish University of Leuven. Janssens's short paper simply described what he had seen under his microscope: chromosomes did some very odd things during meiosis. Before their final reduction into half-sets, they wrapped themselves around each other and appeared to break apart and rejoin. Janssens's phenomenon was christened 'crossing-over'. Morgan and his students suggested that crossing-over explained the occasional breakdown of linkage. They pictured the hereditary factors rather like beads strung along the chromosome. During crossing-over, the two chromosome necklaces – one originally inherited from the father, the other from the mother – broke at random points and then joined again. If the chromosome the fly inherits from its father is pictured as a string of green beads, and that from its mother as a string of red beads, after crossing-over, there might be a green string with a few red beads at one end, while the red string had acquired a few green beads; this was how linked factors sometimes got unlinked. If crossing-over was random it would produce a different mixture of red and green on each of the fly's four pairs of chromosomes. However, the crucial point was that the closer together any two beads were, the less likely it was that a random break would occur in-between them; the further apart they were, the greater the chance of a break. If two of the hereditary beads were close together on the chromosome, they would almost always be inherited together, providing strong linkage, but if they were far apart, they would be separated so often that they almost appeared to be inherited independently of one another. As Morgan wrote, that was why 'we find coupling in certain characters, and little or no evidence at all of coupling in other characters; the difference depending on the linear distance apart of the chromosomal materials that represent the factors'.[20]

Inspired by this idea, Morgan and his students began to look for mutations that were usually linked. As they searched, they discovered that the mutations fell into four distinct 'linkage groups'; the fact that *Drosophila melanogaster* had four pairs of chromosomes made it overwhelmingly likely that each linkage group corresponded to a pair of chromosomes. By 1914, it was clear that the Mendelian theory and the chromosome theory were not rivals, they were one and the same: the Mendelian factors were real and they were located on the chromosomes. As this became increasingly accepted, the carefully neutral language of 'factors', which had been used before anyone knew what they were dealing with, was gradually dropped and a new term, 'gene', came into use. By 1917, Morgan and his group were using it and soon everyone was referring to the study of inheritance as 'genetics' (a term coined by Bateson) and the people doing it as 'geneticists'. Even though no one yet knew how these genes actually worked, the new language reflected geneticists' growing confidence that they were working on a tangible, physical phenomenon whose precise nature would, sooner or later, be fully understood.

Once the connections between linkage, crossing-over and chromosomes were understood, the flies were put to a brand-new use. Sturtevant, still just nineteen-years-old, realized that the frequency with which two genes crossed over could be used to estimate how far apart they were. The fly workers were now going to persuade Drosophila to help them work out precisely where on each chromosome each gene was, to fix the position of each bead on the string. Sturtevant took Morgan's data home one night in winter 1911 and came back to the lab the next morning with the first, basic chromosome map. The following year, Morgan set him and Bridges the task of mapping all the fly's chromosomes.

The principle of mapping was simple: mutations were a tool that allowed the fly boys to see what was happening to each gene in a cross, but to take advantage of them, careful breeding was needed to create a stock of flies that combined a specific pair of mutations. Once that was done, cross-breeding stocks with particular combinations of mutations revealed the frequency of crossing-over – by simply counting the flies' offspring and observing how many times the genes remained linked and how often they were separated. Bridges had unparalleled patience for this kind of work, sitting at a microscope for hours at a time, counting thousands of flies while looking out for new mutants. However, the frequency simply revealed whether the genes were close together or far apart. The next step was to try to translate that into a precise distance: to determine exactly how far apart on their chromosome the factors lay. Doing this required three genes, A, B and C. They were usually inherited together, so they were on the same chromosome, but – thanks to crossing-over – the links sometimes broke down. The data showed that A and B were more closely linked (i.e. closer together on the chromosome) than A and C (because the link between A and B broke less frequently than that between A and C). The same was true for B and C – they were also closer together than A and C. That suggested that B was somewhere in between A and C. To test this, a researcher would carefully measure how often A and B crossed over, and then do the same for B and C (which took two sets of time-consuming, tedious experiments). Adding the frequency from A to B to that from B to C gave a prediction of how often A to C ought to cross over. That estimate could then be checked using a third experiment.

Similar sets of experiments needed to be done for every identifiable factor on each of the four chromosomes, but thanks to the speed with which the flies bred, the relative positions of the genes along each of the chromosomes were gradually worked out. If that all sounds mind-numbingly tedious, it was: the chromosome maps produced between 1919 and 1923 used data from between 13 and 20 million flies, every one of which had to be selected, cross-bred, etherized and counted. Although a single fly is only a couple of millimetres long, if all these flies had been laid end-to-end, they would have formed a line of flies over thirty-five miles long. The efficient, mass-production techniques developed in the fly room were essential; without them, the work could never have been done.

However, even this description of the work does not begin to capture the labour involved. The basic mapping technique assumed that crossing-over was uniform along the chromosome – that it was equally likely to break at any point along its length – but that proved not to be the case. As Sturtevant and Bridges mapped, they kept getting anomalous results. Investigating these resulted in the discovery of many types of genes whose existence no one had suspected: such as genes that had no visible effect on the fly, but reduced the rate of crossing-over. Sometimes a useful, visible mutation occurred too close to one that made the flies weak, or even killed them. Months of frustration could result and many more months would be needed to breed new Drosophila stocks without the troublesome gene.

As the work continued, something remarkable was taking place in the fly room. The wild fruit flies were not merely being domesticated, they were being rebuilt into standardized organisms, as standardized as the Wistar rats or Sellers-thread screws. The flies were 'cleaned' of unhelpful genes that complicated the experiments; once the problematic gene had been detected, a new stock of flies would be created through careful cross-breeding, as it would if a stock were found to be too slow-breeding, or susceptible to disease. Anything that slowed down the work was bred out. In the process, the fly boys were learning a great deal about genes, but they were also building a new kind of fly, one that was an amalgam of genes from many different wild varieties of the insect. The fly room eventually contained nearly 400 unique lab-bred stocks, each cleaned up so that it combined a precisely selected combination of mutations; each stock was both an appealing object of study and also a tool, a genetic probe, that could be used to investigate the genes of an unknown fly.

As the standard fly was constructed, its value as a laboratory tool rose and the fly people began to value and respect it more. In the early days of the research, they often complained of being smothered in pesky flies, but such observations gradually faded; Drosophila was even referred to as 'that noble animal' by one worker.[21] In their early papers, many researchers omitted to mention that they had brought colonies of these rather revolting little insects into their labs, but as the fly showed what it could do, the geneticists who worked on it began to identify themselves as a community, referring to themselves as 'fly people' or 'Drosophilists'.

Have flies, will travel

The unique atmosphere of the fly room attracted much interest, initially within Columbia itself, but eventually all over the world. Everyone who visited commented on the fly room's informality and the air of excitement it generated. Sturtevant later wondered 'how any work got done at all, with the amount of talk that went on'.[22]

Sturtevant was especially interested in theoretical issues, while Bridges was in charge of giving away the carefully constructed stocks of Drosophila. One of the great advantages of working with flies was that they were cheap and portable; they took up so little space that an entire breeding colony could be transported in a bottle, and it was easy to simply give them away to researchers who wanted to do their own experiments. Gifts of flies were an important part of making Drosophila into a standard organism, just like the Sellers-screw thread; a standard is no use unless everyone is using it. It became clear very early on that the flies could produce interesting problems so rapidly that there would be more than enough work to keep Morgan's team busy; giving away flies helped establish a community who would share what they knew and solve problems more rapidly. But it was no good distributing flies if people could not use them, so along with the flies, the Morgan group passed on what they knew about them. Bridges in particular was happy to teach anyone who was interested, all the tricks and techniques he had invented to keep the flies breeding. Eventually a printed newsletter – the *Drosophila Information Service* – was produced to help record and spread the essential fly lore. Freely exchanging flies and information about them became one of the unwritten rules of the fly community; the Drosophilists decided that it was in everyone's interest to share, and researchers who were disinclined to were quietly cut out of the network.

Among the many visitors to the fly room was Hermann Joseph Muller, a masters student from Columbia's physiology department, who started dropping in on Sturtevant and Bridges every Thursday. That was his one free day, since he had to support himself financially by teaching embryology to undergraduates and English to foreigners, and by working as a hotel clerk. He began to attend the fly room's evening reading groups, where new ideas were discussed over cheese and beer at Morgan's house. Muller found the fly group's work thrilling; Sturtevant recalled that the night on which he revealed the first chromosome map, Muller literally jumped with excitement.

Muller was born in New York, the grandson of a migrant metal worker, a bright, hard-working boy who had been top of his high-school class and won a scholarship to Columbia. He was soon fascinated by biology and organized an undergraduate biology club, through which he met Sturtevant and Bridges, who told him about the fly room. Although Muller was eager to join the

Drosophilists, there was initially no space for him in Morgan's lab.

Morgan, Sturtevant and Bridges were easy-going, much given to self-mockery and joking, but Muller took himself very seriously. His personality and late arrival made him feel like an outsider in the fly room, a feeling exacerbated by his tendency towards rigid opinions and ideas. Morgan was a humanitarian middle-of-the-road conservative, horrified by extremism of all kinds, but basically uninterested in politics. By contrast, Muller was a political radical, attracted to Marxism and communist ideas. Both scientifically and politically, Morgan seems to have regarded Muller as a bit of a zealot, while Muller came to feel that his contributions were often overlooked, with others getting more credit. The fly room's informality meant that everyone shared their ideas freely, but formal acknowledgement – in the shape of getting your name on a publication – went only to those who had done the actual experiments. Muller gradually began to suspect that these unwritten rules had been devised by Morgan and Sturtevant to deprive him of his fair share of recognition. However, Muller probably suffered because he was a quick thinker, but a slow worker: he produced ideas much faster than results, working away methodically at incredibly complex, sophisticated experiments that took years to complete. Meanwhile, the others took advantage of his ideas to get their work done first. Muller helped create many of the fly stocks the others used, including some of the trickiest ones, and eventually he came to resent Morgan, in private at least, feeling that the latter had hindered his career.

Morgan, who came from old Southern stock and had a somewhat aristocratic manner, and his golden boy, Sturtevant, who was clearly Morgan's favourite, bore the brunt of Muller's resentment. Calvin Bridges was largely exempt, partly because of their shared political views – even though Bridges' communist sympathies were arguably more a matter of affectation than of deeply felt conviction. Muller also seems to have regarded Bridges as an exploited blue-collar worker on the Drosophila shop-floor: he did the practical work of maintaining the fly stocks, while Sturtevant became more of a theorist because his colour-blindness made him poor at spotting mutants.

In 1917, while the fly work was at its height, the Russian Revolution had created the world's first avowedly socialist country. Like many young Westerners at the time, Muller viewed it as a heroic, idealistic revolt against a corrupt, undemocratic regime; Stalin's gulags and show trials were still years in the future. The revolution enthralled him, and in 1922 he decided to visit the Soviet Union, eager to see the new communist society for himself and to meet Soviet biologists.

The USSR was only five years old at the time, a fragile experiment still struggling to survive after a bloody civil war and a series of devastating famines. There were shortages of almost everything, so Muller – like any polite visitor – brought his hosts a gift, thirty-two small bottles. But instead of duty-free alcohol, they contained live Drosophila, samples of the fly stocks he had helped create. One might imagine that acquiring American fruit flies would not have been high on the Bolshevik government's list of priorities, but Muller's Soviet colleagues were thrilled. Lenin's regime took a great interest in science, especially biology. Inspired by the vision of Marxism as a scientific theory, which promised its adherents the power to reshape the world for the better, the Bolsheviks poured money into sciences such as plant- and animal-breeding, hoping to create better crops and animals, to avert future famines.

Muller met the leading Soviet geneticists, including Nikolai Vavilov, head of Moscow's gigantic Institute of Applied Botany, who had studied genetics in Britain with Bateson. However, it was another Nikolai, Nikolai Kol'tsov, who initally showed the greatest interest in Muller's ideas.

Before the revolution, Kol'tsov had been a liberal critic of the Tsarist government, which had cost him his job at Moscow University. He had managed to persuade a Russian railway millionaire to fund a new Institute of Experimental Biology (which became known as the Kol'tsov Institute), dedicated to Kol'tsov's vision of a new biology, which would bring together the science's great nineteenth-century achievements and combine them with newer ideas of Mendelism, biometrics and chemistry.

Inspired by the laboratory revolution, Kol'tsov encouraged his students to master lab techniques, but – more unusually – he also made sure they went out into the field and observed living creatures in the wild.

Kol'tsov invited Muller to give a talk on the American Drosophila research at his institute, and published a Russian translation of it. When Muller returned to the United States he left a stock of flies so that the Institute could develop its own Drosophila research. But Kol'tsov had a problem – there was no one at the Institute who knew anything about insects. So he turned to an old friend, Sergei Chetverikov, an entomologist who had studied with him at Moscow University.

Chetverikov must have seemed an unlikely figure to head a genetics programme: not only did he know almost nothing about genetics, he lectured on biometrics – most biometricians were hostile to Mendelism – and he was also a field entomologist, not a lab worker. But he knew a lot about insects and that was enough for Kol'tsov, who decided that Chetverikov's interest in biometrics – far from being a disadvantage – was a sign that his friend was open to new ideas.

Chetverikov did indeed prove eager to learn. He and his co-workers decided that they needed to start by learning English, but rather than start with textbooks or novels, they decided to read the latest American genetics papers. They would divide these up between them, take them home, read and translate them, then meet up at each other's apartments in the evenings to discuss what they had learned. They were, in effect, learning two foreign languages at once: English and Mendelian genetics. Inspired, perhaps, by the close-knit groups around Muller and Morgan, the Russian fly workers also became a group, which became known as the *Droz-So-or* (an acronym for *sovmestnoe oranie drozofil'shchikov*, or 'the Combined Cacophony of Drosophilists'), a name that gently mocked the endless proliferating bureaucratic acronyms that adorned the new institutions of Soviet power.

In many ways, the *Droz-So-or* was just like the Morgan group, except that a third of Chetverikov's colleagues were women, in sharp contrast to the 'boys' who dominated the American fly rooms. Like their American colleagues, some of the Russians became interested in studying and mapping chromosomes, but Chetverikov was more interested in studying the genetics of wild populations. Still somewhat sceptical of the relevance of laboratory work to wild populations, he set out to see if the Morgan group's small mutations could also be found in wild populations.

The Combined Cacophony of Drosophilists began to trap wild flies and cross-breed them with Muller's lab-grown flies to discover the hidden genetic make-up of the wild population. The lab flies were rather like chemical reagents, which are used to determine the chemical composition of an unknown substance because they produce predicable reactions. Chetverikov's group knew which genes the American flies contained, so by patiently catching, crossing and counting wild flies, they could deduce which genes were present. The crosses revealed that wild flies varied enormously, carrying all kinds of recessive genes that only became visible when crossed with other recessive stocks.

As we saw with Mendel's peas, if a plant had yellow peas, it could either have two copies of the yellow version of the colour gene, or one yellow and one green (because yellow is dominant while green is recessive). In the early twentieth century, these different versions of a gene became known as alleles (from the Greek word for 'another'). Bateson coined the terms that are still used to describe an organism's genes: those with two copies of the same allele were homozygous; while those with two different alleles are heterozygous. The only way to distinguish the two cases was to cross-breed your pea plant with one that you knew was homozygous (because green is recessive, plants with green peas must be homozygous): the ratio of greens to yellows in the next generation would make it clear whether your original plant had carried two copies of the yellow allele, or one yellow and one green.

Chetverikov's group could use this kind of information by crossing wild flies with laboratory stocks that were known to be homozygous for the recessive gene; the offspring of such a cross would then reveal if the wild fly had also carried the recessive gene. Their experiment convinced them that wild populations contained considerable hidden variation. However, wild populations of flies could not be treated in the same way as laboratory populations in bottles. In the lab, mating could be controlled, so the ancestry of any individual fly could be known and the precise combination of genes it carried could be calculated. A laboratory fly was like one of Sturtevant's father's prize racehorses; its pedigree could be inspected in the stud book. Knowing that all the flies in a bottle shared a pedigree meant they shared a common set of genes. None of this was possible with wild flies, so Chetverikov began to apply some of the mathematical techniques he had learned from the British biometricians for understanding the genetics of wild populations. His knowledge of statistics allowed him to take a sample from the wild, perform experiments in the lab to determine their make-up, and then extrapolate mathematically – applying the results to the world beyond the lab. Chetverikov calculated how frequently different recessive traits might occur in wild Drosophila populations, guessing that when a fly with a particular gene became more or less common, this was presumably the effect of natural selection, since the gene in question had either improved or reduced the fly's survival chances.

Chetverikov published the group's initial findings as *On Certain Features of the Evolutionary Process from the Viewpoint of Modern Genetics* (1926), a prosaic title that concealed something extremely important: for the first time, the biometricians' understanding of Darwinian natural selection and the Morgan group's version of Mendelian genetics, which had been seen as *competing* interpretations of evolution, were presented as *complementary*. Kol'tsov was impressed, and referred to Chetverikov's work as a synthesis that had 'great theoretical interest, in as much as it connects experimental laboratory genetics with the problem of the evolution of organisms in nature'.[23] If Chetverikov was right, he had found what de Vries had been searching for: a way of removing evolution from the realm of speculation and bringing it into the lab. But initially almost no one outside the USSR knew of his work.

The end of *Oenothera*

While the Russians were finding new uses for Drosophila, the Americans were also making further discoveries. In 1918, Morgan wrote to de Vries, asking for his comments on a draft article describing an exciting new discovery – made by Muller – that was to prove crucial in resolving a key aspect of the Oenothera mystery.

The fly boys had discovered that in some cases a gene that normally performed some vital function in the fly could mutate into a form which simply did not work: a heterozygous fly (one that had one working and one non-working version of the gene) would be fine, but it would pass the non-functioning allele on to half its offspring. If any of the unfortunate offspring got two copies of the non-functioning allele (one from each parent), they would die. The story becomes even more complicated in cases where the fly carried two such genes, which is what Muller had discovered. Suppose gene one comes in two versions – A (the working version) and a (the non-working one); and so does gene two – B and b. The only organisms that will survive are those that have at least one copy of both working genes (they would have to be either AA or Aa, combined with either BB or Bb). Any flies that are homozygous for either of the recessive alleles (i.e. that have either the aa combination, or the bb one) died.

The problem was how to detect the lethal recessive genes. Normally, a recessive trait – like the green colour in peas – shows up because some offspring are homozygous for the recessive trait; they have two copies of the green allele and so produce green peas, but any flies with two copies of the recessive lethal allele simply died. When Muller came to count the flies from his carefully constructed cross-breeding experiments, the normal Mendelian ratios that would have told him that his flies carried a hidden recessive gene were all confused. Understandably, it took Muller a massive amount of work to understand what was going on in these complicated cases. Once he had solved the problem, he realized that the same thing must be happening in some of de Vries's Oenotheras. *Oenothera lamarckiana* proved to be a case of what Muller termed 'balanced lethal factors', which meant that the plant looked homozygous (because no recessives showed up in the crosses) when it was in fact heterozygous (carrying the hidden lethal alleles). Such cases are so rare that they had not been detected previously (another effect of the fly room's mass production); de Vries had concluded he had found a new species when what he really had was a very unusual kind of hybrid.

Morgan described Muller's findings and their implications in his letter to de Vries and concluded, 'I venture to think that the *mutation* problem of *Oenothera* may find a very happy solution in the theory of balanced lethal factors.' In the margin of the manuscript, de Vries wrote the single word 'unhappy'.[24]

However, if Muller's discovery made de Vries understandably unhappy, the worst was yet to come. As the fly work unfolded, the idea of big mutations producing a new species in a single leap became increasingly implausible. The gradual unravelling of the mysteries of the chromosome made it increasingly obvious that Oenothera was a freak. Researchers in several countries discovered that the plant's chromosomes behaved in the most unusual way, so that the normal Mendelian rules broke down completely. Among those investigating the plant's chromosomes was Reginald Ruggles Gates, who, despite his enthusiasm for the Mutation Theory, was to play a key part in the theory's undoing. In 1906 he and Anna Mae Lutz discovered that the *gigas* mutant, the large and vigorous 'new species' of Oenothera de Vries had first discovered, had twice as many chromosomes as a normal *Oenothera lamarckiana* (twenty-eight instead of fourteen). Over the next few years this quickly proved to be a common phenomenon among the Oenotheras; many of the mutant forms had unusual numbers of chromosomes.

What had happened in these cases was that Oenothera's reduction division (meiosis) had failed to work properly: the number of chromosomes had not been halved when the plant was producing ova and pollen. Some of the sex cells remained diploid, so when pollination occurred the new plants got extra chromosomes. Such duplications are rare in animals, but more common in plants; the phenomenon is known as polyploidy (meaning 'multiple form').

Oenothera was one of the first polyploid plants to be identified; its fame ensured that a lot of researchers had worked on it, which helped to reveal polyploidy's significance. Recall that when gametes fuse during fertilization, the chromosomes from each parent have to match up to form the usual diploid pairs that are characteristic of the normal body cells. However, if two different species hybridize, their chromosomes do not match and so cannot form pairs; sometimes no offspring result, sometimes they produce sterile progeny. The classic case is of course the mule, the offspring of a donkey (*Equus asinus*, which has sixty-two chromosomes) and a horse (*Equus caballus*, which has sixty-four chromosomes); when a horse and a donkey mate, their chromosomes are unable to pair up and the result is that mules, despite being tough and sturdy, are sterile. However, in a plant where the chromosomes have already been duplicated during the production of pollen or ova – because the reduction division has not worked – there will be two copies of each chromosome in the fertilized ovum, so each chromosome will still be able to find a partner. The result is that the new hybrid plant is fertile, but only when crossed with other polyploid hybrids. If crossed with either of its original parents, sterility results. Since the new polyploid plants often look very different from their parents and will not interbreed with them, they appear to be new species – exactly what de Vries found in some of his Oenotheras. A similar phenomenon occurs in many species of Hieracium, which was the other reason their behaviour was so baffling to Mendel.

Polyploidy also turns out to be relevant to another of the fly workers' discoveries. As we have seen, Mendel's pea experiments had been done with pure-bred strains that possessed very clear-cut, contrasting characteristics, such as green or yellow peas, so the twentieth-century Mendelians tended to think of genes as being like switches – they turned some feature of an organism on or off. However, the fly work produced a much more complex picture. It gradually became clear that genes also affect what other genes do; sometimes, for example, bits of chromosome are duplicated during crossing-over, so that a fly ends up with two copies of a gene, and sometimes two genes produce twice the effect of one. This effect can also occur in polyploid plants, which is why the *gigas* 'mutant', with twice the normal number of chromosomes, was larger and more robust than its parent plants: some of the genes that determined its size had been duplicated and so their effect was doubled. It turns out that the same thing has happened in some of the plants we rely on for food: common wheat, for example, has managed to accumulate no less than six sets of chromosomes; these duplications have produced a plant with much larger, more nutritious seeds – a feature that *Homo sapiens* has understandably found particularly attractive. A comparison of the plump, appetizing grains of wheat with the tiny, tough seeds of most wild grasses reveals the effects of polyploidy in action.

Polyploidy can also explain how bananas became the favourite fruit of toothless babies. Wild bananas are full of hard, black seeds, which make them virtually inedible to humans, but the ones we buy have no seeds at all. This is another effect of polyploidy; somewhere in their evolutionary history, banana chromosomes were duplicated – all modern, cultivated varieties have three sets of chromosomes, instead of two – which produces bigger fruit without seeds (because, like mules, they are sterile). This would appear to be an evolutionary dead end, but bananas are one of many kinds of plants that can reproduce themselves by sending out shoots or roots that can grow into entirely new plants; these are known as suckers – one of the phenomena Henslow lectured on. Luckily, one of our sweet-toothed and sharp-eyed ancestors noticed these seedless bananas and worked out how to grow them by taking advantage of their tendency to produce suckers.

The fly who loved me

Drosophila have continued to prove highly adaptable and they are still exploiting the warm, food-rich ecological niche of the academic research laboratory. But they have nevertheless had to cope with the changing seasons of scientific fashion. For a while, after the Second World War, it seemed as if the fly's days were over, as smaller, simpler organisms were preferred by researchers.

But fly populations recover rapidly after a bad season, and in the 1970s Drosophila became the focus of renewed interest as new kinds of genetic research became possible, as we shall see. Today, countless millions of these tiny flies continue to buzz away in labs all over the world, teaching beginners the essentials of genetics and helping Nobel Prize-winners understand how complex behaviours are controlled by genes. Despite all the extraordinary advances in scientific understanding and technology that have been made since the first flies landed on the first over-ripe banana on a lab windowsill, many of the basic tricks of the fly trade have remained the same. Students still have to learn how to get the flies living and breeding before they can do their experiments. As one recent handbook jokingly informs the would-be Drosophilist, 'the flies frequently require you to do an apprenticeship on any important project – they will not start to perform until they are certain you are serious'.[25]

Thanks to those who took the time to get serious about flies, the list of things they have taught us is almost endless. Resolving the precise connection between chromosomes and inheritance was perhaps the most important, but inspired by the success of the fly workers, who rapidly became scientific stars, researchers began to study chromosomes in all kinds of plants and animals. Gradually, the idea of genes as simple switches, that could only be on or off, began to give way to a more complex picture. One result, as we shall see in the next chapter, was that the seemingly sharp distinction between smooth continuous variation and abrupt, jumpy changes began to break down.

In 1915 Morgan and his students published *The Mechanism of Mendelian Heredity*, which summarized many of their discoveries. Thanks to the pace at which the flies bred, it had taken only a few years to assemble a mass of evidence to support their claim that the hereditary particles were on the chromosomes. They were not the first to suggest this, but they were able to provide better evidence than ever before, evidence that the Mendelian factors were at specific physical locations on specific chromosomes. The fact that their results came from laboratory experiments made them even more persuasive, as did the fact that Morgan's team would happily send stocks of the relevant flies to anyone who wanted to check the results for themselves. But while the fascination with flies spread rapidly, not everyone was fully persuaded by the new ideas. In Britain, Bateson had his own theories about inheritance and resisted the American ideas for many years; his considerable reputation helped to slow the fly's British advance for many years.

However, the strongest opposition to the Mendelian chromosome theory came, as one might expect, from a few of the naturalists and field workers, especially from the biometricians. They complained that breeding flies in milk bottles and adjusting the temperature so that they bred continuously created entirely unnatural conditions; of course they mutated under such stresses, and so the results obtained had no bearing on wild flies. Morgan was, understandably, contemptuous of these criticisms. He observed that critics implied 'that results obtained from the breeding pen, the seed pan, the flower pot and the milk bottle do not apply to evolution in the "open", nature at large or to wild types'. However, if biologists were to give up these experiments, chemists and physicists should give up using spectroscopes, test tubes and galvanometers, since these were all 'unnatural instruments'. Morgan argued that 'the real antithesis is not between unnatural and natural treatment of Nature, but rather between controlled or verifiable data on the one hand, and unrestrained generalizations on the other'.[26]

Yet even the most sympathetic naturalists found it hard to see how the discoveries of the fly room could be applied to evolution in the wild. The fly experiments depended on creating pure-bred flies with known pedigrees; that seemed to be the only way experimenters could know what genes they were dealing with. Of course, had the naturalists known of Chetverikov's work, they might have understood its implications sooner, but Chetverikov never got to complete the experiments necessary to develop his ideas. His career was cut short after Stalin came to power and the USSR became an increasingly repressive society. Among the millions arrested during the crack-downs and purges of Stalinist Russia was Chetverikov; in 1929 he was sent into internal exile, forbidden to visit Moscow or Leningrad, and forced to work as a schoolteacher. He was lucky, in that he survived and eventually died of natural causes, but he was unable to publish anything further on genetics.

For a while, Chetverikov's students continued his work, but the Combined Cacophony was eventually devastated by Trofim Lysenko, who gained control over Soviet biology during Stalin's regime. Lysenko rejected orthodox genetics in favour of a form of Lamarckism which claimed to be able to evolve plants much more quickly than Mendelian methods. Lysenko claimed his views were more Marxist than those he called bourgeois 'fly lovers', whose links with US genetics were well known. Lysenko instituted an assault on orthodox genetics and the country's political leaders began to support him during the new famines of the 1930s. He promised a rapid solution to famine and, with Stalin's support, he eventually acquired the power to virtually outlaw Mendelian genetics. Many geneticists disappeared, were arrested and, in some cases, were even executed.

After Chetverikov's arrest and the dispersal of the *Droz-So-or*, their work might have remained almost unknown outside the USSR, but for a handful of people who publicized it in the West. Among them was the British biologist J.B.S. Haldane, another left-wing sympathizer who visited the USSR in the 1920s and came away deeply impressed by the level of government support for science (still unheard of in the West). A few years later, Haldane met Chetverikov at an international genetics congress, arranged for some of the Russian work to be translated, and encouraged his British students to read it. Haldane became intrigued by the possibility of applying biometric tools to genetics, but working out how to do this was to require a lot of maths – and a lot of guinea pigs.

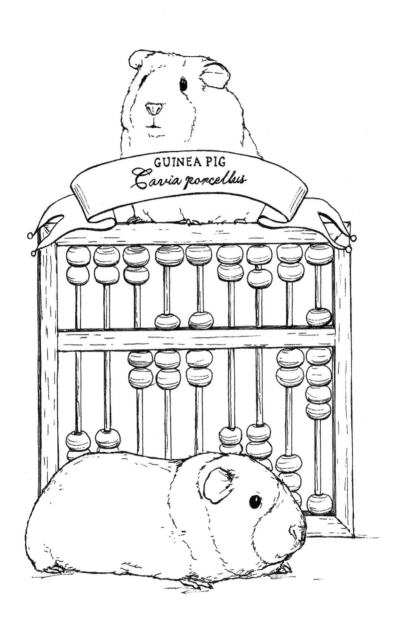

GUINEA PIG
Cavia porcellus

Chapter 7
Cavia porcellus:
Mathematical guinea pigs

Within the seemingly innocuous dark blue covers of the twenty volumes of the *Oxford English Dictionary* lies a record of invasion and colonization. The story of English is one of wholesale theft – of countries, their animals, their plants and, most noticeably in this context, their words. The Latin roots of some English words are, in part, a record of the vocabulary the Romans imposed on the British, in an effort to educate and civilize them. When we do manual work for which we are remunerated, we are doubly commemorating our one-time Roman overlords – both 'manual' and 'remunerate' come from the Latin *manus*, a hand. And our salaries derive from the money paid to Roman soldiers to allow them to buy salt: from the Latin *salarius*, or 'salt-related'. Similarly, the fact that we call certain animals 'cows' when they are in a field, but 'beef' when they appear on our dinner plates, is a relic of 1066 and all that: the Norman lord called the animal *boeuf* when eating it, while the conquered Saxon peasant, who actually herded it, still referred to it by the old English word, *cú* (cow), which shares common Indo-European roots with a dozen similar words across northern Europe.

The kind of person who puts on jodhpurs (possibly khaki in colour) to go to a gymkhana and afterwards sit on the veranda of their bungalow is commemorating the impact on English of the British Raj: *khaki* is Urdu for dust; *Jodhpur* is a city in Rajasthan; gymkhana and bungalow both come from Hindustani – the first is a modification of Hindustani *gend-khna*, 'ball-house', a racquet-court, while bungalow derives from *bangla*, meaning from Bengal. However, veranda, which is often thought to be an Indian word, is in fact one that travelled the other way: it comes from the Portuguese and older Spanish *varanda* (or *baranda*) meaning a railing or balcony; Europeans took it to India and it was then adopted into Hindi and Bengali.

Linguistic imperialism is also evident in the names of many of the plants and animals of the New World; as we have seen, most – like Oenothera – lost their indigenous names when Europeans renamed them, but a few native names survived. On occasion, the colonists simply failed to come up with names for the new species and borrowed the indigenous terms, which is why English has ended up with numerous borrowings from Quechua, the language of the Incas.

Quechua is still the most widely spoken Amerindian language and has given English dozens of words: guano, the once invaluable bird droppings used as fertilizer, comes from the Quechua word for dung, *huanu*. The beef 'jerky' that cowboys eat in movies derives its name from *ccharquini*, 'to prepare dried meat'. But perhaps the most important Quechua borrowing came from a tree which the indigenous people called *kina-kina*. Early Spanish explorers learned from the locals that the tree's bark had an almost miraculous ability to cure fever, even the lethal malarial fevers that defended the tropics from European invasion. In Europe, the mysterious bark became a Spanish monopoly, allowing them to successfully go where few white people had gone before. It thus became a key weapon for the invaders against the very people who had taught them its secrets. The British referred to the mysterious substance as 'Jesuit's bark', and put great effort into 'acquiring' seeds so that they could grow it in India and break the Spanish monopoly. The Spanish spelled the bark's name *quina-quina*, from which we get the drug's modern name, quinine.

Because there are often many indigenous terms for a species, or several species that bear the same name, native names are seldom retained as scientific ones, but the scientific name of the guinea pig is an exception. The genus's scientific name, *Cavia*, derives from the Quechua *cui* or *cuy*, which may have echoed the squeaking *kwee-kwee* sound the animals make. The *cuy* is one of only four mammals to have been domesticated in the Americas – the alpaca, guanaco and llama being the others. The latter trio were domesticated for their wool or as beasts of burden, but *cuys* were mainly a source of meat. Several thousand years ago, perhaps as early as 7,000 BCE, the people of the Andes started to keep *cuys* as livestock. They were probably treated much as they are today; most rural families in the Andes have a dozen *cuys* living in their house – they usually live in hutches in the kitchen and feed on scraps. Once they are fat enough, they are killed and cooked. Among the Andean people they still have other uses, as gifts or as a form of pocket money for children. They play a part in traditional healing ceremonies, during which live *cuys* are sometimes still rubbed on the affected parts of people who are ill, and then sacrificed. *Cuys* are found at the heart of all kinds of Andean rituals, from birthdays to weddings; their ritual significance survived the transition to Christianity – on All Souls' Day, the dead are offered a portion of *cuy* meat.

As with any domesticated animal, the Incas began a haphazard kind of selective breeding of the *cuy*, almost without realizing it, as soon as they took an interest in the species. *Cuys* that were too fast to catch, too aggressive, or too skinny to eat simply did not find a home in Andean hutches. As with other domesticated species, features of the animal that would have been a distinct drawback in the wild became a plus in the domesticated breed; natural selection is unlikely to favour an animal that resembles a placid, slow-moving, substantial meal, but artificial selection favours precisely those traits. As the *cuy* changed its shape and behaviour to suit its new habitat, humans must have noticed what was happening and began to breed them deliberately, probably about 3,000 years ago. By the time the Spanish invaded South America in the sixteenth century, the *cuy* was fully domesticated and was used both for food and in Inca religious ceremonies. Mummified *cuys* have been found in Inca tombs, along with terracotta statues representing them. Modern guinea pigs have been steadily adapted to human needs ever since, so much so that they are now classed as a separate species, *Cavia porcellus*, whose precise relationship with the wild species of the Andes is no longer clear.

The *cuy* made its debut in print in 1547, when Gonzalo Fernández de Oviedo y Valdés (usually known simply as Oviedo) published his *Historia general y natural de las Indias* ('General and Natural History of the Indies'), the earliest illustrated natural history printed in Spain. Oviedo was an official imperial chronicler for the Spanish court, and accompanied the conquistador Francisco Pizarro to make an inventory of Spain's new possessions. The immensity of the New World's endless, nameless jungles threatened to make Oviedo's work impossible; he wrote that 'although it is visible, we ignore most of it, since we do not yet know either the names nor the properties of such trees'.[1] His interest in naming the contents of these new territories was very pragmatic; like most Europeans of the time, he saw the New World in terms of resources and named them so that they would be easier to exploit. When he preserved indigenous names it was for precision, to make the identity of a resource obvious. He saw his first *cuy* in Santo Domingo, but since they are not indigenous to any part of central America, these would have been domesticated ones the Spanish had brought there. Oviedo renamed them the '*chanchito de la India*', or little pig of the Indies. His decision to dub the animals 'pigs' may have been prompted by their squeaks and squeals, but the name seems more likely to refer to the fact that European pigs were often kept in much the same way: allowed to wander about the homestead, eating scraps until they were fat enough to eat.

As we have seen, Europe's naturalists were enthusiastically creating comprehensive catalogues of everything the world contained, often by copying and compiling each others' work, so once Oviedo described the *chanchito de la India* it became a feature of natural history books. The Swiss naturalist Konrad Gesner mentioned it in his *Icones animalium quadrupedum* (1553), giving it the name *Cuniculus indus*, or rabbit of the Indies – perhaps because of its size and habit of burrowing. It made its debut in English in Edward Topsell's *The Historie of Foure-Footed Beastes*, which appeared in 1607. Topsell was a clergyman with no pretensions to be a naturalist – most of his animal stories served to illustrate moral points – and his book is largely a reworking of Gesner's. Topsell knew that the animal Gesner compared to a rabbit had also been likened to a pig, so he compromised by giving it the name 'Indian little pig coney', coney being a common word for a rabbit.

Sometime in the seventeenth century, English speakers started calling these animals guinea pigs. It remains unclear exactly when or why, but in 1664 the English natural philosopher Henry Power was confident that they were familiar enough to be used to provide a comparison for the distinctly unfamiliar cheese mites he had observed under his microscope, which he described as looking 'like so many Ginny-Pigs, munching and chewing the cud'.[2] The origins of 'guinea' remain a mystery. All kinds of implausible suggestions have been made, including the idea that these pets sold for a guinea (21 shillings, which would be £100 in today's money – rather a lot for a small pet). It has also been suggested that the ships bringing the animals from South America called in at Guinea in West Africa on their way to Europe, but there is no evidence for this, nor for the more plausible confusion of Guinea with the South American country of Guyana. It is more likely that the British used 'Guinea' in a very loose sense to mean any far-off, exotic country – somewhere so distant and foreign that no one knew (or perhaps cared) exactly where it was.

The animals were formally given the generic name *Cavia* by the German naturalist Peter Simon Pallas, in his *Miscellanea zoologica* (1766); *Cavia* being a Latinized version of *cuy*. The species name, *porcellus*, or 'little pig', was conferred by Linnaeus. The British naturalist Thomas Pennant seems – understandably – to have felt that the common English name 'guinea pig' was inappropriate for an animal that was neither a pig nor from Guinea, and so he added another Quechua name to the English dictionary, borrowing *cuy* via *cavia*, and re-christening Topsell's little pig coney as the 'restless cavy', in 1781.

However, 'restless' was a singularly inappropriate adjective for domesticated cavies, which had been selected over thousands of years for placid temperaments. Noticing that they rarely bit people, Dutch sailors were bringing them to Europe as pets for their children not long after Oviedo first described them. The English took to the guinea pig particularly enthusiastically and they soon became popular pets; even Queen Elizabeth I had one. Their good natures endeared them to the ladies of the Court, who were often to be seen accompanied by a servant carrying a pet guinea pig on a silk pillow. By Victorian times they were so popular that everyone in Britain could be assumed to know what a guinea pig looked and sounded like, to judge by the regularity with which they were used metaphorically. George Eliot mentions them several times; for example, describing a character in *Daniel Deronda* as possessing 'a pair of glistening eyes that suggested a miraculous guinea-pig'.[3] They also became part of the language in several common expressions: those who took company director-ships only for the sake of the fees (paid in guineas) became known as 'guinea pigs', as did clergymen who were paid to give sermons on behalf of their wealthier, but more indolent, colleagues.

In nineteenth-century Britain, pet-keeping became an increasingly competitive business as dog shows and pigeon-fancying clubs became common. The guinea pig was soon attracting its own 'fancy', largely thanks to one man, Charles Cumberland, a writer and Fellow of the Zoological Society, who published *The Guinea Pig or Domestic Cavy* in 1886. Cumberland proffered his claim of having 'something new to tell' as 'my chief excuse for laying this brief treatise before the public', in the hope that its defects, 'of which I am conscious, may be atoned for by its novelty, and the value of its facts'.[4]

Cumberland's claim to novelty was not immediately apparent, since he began with a recapitulation of the relevant writings of Gesner, Topsell and the French naturalist, Buffon, but Cumberland claimed to write from first-hand experience, asserting that many of his predecessors had simply copied from other writers without checking their facts. As a result, basic information, like the number of the animal's toes, had been continuously misreported for centuries. Yet despite his desire to sweep away misinformation with science, he added to the confusing legends about the animal's common name by suggesting that it probably came from the Spanish having first encountered them on sale in markets, prepared by 'scalding and scraping them in the same manner as we should treat a pig'; hence they looked rather like small suckling pigs. Cumberland lists their numerous European names – *Cochon d'Inde* (Indian pigs, in France); *Cochinillo das Indias* (Indian pigs again, in Spain); and *Meerschweinchen* ('little sea pigs' in Germany) – and suggests that they would be best known as domestic cavies, noting that 'cavy' is close to the Peruvian Indian words *Coüi* or *Coüy* (now spelt *cuy*).[5]

Cumberland relied on the libraries of the Zoological Society and the British Museum for the historical sections of his book, but felt that the days of compiling from other writers were over. Gripped by the prevailing spirit of empiricism, he proudly claimed that 'the observations upon the management of Cavies are based upon experiments, conducted by myself, on a considerable scale, during a period extending over more than five years'. Cumberland kept his guinea pigs much as the Andeans kept theirs; he describes how one of his males, 'Bobby', was allowed to run 'loose about a kitchen, and was much petted'. Once you got to know your cavy, Cumberland suggested, you would soon learn to recognize the 'little grumbling note, by which it appears to express satisfaction or affection' and to distinguish that sound from the call 'with which it greets the sound of the well-known step of its feeder or owner'. Those 'who are intimate with the animal, will . . . find many gratifying marks of intelligence and affection'.[6]

Cumberland's book was intended to promote fancy guinea pig shows, so he urged his readers to keep and breed cavies, to join or form cavy clubs, so as to exchange breeding stock and thus 'avoid the evil effects' of inbreeding. The would-be breeder nevertheless faced the problem of what to do with the 'weeds', the substandard members of any litter, which are useless for breeding and cannot be sold. 'This difficulty,' Cumberland wrote, 'may be removed by sending the useless Cavies to table, for which purpose they were, probably, in the first instance domesticated.' Indeed, not only does he present the eating of cavies as a necessary by-product of breeding them, but – as Cumberland acknowledged – persuading his readers to think of them as edible was 'the principal object I had in view when I began the cultivation of the cavy'. (His book, which went into several editions, bore the subtitle 'for Food, Fur and Fancy' – although he had to admit that their fur is not much use for anything.)

Cumberland gave detailed instructions on how to kill and clean your guinea pigs, but added that he looked forward 'to the time when Cavies will be bought up for market purposes by people who will make a business of fatting, killing, and preparing for cooking'. And just to ensure everyone got his point, Cumberland added a few recipes to his book, including curried cavy, *Cavy aux Fines Herbes* and '*Cavy en Gibelotte*' (sautéd cavy served with an eel in a white sauce, a few mushrooms, some white wine, 'and season with salt, pepper and, a bouquet of parsley, thyme, and little green onion'). Adding:

> I do not wish it to be supposed that I recommend Cavy as a cheap food, but rather for its delicious flavour and *recherché* quality. It may, no doubt, be sometimes grown at small cost; but I look upon it as being so excellent for the table as to be worthy both of trouble and expense in its cultivation. Think of its value in the game course when game is out of season.[7]

Into the lab

If being served sautéd with eel seems like a grim fate for a guinea pig, worse lay in store for *Cavia porcellus* when men of science started to take a serious interest in the poor creatures. Cumberland noted that the animals had first entered a lab in about 1780, when the pioneering French chemist Antoine Lavoisier used guinea pigs to measure the amount of oxygen consumed and carbon dioxide produced during respiration. The very qualities that made guinea pigs into ideal pets – their being small, docile and easy to look after – also made them ideal laboratory animals. And they bred comparatively rapidly; females can become pregnant when they are just three months old and are fertile every two to three months thereafter. They usually have two to four young at a time, but litters of up to eight are not uncommon.

By the time Cumberland was researching his book, guinea pigs were to be found in scientific laboratories all over Europe. In Germany, Robert Koch had used guinea pigs to persuade doctors of the truth of the still new 'germ theory', that diseases were spread by newly discovered minute creatures called microbes. Koch pioneered new techniques for identifying microbes under a microscope and proved that each disease was caused by a different germ; he and his co-workers identified the microbe that caused tuberculosis in 1882 and that for cholera in 1883. A few years later, an American journal was able to inform its readers of Koch's latest breakthrough: 'he had found the means of arresting the development of tuberculosis', as a result of experiments done on guinea pigs, which were 'even more sensitive than man' to the microbe that caused the disease.[8] A few years later, *Harper's* reported that each microbe had its own anti-toxin, and described experiments that concluded with one cage containing 'a dead guinea-pig, inoculated with diphtheritic poison, while its companion, inoculated . . . with the same poison and also with its correspondent antitoxine [*sic*], seemed to be a little ragged and under the weather, but otherwise in cheerful spirits and condition'.[9]

Among the many medical and scientific men who found guinea pigs invaluable for their research was the American-born physiologist Charles Édouard Brown-Séquard, who used them in his research on epilepsy, discovering that guinea pigs could be 'rendered epileptic in consequence of an injury to the spinal cord'.[10] Some of his contemporaries, including Charles Darwin, were intrigued by his results and defended animal experiments as vital to scientific progress, but others were deeply concerned by

Brown-Séquard's work. When he had referred to the guinea pig's having suffered 'certain injuries to the spinal cord' he was – perhaps not entirely accidentally – glossing over a significant point: he had caused those injuries. In an effort to discover what role the nerves and the signals they carried played in epilepsy, Brown-Séquard had deliberately severed the animal's major sciatic nerve, which joins the spinal cord to the leg and foot muscles. He had done the experiments for what he had felt were very good reasons: it had long been known that some human epileptics could be cured through having their toes flexed or immobilized.

Yet, however justified physiologists like Brown-Séquard felt they were in experimenting on animals, an increasingly vociferous and influential section of the public disagreed with them. Gradually, Brown-Séquard's fame took on a new quality: as an American doctor writing in the popular magazine *Scribner's* observed, he 'has probably inflicted more animal suffering than any other man in his time'. The article described a visit to Paris to observe Brown-Séquard at work: 'a Guinea-pig was produced – a little creature, about the size of a half- grown kitten – and the operation was effected, accompanied by a series of piercing little squeaks', after which the guinea pig ran in desperate circles, the injury to its brain leaving it unable to walk in a straight line. 'This experiment,' the writer argued, 'had not the slightest relation what-ever to the cure of disease.'[11] Why then were animals made to suffer? Partly as a result of such publicity, Brown-Séquard found that he was sometimes unable to get to scientific meetings because of the threat of demonstrations by anti-vivisectionists.

However, in the following issue of *Scribner's*, another doctor sprang to vivisection's defence, arguing that Brown-Séquard had not just induced epilepsy in guinea pigs, but had also discovered that 'if a certain region of the skin of the face is cut out, the animal gets well'. This had direct implications for human health:

Some time since, a boy was struck on the head with a brick; epilepsy followed, and two years of complete wreck of health,

threatening idiocy. A vivisector was at last called in con-
sultation, and, bearing in mind Brown-Sequard's experiments,
had the scar on the head cut out. Result – cure. A considerable
gain, that, to one young life.[12]

For some doctors, such cures were sufficient to justify vivisection.
They also argued that the pain they caused, which 'accomplishes
so much for the human race', was 'inconceivably minute' when
compared to that which nature inflicted through disease,
predation and parasites. The clinching argument offered to the
magazine's readers was that the regulation of vivisection would be
expensive: the writer asked rhetorically if it was really appropriate
'that the population shall be taxed' even more heavily, simply 'to
render more irksome and laborious that progress in the divine art
of healing'?[13]

Because of their popularity as pets, guinea pigs were often
prominent in these debates over vivisection and gradually 'guinea
pig' came to be synonymous with 'experimental organism'.
Cavies made one of their stranger cultural appearances in a short
story, 'A Point in Morals', by the American novelist, Ellen
Glasgow. In her tale, several characters discuss whether human
life has become over-valued until everyone is being kept alive and
'the survival of the fittest is checkmated'. (Her story first appeared
in 1895, just as interest in eugenics was beginning to revive on
both sides of the Atlantic.) One of the characters, 'a well-known
alienist [psychiatrist] on his way to a convention in Vienna',
describes meeting a murderer on a train. The man admits his
crime and claims to have no regrets: 'I was ridding the world of a
damned traitor,' he argues. But he has decided to kill himself, so
as to spare his wife and family the pain of seeing him tried and
executed. He carries a vial of carbolic acid for the purpose, but has
realized that the alienist can grant him a much less painful death
– thanks to a large quantity of morphine the latter happens to
have in his bag – and the murderer begs for it. The alienist
hesitates over the morality of assisting a suicide and helping a
murderer escape justice, but remembers 'that I had once seen a

guinea-pig die from the effects of carbolic acid, and the remembrance sickened me suddenly'. Should he spare the guilty murderer the agony that the innocent animal suffered? He wonders what his favourite philosophers would advise him and imagines the man's 'broad-faced Irish wife and the two children' and the misery and disgrace they would otherwise face. And then 'I thought of the dying guinea-pig', and as the train pulls into the station where the alienist is to get out, 'I stooped, opened my bag, and laid the chemist's package upon the seat. Then I stepped out, closing the door after me.' When he reads of the man's death in the paper the following day, he admits that he feels like a murderer himself, but 'a conscientious murderer'.[14] By the end of the century, thanks to the anti-vivisectionists, guinea pigs were playing a rather different role in fiction from George Eliot's glistening eyes of twenty years earlier.

Anti-vivisectionists often protested that not only were animal experiments cruel, but nothing of value was learned from them. The increasing cost of modern science and the growing expectation that government should finance it brought guinea pigs into wider debates over who should pay for science. One magazine writer satirized the scientist's demands: 'Give me a thousand or fifteen hundred a year,' the fictitious physiologist asks the State, and 'In return I will give you some new facts about . . . the length of time a new poison takes to kill a guinea-pig.'[15]

Fortunately for the scientists – if not for the guinea pigs – cavies proved susceptible to scurvy, a fact that was to do much to redeem the public image of laboratory science. For centuries, scurvy had caused sailors more suffering than storms and pirates combined. After long voyages, mariners returned home covered in bruises, their mouths bleeding and their teeth falling out; left untreated, they eventually suffered internal bleeding and died of the disease. In the mid-eighteenth century a British doctor, James Lind, had discovered that fresh oranges prevented and cured scurvy. From then on, British navy ships always carried fruit, usually lemon or lime juice. After 1844 it became law for all British merchant vessels to do the same, so that British sailors and ships' passengers became known in Australia and America as 'lime-juicers', or simply limeys. But, mysteriously, sometimes the lime juice failed and entire crews who had been drinking it regularly nevertheless became sick.

Doctors still did not know exactly what scurvy was, nor why lime juice sometimes cured it – and sometimes failed to. The disease's symptoms were similar to those of an even more painful disease, rickets, that regularly afflicted the children of Britain's seething slums in the late nineteenth century. In many cases, children moaning in agony were admitted to hospital and were found to be suffering from both diseases. Doctors found that, like scurvy, rickets could be cured if they were quick enough, by feeding children fresh milk, fruit and vegetables, which slum children rarely saw. Even children from prosperous homes suffered from the disease, because some misguided middle-class parents thought strongly flavoured foods, such as fruit and vegetables, were unsuitable for a child's diet. Another similar illness, known as 'ship's beriberi', seemed to afflict sailors on long voyages from the East, even though they had meat and vegetables on board. Some doctors wondered if the relatively new process of canning vegetables led to contamination and although no evidence of poisons could be found in the cans, it was clear that for some reason tinned or dried vegetables could not protect against scurvy-like diseases – fresh vegetables possessed some virtue that was lost when they were preserved.

Impressed by the success of microbe hunters like Koch, doctors turned to the new laboratories to try to work out exactly what caused the various scurvy-like diseases. Some assumed infectious microbes were at fault, others suspected poisons in the often badly-preserved foods on ships, while a third group assumed that diet alone was responsible. Initially, experiments were done with pigeons, but as it became clear that pigeons remained healthy on diets that made people sick, researchers started looking for a mammal to experiment on – in the hope that its reactions would be more comparable to human's. Guinea pigs, by this time common in laboratories, were an obvious candidate. Two

Norwegian doctors, Axel Holst (who had worked in Koch's lab) and Theodor Frölich, fed guinea pigs on a variety of diets and proved that while they remained healthy on fresh potatoes, they died if they were only fed dried ones. They also showed that well-known anti-scurvy treatments, like cabbage, became less effective or even worthless the longer they were cooked. When they published their results in 1907, they had found no evidence for either microbes or poisons – poor diet alone caused scurvy. Unfortunately, the chronic lack of funding for Norwegian scientific research at the time made it almost impossible for them to carry on with their work, which was largely ignored.

Despite the growing acceptance that the scurvy-like diseases (which, in addition to rickets and beriberi, also included pellagra) were all caused by dietary deficiencies, it was clear that the British Army had not learned the lessons their naval colleagues had long understood. During the First World War, scurvy became wide-spread among the troops; many of those evacuated from the disaster at Gallipoli were found to be suffering from diseases like beriberi and scurvy. More experiments, with more guinea pigs, were performed at London's Lister Institute, where a few years earlier a Polish chemist named Casimir Funk had demonstrated that an extract of rice could cure beriberi. Because the extract contained a chemical known as an amine (a derivative of ammonia), Funk decided he had discovered a new group of chemicals which he called 'vital amines', or vitamines.[16] By 1915, Funk had emigrated to America and nearly all the Lister Institute's other male scientists were away fighting in the war, so their colleagues Harriette Chick and Margaret Hume led a team of women who were given the task of identifying foods that could be easily transported to the troops to keep them healthy. They discovered that guinea pigs which were fed nothing but oats, bran and water quickly contracted scurvy, so they tried supplementing their diets, adding one extra food at a time, to discover what worked best to prevent the disease. Among other things, they discovered that not all limes are created equal: the West Indian sour lime, which had been the Royal Navy's main source of lime juice since the late nineteenth century, turned out to be much less effective than the Mediterranean sweet lime it had replaced. Even worse, its effectiveness against scurvy fell rapidly when it was preserved in alcohol, as it had to be before shipboard refrigeration.

Meanwhile, on the other side of the Atlantic, rats were overcoming decades of prejudice to find their way into biology labs. They had, of course, long been regarded as aggressive, disease-carrying vermin, but were finally becoming common laboratory animals, as we saw with the Wistar rats in the previous chapter. Elmer V. McCollum, who led a research team at the University of Wisconsin, was particularly keen to promote the rat as a standard animal ('McCollum rats' are still in use laboratories today), not least because they bred much faster than guinea pigs. However, when he first suggested them to his boss, he was told that if anyone discovered his team were 'using federal and state funds to feed rats we should be in disgrace and could never live it down'.[17] Despite this reaction, McCollum's team managed to acquire some taxpayer-funded rats and used them to investigate dietary diseases. In the process, they discovered that each was caused by a different deficiency and christened the mystery factors 'A' (whose absence caused childhood blindness and reduced immunity to other illnesses) and 'B' (where deficiency caused beriberi). The Americans also repeated the British guinea-pig experiments, but found that fresh milk did not appear to prevent scurvy as the British had claimed. Since McCollum's team had been unable to link scurvy to either factor A or factor B, they suggested it might not be a deficiency disease at all. Back in Britain, Chick and Hume responded by repeating their initial experiment and finding that fresh milk *did* prevent scurvy.

Why did the experiments on opposite sides of the Atlantic have different outcomes? Was there something in British milk that its US equivalent lacked? A chemical analysis of the different diets eventually identified a third component, originally called 'accessory food factor C', a name that – not surprisingly – did not catch on. The public had grown used to talk of 'vitamines' so the scientists ended up adopting the name, despite the fact that amines turned out not to be important. As a concession to scientific accuracy, the final 'e' was dropped and vitamin C was born.

Fresh milk is a good source of vitamin C, so the failure of the US experiments was puzzling, but was explained by one of Chick and Hume's colleagues, who observed that 'the one thing which is fatal to nutritional work is, to send the animals away to an animal-house to be looked after by someone else', which is what the Americans had done. In the animal house, no one had checked to see whether or not the guinea pigs actually drank their milk; by contrast, Chick and Hume's guinea pigs 'have been tended and nursed and fed by the observers themselves'.[18] As long as the guinea pigs actually drank their milk, they stayed healthy, but another mystery remained, which was why guinea pigs developed scurvy on diets that were evidently adequate for rats. This confusion turned out to be a result of the fact that guinea pigs and humans cannot make vitamin C in their bodies and so have to consume it if they are to stay healthy; however, rats and pigeons produce their own vitamin C if their diets are deficient. Once it was clear that the guinea pig was the appropriate animal to work with, research began to find out exactly what vitamin C was and eventually – once the chemical structure was discovered – the first synthetic vitamin C was created in 1933 and successfully tested on guinea pigs.

The conquest of diseases like scurvy became one of the first big success stories for a new science, biochemistry, which used the vitamin story to promote the idea that it was a science of life, at a time when the older physical sciences were increasingly being used to create new and more dangerous weapons – bombs and poison gas. The massively influential and wealthy Rockefeller Foundation was persuaded by these arguments to switch much of its funding from physics to biological research in the 1930s. Thanks in part to guinea pigs, the newly wealthy life sciences began to attract bright young scientists who set to work to analyse the processes that sustain life, from fermentation and photo-synthesis, to respiration and digestion. Understanding the chemistry of life was the first stage in a long boom for biology, until it eventually threatened to displace physics as the queen of the sciences.

From agriculture to jazz

Ironically, the Norwegians Holst and Frölich were not the first to produce scurvy in guinea pigs: a group at the US Department of Agriculture's (USDA) Bureau of Animal Industry had accidentally done the same thing a decade earlier, when the supply of fresh grass ran out and the animal's keeper forgot to give the animals fresh vegetables instead. The Bureau's annual report related the details of the error rather apologetically and – unfortunately as it turned out – did not publicize their accident too widely; the story of vitamin C might have been very different had they done so.

The USDA had been established in 1862 by Abraham Lincoln. Coming from a farming background himself, Lincoln knew that the majority of his countrymen were farmers who needed the best seeds, crops and advice that science could provide. To head the new department, Lincoln appointed a man called Isaac Newton (no relation), a successful farmer from Pennsylvania who had previously been in charge of the agricultural section of the Patent Office. In his first report on the new department, Newton identified its main goals as: publishing useful agricultural information; introducing valuable plants and animals; responding to inquiries from farmers; and testing new agricultural machinery and inventions.

Guinea pig keeping was, unsurprisingly, not on Newton's list, yet within a few decades the USDA had a substantial colony, used to test vaccines. A dozen years later the guinea pigs were being bred on a large scale, as a substitute for traditional farm animals; but unlike Charles Cumberland, the Bureau was not breeding guinea pigs as a replacement for 'the game course when game is out of season', nor for their '*recherché*' flavour. The USDA wanted to investigate the breeding of livestock, but had no space for thousands of pigs or sheep. Since guinea pigs had proved a valuable alternative to humans in testing vaccines and vitamins, they were now becoming a substitute for farm animals in a large-scale investigation of the effects of inbreeding. In 1906, George M. Rommel, head of the Bureau's Division of Animal Husbandry, had decided to investigate inbreeding because it remained common – if controversial – among commercial animal breeders. As we have seen, some argued that this was the best way to 'fix' a desirable trait, while others argued that such incestuous couplings were invariably harmful. Rommel instituted controlled experiments to answer the question and, since the Bureau already had a colony of guinea pigs, he opted to use them.

A guinea pig's life at the Bureau's experimental farm in Maryland was more attractive than those of its colleagues in other labs; all the US government wanted of them was that they have lots of sex, albeit with their own siblings. The Bureau's researchers crossed brother and sister guinea pigs for more than two dozen generations, in order to create a number of heavily inbred families, but having done so, they had a problem. By 1915, they had accumulated data from tens of thousands of guinea pigs, but were unsure of the best approach to take in analysing it. Clearly, they needed a clever young geneticist who had learned the latest techniques, so Rommel approached his friend William E. Castle at Harvard to see if he could recommend one.

As we have seen, Castle was also interested in inbreeding, which was why he had first brought fruit flies into his lab, but his real interest was in mammals. He had been born on a farm in Ohio and studied biology at Harvard in the early 1890s. Among his teachers was Charles Davenport, one of the pioneers of laboratory biology in America. Castle stayed on as Davenport's assistant, then taught zoology back in the Midwest for a couple of years before returning to Harvard, where he began rearing guinea pigs and mice for experiments. When the 'rediscovery' of Mendel's work was announced, Castle was intrigued and started using his mammal colonies for inheritance research, becoming one of the first mammalian Mendelians.

Within a few years of his return to Harvard, Castle had bred over 1,500 rabbits, 4,000 rats and 11,000 guinea pigs. He was getting desperately short of space when he heard that Harvard's Bussey Institution – which had originally been founded to teach agriculture – was about to be closed down. Castle and others successfully lobbied to have it turned into a research facility for the biological sciences and persuaded the Carnegie Institution to contribute to the cost.

With the money and facilities secured to continue his mammal work, Castle set out to test de Vries's Mutation Theory. He had originally been largely convinced by it – and had been one of the scientists invited by Davenport to address the 1905 Philadelphia meeting of the American Society of Naturalists on the subject – but had since begun to have doubts. With the help of one of his students, Hansford MacCurdy, Castle set out to test the power of selection; he had told the Philadelphia meeting that 'the formation of new breeds begins with the discovery of an exceptional individual', and that 'such exceptional individuals are mutations'.[19] The question was whether, having found such an individual, it was possible through selective breeding to create not merely a new variety, but a new species. As Castle and MacCurdy put it in the published account of their experiments, a great deal had been written both for and against de Vries's theory, 'but discussion is at present less needed than experimental tests of the views outlined'.[20]

For their experiment, Castle and MacGurdy chose hooded rats (so called because they are white with black heads). The rats presented what looked like a classic case of continuous variation: some had very little black fur (giving small 'hoods') while others were predominantly black (large hoods), with most of the rats lying somewhere between these extremes. Since the amount of black varied smoothly, with no jumps or breaks in the sequence, it seemed unlikely that the pigmentation was controlled by a simple on/off Mendelian factor. Castle and MacCurdy mated large-hooded rats with each other, and did the same with the small-hooded rats. They expected to find, as plant breeders had done earlier, that selection could only shift a species so far; the amount of black could be increased or reduced up to a certain point, but eventually what was called a 'pure line' would be created and no further change would be possible unless a fresh mutation occurred.[21] They also assumed that once they stopped selecting, put their large- and small-hooded varieties back together and let them once more mate at will, later generations of rats would rapidly revert to type and show the original variation along a continuum.

To Castle and MacCurdy's surprise their rats did not revert; they appeared to have created some lasting genetic change. As they noted, these were results that 'support the Darwinian view rather than that of De Vries'.[22] Castle became a convinced Darwinian and argued that – despite appearances – there must indeed be a Mendelian factor for 'hoodedness', but that the factor was in some way altered by selection so as to produce discontinuous variations. However, his colleagues were unconvinced by this suggestion and argued that cases such as the hooded rats were better explained by assuming that several Mendelian factors were at work, which affected each other but did not actually change. There might, for example, be a basic pair of genes – large-hood and small-hood – but also several other 'modifier genes' that increased or reduced the effect of the basic pair. By removing the rats with intermediate amounts of black, Castle and MacGurdy had been removing the modifier genes from their breeding stock, until the animals that remained possessed only either the large-hood or small-hood gene. When these stocks were allowed to interbreed again, the original smoothly graduated range of blackness did not reappear, because no modifier genes remained. After several years of debate, it was this latter view that became widely accepted, largely because it was born out by the fly boys' work, which – as we have seen – showed that genes were not the simple switches they had originally been assumed to be. As a result, the long-standing distinction between continuous and discontinuous variation – and between blending and non-blending inheritance – began to be abandoned. 'It is impossible,' Castle said, 'to make a sharp distinction between continuous and discontinuous variation'; they were both controlled by genes.

Castle continued that it therefore seemed misleading 'to assign all evolutionary progress to one sort of variation or to one sort of inheritance'.[23] Between them, the rats and flies had finally persuaded most biologists that all inheritance was Mendelian – everything was in the genes.

Castle also had the rats to thank for the fact that he became one of the most influential of the early American geneticists. He travelled, lectured and wrote, explaining how the Mendelians were 'able to predict the production of new varieties, and to produce them'.[24] Prediction was one of the highest goals of any science: ideally, a scientific theory allows you to calculate the outcome of an experiment in advance – that way, actually performing the experiment allows you to test whether or not your theory was right. Castle's audiences would have grasped his implication: the biometricians – for all their complex mathematical tools – could not predict the outcome of a specific mating, while the Mendelians could. In the spring of 1912, Castle visited the University of Illinois to lecture on hooded rats and Mendelian genetics; in his audience was a young graduate student, Sewall Wright, who was excited by what he heard. Wright went up to Castle afterwards, to tell him 'that I was very much interested in genetics but that no course was given in it at Illinois', and asked if it might be possible to do research with Castle.[25] Wright had taught himself genetics from an article on Mendelism in the *Encyclopaedia Britannica*; its description of the simple mathematical rules that underpinned inheritance greatly appealed to him. Castle was impressed by the young man's intelligence and enthusiasm – he invited Wright to come and work with him.

When he arrived at Harvard, Wright found that he was not going to be working on flies or rats, but on guinea pigs. Castle was about to lose his existing guinea pig keeper, John Detlefsen, who had finished his graduate work; Wright was to be his replacement. Before he went off to start his first job, Detlefsen introduced Wright to the guinea pigs who, unbeknownst to the newcomer, were to be the focus of the rest of his working life. As well as teaching him the basics of guinea pig care and feeding, Detlefsen also showed Wright his data on six generations of guinea pigs; crosses between the common lab guinea pig *Cavia porcellus* and wild Peruvian cavies, *Cavia rufescens*, which Castle had collected a few years earlier. These crosses showed that while the first generation were almost all sterile, the population's fertility gradually recovered over successive generations, as the percentage of the other species' genes declined. However, Detlefsen could not produce a more precise analysis of his data. Wright took one look at it and did a quick calculation on the assumption that there might be several Mendelian factors operating (as with the hooded rats), each of which contributed equally to sterility. He immediately realized that if there were eight factors his calculation gave theoretical percentages that closely matched those Detlefsen had actually observed. The entire exercise took Wright only a few minutes and he was surprised by how impressed Detlefsen was; both the theory and the calculation had seemed quite obvious to him.

As the speed at which Wright had done his calculation suggests, he was fascinated by mathematics and had been from an early age. When he began school, just before his eighth birthday, he had been asked to demonstrate his mathematical skills, to enable his teachers to assign him to a grade. He remembers volunteering 'that I could extract square and cube roots', ruefully adding that, in retrospect, he would not have done so if he had had a little more experience of school life. He was taken down to the eighth grade's classroom and despite being almost unable to reach the blackboard, successfully extracted the cube root of a number. 'It must,' Wright later recalled, 'have been a most disgusting spectacle to the students.'[26] Being younger and shorter than most of his schoolfellows left Wright feeling 'much out of place', but his enthusiasm for maths was undiminished. He later wished that he had acquired a more thorough mathematical education, but nevertheless – as he modestly put it – he 'acquired some facility in translating questions into mathematical symbolism and solving [them] as best as I could'.[27]

Wright's modesty about his abilities was entirely characteristic: his contributions to the mathematical understanding of inheritance were to make him famous; he helped create a new mathematical approach to genetics that ended the long running battle between the Mendelians and biometricians. This mathematical treatment of evolution – along with his guinea pigs – also brought Wright into contact with one of the more extraordinary figures of early twentieth-century-British biology, John Burdon Sanderson Haldane.

Bombs, biochemistry and beanbags

Haldane and Wright could hardly have been more different: while Wright came from a fairly ordinary middle-class American family, Haldane was an aristocrat in two senses: his family could trace its lineage back to the ancient Scottish nobility, but – more relevantly – he was born into Britain's scientific aristocracy. His father was a distinguished physiology professor at Oxford, an expert on respiration who was frequently called in by the government to advise on such matters as safety issues in the mines. From the age of four, young JBS (as he was almost always known) was fascinated by what he referred to as his father's 'labertree' and the interesting game of 'experiments' that he played there. A precocious child, JBS could read by the time he was three; by the age of five he had learned enough German from his nurse to leave her little notes that read 'I hate you.'[28]

Haldane and Wright probably first heard of each other in 1915, when they each published papers demonstrating genetic linkage in mammals for the first time. Demonstrating that specific Mendelian factors were invariably inherited together in mammals – because they occurred on the same chromosome, as had been shown in Drosophila – was an important step to establishing that the connection between chromosomes and inheritance was universal; linkage had by now been found in plants too, so it seemed that every living thing shared the same machinery for passing on its variations. Wright's work grew out of his graduate studies at Harvard, which were in turn a continuation of Castle's

231

rat work, but Haldane's had rather more unusual roots. In 1901 his father had taken his eight-year-old son to the Oxford Junior Scientific Club, to hear a lecture by Arthur Darbishire on the newly rediscovered principles of Mendelism. A few years later, his sister, Naomi (who would later become a celebrated novelist under her married name, Naomi Mitchison), developed an allergy to the horses she had loved and took up keeping guinea pigs instead. She loved the animals and knew many of them by name; she could impersonate their squeaks and grunts so well that they would answer her. When her elder brother came home from Eton for the school holidays and discovered her new pets, he 'suggested that we should try out what was then called Mendelism on them'. She agreed, deciding that 'Mendelism seemed quite within my intellectual grasp', and so her pet population began to expand. Well-known scientists, including pioneers of genetics, were familiar figures in the Haldane household – Naomi named one of her guinea pigs Bateson in honour of one of her father's visitors. One of JBS's friends remembered that in 1908 the lawn of the Haldanes' house was entirely free from the usual upper-class clutter of croquet hoops and tennis nets; instead, 'behind the wire fencing, were 300 guinea-pigs'. Naomi looked after them during term-time, and though five years younger than JBS, she became deeply interested in genetics, later recalling the 'the terrific thrill I got out of Morgan's great book [*The Mechanism of Mendelian Heredity*], reading it eagerly, curled up in a corner of the school room sofa, seeing it make sense of our puzzles'.[29]

As a result of these back-garden experiments, JBS and Naomi found they had, as she put it, 'tumbled onto what was then called linkage' (it was in fact often called 'reduplication' at the time, but 'linkage' became the accepted term soon after). JBS read all the papers on Mendelism then available to try to make sense of their results. In a recent scientific paper by Darbishire, Haldane noticed evidence of linkage, which the paper's author had overlooked; he and Naomi tried to prove that it existed using their pets. 'The guinea pigs were a mine of information,' Naomi recalled, 'we had to arrange marriages, which sometimes went against the apparent inclinations of the partners, though I rather enjoyed exercising power over them.'[30] But tragedy intervened. One of JBS's school friends, Cedric Davidson, recollected that the experiments 'necessitated the breeding of many generations of guinea-pigs, and our wretched little fox terrier . . . Billy . . . crawled over your front gate . . . goodness only knows how . . . promptly jumped on the cages in which were your g-pigs and they one and all died of fright', a double tragedy since 'they were the penultimate generation which were to prove your theory!!' Davidson was appalled; forty years later he could still remember how upset the Haldanes' mother had been as she waited for JBS to return from Eton that afternoon and discover what had happened. But Davidson recalled that when JBS got back, 'you came up to see us & told us not to worry, that you yourself were quite satisfied that your theory was correct, but admitted it only required one further generation for you to submit as a scientific fact'. Long afterwards, when Davidson wrote to his now famous friend to remind him of the disaster, he observed that 'I thought then and still think you lied and lied most nobly. I have never forgotten it and never shall.'[31] Nor, it seemed, had Haldane: his reply to Davidson's long, chatty letter was noticeably terse.

Thanks to Billy the fox terrier, Haldane's announcement of linkage in mammals had to wait until 1912, when he presented his analysis of Darbishire's mouse data to an undergraduate seminar at Oxford, where he studied first mathematics and then classics. He was advised to gather his own data before publishing and opted to work with mice, perhaps still mourning the fate of the original guinea pigs. Naomi and another friend, A.D. Sprunt, helped with the work, but before it could be published the First World War had broken out – by the time the paper appeared in 1915, Sprunt was already dead and Haldane was in the trenches.

Haldane had joined up immediately and served with considerable courage (and not a little foolhardiness) as a bombardier in the Black Watch. He briefly left the front after the first German poison gas attacks of the war: his father had been asked to help devise defences against the gas and had requested his son's assistance. As with many earlier experiments, the Haldanes, father and son, used themselves as guinea pigs (the Haldane family motto was the single word, 'Suffer'). They tested gas masks by entering a sealed chamber full of hazardous gases and then seeing how well they could walk, run or recite poetry, with and without the masks. Their work resulted in improved respirators that saved thousands of lives. Once it was done, JBS went back to fight, to discover that most of his fellow officers in the Third Battalion of the Black Watch had been killed during his absence.

After the war Haldane returned to Oxford, where he started teaching physiology, despite having no qualifications in the subject other than his famous father (JBS never in fact took a science degree of any kind); all he had, he claimed, was 'about six weeks' start on my future pupils' – but it proved to be enough. He was still interested in genetics, however, and discovering that Wright and Castle had published a paper on linkage in mammals, Haldane sent Wright a copy of his own paper on mice. He worried that 'it is not very intelligible', but explained that 'I was wounded at the time I wrote it, and thought I had better publish as quickly as possible'. [32]

The two geneticists must have rapidly discovered their common interest in guinea pigs. Wright's doctoral work had involved searching for a Mendelian explanation of the guinea pig's continuously varying coat colour. So when George Rommel from the USDA approached Castle for a Mendelian with expertise in guinea pigs, Wright – who had just completed his PhD – was the obvious choice and he became 'senior animal husbandman' for the Animal Husbandry Division of the Bureau of Animal Industry.

As well as doing research, Wright was still expected to keep up the long-standing tradition of the USDA and answer questions from farmers, amateur and professional breeders and, indeed, any random crackpot who felt like writing to him, including the secretary of the Illinois Vigilance Association. The Association, as its letterhead proclaimed, was 'organized for the purpose of suppressing the traffic in women and girls and the conditions which make that traffic possible'. Among other things it campaigned against the evil effects of Chicago's flourishing jazz scene. 'Moral disaster is coming to hundreds of young American girls,' wrote a journalist in the *New York American,* 'through the pathological, nerve-irritating, sex-exciting music of jazz orchestras.' He added that 'according to the Illinois Vigilance Association, in Chicago alone the association's representatives have traced the fall of 1,000 girls in the last two years to jazz music'.[33] The Association's secretary wrote to ask Wright if suppressing the 'sex instinct' could be scientifically proven to benefit the species. Wright's characteristically polite and careful reply simply observed that he knew of no evidence of ill effects from 'infrequent breeding of domestic or wild animals' apart from 'temporary sterility in the male'. The Association's secretary requested that Wright provide him with any future evidence on 'the benefits of the restraint of the sex instinct', if he should come across any.[34] Wright was exceptionally generous with his time and knowledge, feeling that as a public servant he was duty bound to answer every question as fully as he could. It was a habit that never left him; the immense influence he would eventually have in the scientific world was, in part, due to his generosity in answering his colleagues' questions.

Wright was too modest and shy to enjoy publicity, but Haldane revelled in it. He wrote a regular column for the British Communist Party's paper, the *Daily Worker,* as well as numerous popular books and a collection of children's stories, *My Friend Mr Leakey,* which remains in print almost eighty years after it first appeared. He also broadcast regularly for the BBC and proved to be an editor's dream, invariably available whenever an opinion, preferably a controversial one, was needed on any scientific topic. Haldane shared Wright's willingness to answer letters from the public, so while Wright was patiently answering questions about prize bulls and fallen women, Haldane's letters covered everything from the advisability of cousin marriage to the possibility of life on other planets; from offers to send him interesting cats to invitations to address student socialist societies; and explanations of original proofs of the mathematical Four-colour-map theorem.[35]

Haldane's knowledge of genetics inevitably also led to discreet enquiries about eugenic matters, especially whether it was advisable to have children if either spouse suffered from a particular illness or disability. On one occasion, a correspondent who 'suffered considerably from defects which I would rather not risk transmitting' asked JBS to help him locate a sperm donor, preferably 'an "allrounder" mentally and physically' – someone like JBS himself, the writer seemed to be hinting. Haldane pencilled 'really can't' in the margin of the letter and left it to his secretary to break the bad news.[36]

In 1923 Haldane defected twice: from Oxford to Cambridge and from physiology to biochemistry. Attracted by the power and promise of the new science of life, he accepted a new job working alongside Gowland Hopkins, another of the discoverers of vitamins (for which the latter won the Nobel Prize in 1929 – one of twenty-three that guinea pigs have helped win). Ten years after that, Haldane moved again – to University College London – initially as a professor of genetics and then of biometrics. These changes of location, specialization and interests were not as unusual then as they would be now – biology had not yet become quite as specialized as it now is – but they were nevertheless an indication of Haldane's restless intellectual energy, as well as his enormous talent for quarrelling with those around him. His contributions to any conversation were 'frequently caustic, at times vehement, but always profoundly human'.[37] Again, the contrast with Wright is striking: the American worked quietly, checking and rechecking his results before publishing them. Unusually for a geneticist of his generation he said little, in public at least, about eugenics, a subject – like most subjects – on which Haldane, over the course of his life, had numerous opinions, each of which was liable to be loudly discarded in favour of an equally strenuously maintained, but entirely contradictory, position within a few years.

While Haldane was busy shifting his stances and sciences, Wright stuck to his guinea pigs. He arrived in Washington to discover the USDA had kept meticulous, detailed records of over

34,000 matings; the results of each cross having been carefully recorded using a rubber stamp outline of a guinea pig, coloured in to show what the offspring looked like. Impressed by the accuracy of the records, Wright began analysing them with the techniques he had devised at Harvard. He discovered that, despite twenty generations of brother-sister matings, there were no signs of major problems, such as deformities. However, closer inspection showed that litters were becoming smaller and less frequent, the guinea pigs' birth weight was down, as was their disease resistance, and they were dying sooner. Inbreeding was clearly not good for them, a result that Wright must have felt some personal interest in, since his parents – like Charles and Emma Darwin – were first cousins.

Wright also found that the inbred guinea pig families looked very different from one another, despite having all descended from the same original stock. Just as Sergei Chetverikov had found with his wild Drosophila, there could be a lot of variation hidden in a population that only became visible slowly over several generations of inbreeding; this was the effect of recessive genes, which only became visible when an animal had two copies of them. Inbreeding made what were normally rare combinations – many of which were bad for the animal's health – much more common. However, it was also noticeable that when the inbred families were crossed with each other, much healthier, more vigorous animals immediately resulted, a phenomenon that became known as hybrid vigour (and which we will return to). Wright concluded that the best way to improve a breed was not continued selection within the breed as a whole, but the production of heavily inbred lines which might show up rare but useful traits. These could then be crossed to combine the desirable traits while restoring the breed's lost vigour.

Wright must have known of Haldane's interest in guinea pigs, since he offered to send him some of the USDA's stock. Haldane thanked him, but suggested that 'if you are testing this linkage on a big scale, there is no need for me to butt in. So please do not send them – if it is of any great trouble to you.' JBS had already taken on some rabbits from a co-worker and 'In consequence of this I shall have less space than I thought, so I should not be able to keep as many guinea pigs as I had hoped.'[38] A lack of space may have prevented Haldane from working regularly on guinea pigs, but so did his intellectual restlessness; he simply was not temperamentally suited to years of patient attention to detail in the way that Wright was. The practical side of biology also required other skills that JBS lacked; one of his former students remembered that 'he was not himself a good observer – and he was a terrifyingly bad experimenter'.[39]

However, one area in which Haldane and Wright were well matched was mathematics. Haldane had won a mathematical scholarship to Oxford, where he had achieved a First in the subject. In 1924, just after he had switched from physiology to biochemistry, Haldane published his first major genetics paper. Just as Wright had turned his mathematical mind to guinea pigs, Haldane turned his to moths, particularly the peppered moth (then known as *Amphidasys betularia*, now renamed as *Biston betularia*). This was a famous case of natural selection in action: in the mid-nineteenth century, as industrial pollution started to turn much of Britain's landscape black, entomologists noticed that among these greyish-white moths there were increasing numbers of the normally very rare dark form of the insect. The moths usually rested on the pale bark of birch trees, where the occasional dark moth stuck out and soon fell prey to hungry birds. However, in some areas of the country, industrial soot and smoke had darkened the tree bark so much that it was the more common pale moths who were the more exposed to predation, while the dark ones were camouflaged. In polluted areas, the dark moths gradually became more common as the numbers of light ones declined. This was widely accepted in Haldane's day, but he was interested in trying to calculate exactly how much of an advantage the 'dark' version of the moth colour gene would have to have over the 'light' one, in order for the populations to change in the way naturalists were observing. The dark-coloured form had first been recorded in 1848 and fifty years later had become dominant in polluted areas: using his unusual combination of mathematical and genetic expertise, Haldane calculated that the dark form must be producing 50 per cent more surviving offspring than its pale-coloured rivals.

Such topics filled Haldane and Wright's correspondence for many decades; their letters were often covered in calculations and formulae, clarifying and criticizing each other's ideas, but while Haldane had moved from mice and guinea pigs – via moths and horses – to newts, Wright had never taken his eyes away from his guinea pigs. He wrote to Haldane to say that he would not be able to meet him at that summer's major international genetics congress because he needed to use his summer vacation 'to analyze a mass of accumulated data on my guinea pig colony', while his graduate students were out of his way.[40]

Haldane and Wright were not alone in attempting to solve biological problems mathematically. Forty miles from Cambridge were Britain's oldest agricultural research centre, the Rothamsted Experimental Station, and Ronald Aylmer Fisher, whose Cambridge-trained mathematical mind dwarfed even Wright's and Haldane's. His job at Rothamsted involved analysing the results of the station's plant-breeding experiments, such as trials of the effectiveness of different chemical fertilizers, to calculate precisely how much of the difference in yield between two crops was caused by the growing conditions and how much was due to the genetic superiority of one variety over another. The mathematical analysis of the relative contributions of nature and nurture was a subject that was important to Fisher; he was a committed eugenicist. He had helped found the Cambridge University Eugenic Society as a student and had become a close friend of Charles Darwin's son Leonard, president of the Eugenics Education Society.

Nevertheless, it was statistics rather than biology that had brought Fisher into contact with biometrics and Karl Pearson. Even after Fisher became convinced by Mendelism (which led to a falling out with Pearson), he remained interested in the biometrician's techniques. Fisher created a mathematical model of a population of hypothetical organisms, which used subtle and complex mathematical techniques to demonstrate how favourable genes spread through the population. He also showed that while the incidence of unfavourable mutations would decline, they would not necessarily be eliminated if they were recessive, but would survive, as Chetverikov had found among his wild flies. Fisher's mathematical demonstration was important, since some interpreted the biometricians' arguments as implying that natural selection must eventually use up all the variation in a population. If the fittest survived and the unfit died out, would not every organism in time simply have the best possible genes? The resulting genetically homogenous population would be perfectly adapted to its environment, but if and when that environment changed, it would be unable to adapt. It seemed as though perfection must guarantee extinction for every species, including our own. This was an especially worrying prospect for eugenicists, since their selective breeding programmes were intended to achieve perfection even faster than natural selection could. But Fisher showed how the genetic diversity of a population was maintained, in part because organisms with two different versions of a gene (heterozygotes) were sometimes fitter than either of the homozygous forms. The classic example of this process is sickle cell anaemia: one copy of the sickle cell gene confers some resistance to malaria, but two copies cause the painful and some-times life-threatening disease. Nevertheless, the benefit conferred by the gene, the malaria resistance, has been enough to ensure that natural selection did not eliminate it from the population. To biologists, Fisher's calculations, if they were able to follow his complex maths (which many could not), revealed a picture of continuous, gradual evolution – and of evolution without end.

Despite their many differences (political as well as scientific), Wright, Haldane and Fisher had independently arrived at Chetverikov's idea of studying evolution mathematically. Each had recognized the predictive power of Mendelian genetics while realizing that it could not be applied to wild populations using the standard technique of controlled experiments, given the impos-sibility of recording each wild organism's pedigree. Ironically, it was the Mendelians' scientific opponents, the biometricians, who had supplied the solution: their statistical tools made it possible to take a sample, assess its genes experimentally, and extrapolate the results to a whole population. Equations revealed the genes of thousands of wild plants or animals to the laboratory scientist; plotted on graph paper, changes in the genes of an entire population could be observed. They made it possible to extend the power of genetics from the laboratory into nature.

'Just another geneticist'

In 1920 Wright had spent his thirty days' annual leave at Cold Spring Harbor Laboratory, using his summer holiday to do more guinea pig experiments. There he had met a young woman called Louise Williams, a fellow biologist who was at the lab to take care of some rabbits that were being studied by her former teacher. The animal caretaker who had been cleaning her rabbits' cages had recently quit; when Wright arrived she had hoped he was the new animal caretaker. He later recalled that 'the first thing that she had said to me . . . was how badly she had been disappointed in me' when he proved to be 'just another geneticist'. Both chagrined and charmed, Wright immediately offered to help her with the cages. They began to take walks together after dinner and were always seen sitting with each other at the lab's regular picnics. Wright later wrote that 'I was so much in love with her at the end of my vacation and so dubious about merely starting a correspondence that I proposed on the last evening. She demurred at first because of the shortness of our acquaintance and because I had not met her parents or she mine but finally agreed to consider us engaged.' They were married the following year.[41]

A couple of years later, Wright was discreetly approached by the University of Chicago and offered an academic job. Although he and Louise Wright, as she now was, would have preferred to raise their children nearer to open countryside, the prospect of better facilities, a community of other geneticists (Wright felt somewhat isolated in Washington) and a salary increase finally persuaded him to accept the job and the couple moved. Wright briefly considered switching to another experimental animal, but realized that he had invested too much time and energy in guinea pigs to abandon them now; one of his conditions for the new job was that Chicago build him 120 custom-built pens to house the guinea pigs he brought with him from Washington.

At Chicago Wright discovered that he was not a natural teacher, too shy and nervous to make off-the-cuff remarks or banter with the students. Instead, he prepared every class in great detail and always gave formal lectures, even to the smallest groups. Teaching forced Wright to be unfaithful to his guinea pigs – they bred much too slowly to be used for experiments in a ten-week course – so he acquired Drosophila stocks from Morgan's lab at Columbia. One of his students remembered Wright as 'a very mild mannered gentleman who would stand there looking hurt and blinking when his students would not come to the obvious conclusions – obvious to him from the analysis that he put on the blackboard'. Wright would scribble continuously as he talked; 'he would cover the entire blackboard of a 30 foot wide classroom, 3 times a lecture'. Despite the necessity of using Drosophila in experiments, Wright would bring guinea pigs into the classroom whenever he could justify doing so; on one occasion he brought one in to show his class some interesting variations in its coat colour. 'This particular guinea pig was somewhat more fractious than usual and was scurrying around on the desk and was not about to be quiet,' a student recalled, so Wright picked up the restless cavy and tucked it under his armpit, where he usually kept his blackboard eraser. A few minutes later, running out of space for the next equation, he reached for his eraser 'and started to erase the blackboard with a squeaking guinea pig'.[42]

With the (sometimes reluctant) help of his guinea pigs, Wright had devised mathematical tools for making sense of biological problems. Fisher and Haldane had come up with similar techniques for addressing comparable problems; in the process, the three of them found they had invented a new science – population genetics. But there were important differences between them: Haldane and Fisher's work was more mathematically sophisticated than Wright's, but also more abstract. They were forced to treat each gene separately, assigning it an adaptive value and then predicting how it would spread through a hypothetical population. This abstract modelling of the way isolated genes behaved became known, not always politely, as 'bean-bag' genetics, because genes were treated as if they were picked randomly and independently out of a bag, rather than existing alongside each other – and interacting with each other – as in living organisms.

Haldane and Fisher understood that to make their calculations possible (especially in the days before electronic computers) they had to simplify their models, to reduce the number of variables they had to calculate. This led to what they realized were unrealistic assumptions: for example, in a real population of moths – or any other organism – mating is never completely random; apart from anything else, moths (like the rest of us) are much more likely to find potential mates close by. In a small population of moths, it is possible that there might be more dark than light moths purely by chance, so the number of dark offspring could increase without natural selection having anything to do with it. Castle's hooded rats had demonstrated a similar effect – that continued selection within a small population created unusual genetic combinations that might never exist in a larger population; paradoxically, that meant that a small inbred population might actually vary more than a large, freely interbreeding one. To avoid having to calculate the effects of this kind of random factor, Fisher and Haldane assumed that each of their mathematical species had an infinitely large population; all the potential complications that might arise in small isolated groups were simply ignored. Wright's work was subtly different; partly because of his abiding affection for his guinea pigs, his maths also accounted for factors like inbreeding.

One result of the mathematical trio's work was to persuade biologists of the power of natural selection – that it alone was enough to achieve changes such as had been observed in the peppered moth. As a result, some older ideas about inheritance, especially the possibility of Lamarckian inheritance, were finally

abandoned. Haldane had never believed in Lamarckian inheritance, but – in one of history's more bitter ironies – he ended up
defending it for political reasons, because the inheritance of
acquired characteristics was at the heart of the style of genetics
being promoted in the Soviet Union by Lysenko. Lamarckian
ideas had often attracted those on the political left, because they
seemed to suggest that better living conditions would produce
better people. By contrast, many felt that the growing Mendelian
orthodoxy, that genes were a stable given, was a reactionary view,
suggesting that the poor were poor because they were biologically
inferior and only eugenics could cure poverty, by getting rid of
the genes that that 'caused' poverty – and those that carried
them. Certainly many right-wing eugenicists chose to interpret
Mendelism this way, including the Nazis, a fact that Lysenko
seized on to accuse Soviet Mendelians of being in league with
fascism.

Although Haldane had no sympathy for the idea of the inheritance of acquired characteristics, he had plenty for Marxism, first
supporting and then eventually joining the British Communist
Party during exactly the period – from the mid-1930s to the late
1940s – when Lysenko's views were becoming the Soviet
biological orthodoxy. As the nature of Nazism became clearer
during the 1930s, even non-communists began to see the USSR
as the only barrier to the rise of fascism, not least because demo-
cratic governments stood by and did nothing while republican
Spain fought for its survival against fascist forces backed by Hitler
and Mussolini. Haldane was one of hundreds of British communists who went to Spain to fight fascism; he offered his scientific
expertise and made several visits to Madrid to help the republican
government plan for air raids and possible poison gas attacks. The
threat of fascism made support for the USSR an even more
pressing duty for British communists; in these circumstances, any
criticism of Soviet policy was seen as disloyal.

An additional factor that shaped Haldane's views was that, as
we have seen, he had visited the USSR and – like Hermann
Muller before him – had been impressed by the level of state

funding for science, which dwarfed the meagre funds made available to scientists by the British government. Haldane was one of many scientists who were drifting to the left in the 1930s, and the British Communist Party actively and successfully recruited them, promising them a future in which science would be unconstrained by limits on either money or freedom. Such promises began to look increasingly unconvincing as in the Soviet Union, Lysenko manoeuvred to ensure that those who disagreed with him lost their jobs, were arrested or simply disappeared. As we have seen, Chetverikov, his friends and colleagues were all victims of Stalin and Lysenko's terror, but Haldane – who had done so much to publicize and promote Chetverikov's work in the West – stood by and watched, unable or unwilling to speak out in defence of his Russian colleagues. For many years Haldane maintained an uneasy silence, unwilling to say anything that could be useful to anti-communist propagandists. In private, especially at the British party's scientific debates, he was critical of many of Lysenko's claims, recognizing them as unscientific nonsense. But gradually his political loyalties forced him to defend Lysenko and the party line in terms that, if they did not involve outright lies, certainly fell far short of the truth. In the late 1940s, he even began to suggest that there might be some scientific evidence for Lamarckian inheritance, but eventually found he could not live with his divided loyalties any longer. In 1950 he quit the Communist Party, feeling unable to reconcile his political loyalties with his obligation as a scientist to pursue the truth as he saw it. A belated recognition, but he was virtually the only leading British Communist scientist to quit the party over Lysenkoism.

From Russia with flies

Perhaps the only positive product of Lysenko's reign was Theodosius Dobzhansky, possibly the most brilliant biologist of the twentieth century, who was to complete the rapprochement between the Mendelians and biometricians that Haldane, Fisher and Wright had begun. Dobzhansky came to the United States in 1927, to work with Morgan's group at Columbia. He had originally planned to return to Russia after his studies, but when he realized how hostile the political climate in the Soviet Union had become to Mendelians, he decided to stay in America, eventually becoming a US citizen.

Dobzhansky had arrived steeped in the Russian tradition of field work, used to working with wild flies. Although he had not been one of Chetverikov's students, Dobzhansky knew all about his statistical methods and how to apply them to understanding evolution outside the lab. At Columbia, he learned all the lab techniques of Morgan's fly boys and decided to try and combine the Soviet and American approaches; he took his lab on the road. In the late 1930s, he began to drive up into southern California's San Jacinto mountains, his car packed with bottles, microscopes and – most important of all – lab-bred flies. Over the following years, Dobzhansky travelled all over the southern USA, from California to Texas, catching wild Drosophila. Early each morning or late in the evening, he would put some fermenting mashed banana into a half pint milk bottle and wait for the flies to catch the scent and enter the bottle; then he and his students would trap them. Once caught, the flies were crossed with the cleaned-up lab flies, whose genetic make-up was by now well known. Breeding experiments allowed Dobzhansky to use the lab flies as Chetverikov had done – as a probe to reveal the unknown genes of wild Drosophila.

Dobzhansky discovered that the genetics of the wild populations consisted of several distinctive sets of genes, or genotypes, and – more unexpectedly – he found that the percentage of each genotype varied according to the season. One type dominated in the summer, but was much rarer in winter-caught flies. Like most biologists, he had always assumed that natural selection was much too slow for humans to observe in a lifetime, but as he trapped and bred his flies, Dobzhansky realized that only natural selection could account for the changing frequencies of the different genotypes. One type was better adapted to cold weather, another to drought, and so on. He tested this idea by raising mixed populations in large cages, built from fine mesh screens. Some cages were kept wet and cold, others hot and dry, to simulate the seasons; he found that one particular type of fly became dominant in each cage: in one case, a genotype that made up just 10 per cent of the initial population had increased to 70 per cent in just ten months. Thanks to Drosophila's rapid breeding, he was watching evolution in action.

By cross-breeding his lab flies with wild ones, Dobzhansky was completing the transformation Chetverikov had begun; bringing together lab and field. As he proudly wrote, 'Controlled experiments can now take the place of speculation as to what natural selection is or is not able to accomplish.' For the first time in the history of biology, the raw material of evolution, the endless variability of living things, 'now lies within the reach of the experimental method'.[43]

Dobzhansky met Wright in 1932 at an international genetics congress, where – in Dobzhansky's words – he 'fell in love' with Wright; it was the beginning of a long and highly productive friendship. Despite knowing little maths, Dobzhansky absorbed all of Wright's papers by carefully reading their introductions, skimming the central mathematical problems, and carefully reading the conclusions. When he needed help analysing his results, it was Wright to whom he turned, and they collaborated closely on Dobzhansky's enormously influential series of papers that described and analysed his work with the wild flies. Dobzhansky got into the habit of consulting Wright when planning new experiments, and Wright often checked and commented on Dobzhansky's work before it was published.

Dobzhansky and Wright's work was the last, decisive step in one of the most momentous achievements of twentieth-century biology: the creation of the modern theory of evolution. It has become known as the modern synthesis, because – as biologists still like to present it – it combined Darwin's ideas and Mendel's. But describing the synthesis in this way omits all the practical detail of how the story actually happened: the hours Weldon spent counting shrimp, which helped transform Darwinism into biometrics; or the patient fly breeding that turned Mendel's mysterious *Anlagen* into genes on chromosomes. Despite the failure of the Mutation Theory, Oenothera established that evolution could be investigated in laboratories, a point Castle acknowledged when he described his use of rats to test natural selection: 'to De Vries we owe much for showing that such tests are possible'.[44] Though the modern synthesis is often described as the marriage of Darwin and Mendel, it really involved cross-breeding Soviet and American flies with guinea pigs, hybridizing scientific practices and disciplines and forcing biometricians to work with Mendelians so that they could mate maths with moths, to give birth to genes on graph paper.

Wright never really chose to work with guinea pigs; in a sense, they chose him, since his position at Harvard was dependent on his looking after the existing colony. There were good reasons why he might have wanted to work on another species: they were much harder to keep than Drosophila; they bred much slower; they were prone to disease; they have far more chromosomes than the flies (thirty-two, compared to four, although the guinea pigs' number was not even known until long after Wright graduated); and guinea pig chromosomes are small and hard to identify under a microscope. While Morgan's fly team had hundreds of mutations to work with, hardly any were known in guinea pigs (since only two examples of linkage in guinea pigs had been demonstrated, Wright once complained that 'everything in the guinea pig seems to be independent of everything else'), so chromosome mapping was impossible.[45] Worst of all, there was no guinea pig community for Wright to share ideas with; at the time, there were only a couple of other scientists using them for genetic research. Yet Wright seems to have taken to guinea pigs from the outset and never regretted the fact that they chose him. Although few guinea pig genes were known at the time, there were about 3.5 million possible combinations of the genes that were known, so Wright studied the problem of gene interactions – how the genes affected each other – for his entire career.

Because they forced Wright to concentrate on the interaction between genes, the guinea pigs prevented him making some of the elegant but unrealistic simplifications that Haldane and Fisher's work relied on. As a result, his work was more accessible – and more useful – to working field biologists than theirs. Thanks, in part, to Castle's influence, but also through working with the inbred colony in Washington, Wright had to develop mathematical tools with which to analyse the effects of inbreeding. As he struggled to understand the odd-looking inbred families that had been produced from the USDA's single original stock, he was constantly reminded that genes interacted with each other to produce variation, the raw material of evolution. As the organism's original hand of genes was shuffled and redealt with each new mating, new combinations arose, giving the offspring new strengths – or weaknesses. He concluded that evolution was most likely to take place in small, isolated populations – populations that were nothing like the infinitely large ones hypothesized by Haldane and Fisher.

Given the comparative ease with which Wright's insights could be applied to real population studies, it is no surprise that, of the three great mathematical population geneticists, he was the one who most influenced Dobzhansky, as he did many other field biologists, not many of whom could follow the complex mathematics used by the pioneers of population genetics. Wright's ideas were the easiest and most useful for them to apply to their own work, particularly because the importance of treating a species as a series of small, largely isolated sub-populations had emerged quite independently from field studies of wild populations. Thanks in large measure to Dobzhansky's book *Genetics and the Origin of Species* (1937), Wright's work became vital to successive generations of biologists.

Yet despite the huge affection and respect in which Wright was held by his colleagues, his guinea pig work at Chicago produced one major disappointment. He had hoped to connect genetics with development, to show how genes actually turned a fertilized egg into an adult guinea pig. Like many of his contemporaries, Wright suspected a chemical connection, that specific genes produced specific chemicals, probably the enzymes whose workings were then being unravelled by the biochemists (Haldane had played a useful part in this work before he defected to genetics). Yet despite years of hard work, Wright never made any real headway with this problem. Understanding how genes actually worked in practice would require a new science, molecular biology, which was to be built on work with an organism that could not have been more different from the guinea pig – a virus called bacteriophage.

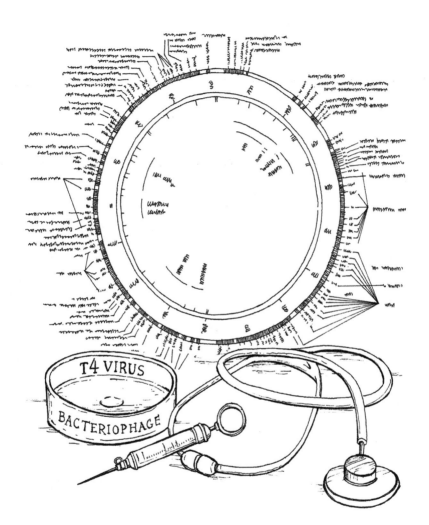

Chapter 8

Bacteriophage: The virus that revealed DNA

'Professor Max Gottlieb was about to assassinate a guinea pig with anthrax germs, and the bacteriology class were nervous. They had studied the forms of bacteria, they had handled Petri dishes and platinum loops . . . and they had now come to . . . the inoculation of a living animal with a swift disease. These two beady-eyed guinea pigs, chittering in a battery jar, would in two days be stiff and dead.'

This scene, from Sinclair Lewis's 1925 novel *Arrowsmith*, dramatizes the predicament facing the medical profession in the early twentieth century; in all too many cases, they were still better at killing than at curing. Thanks to Louis Pasteur, Robert Koch (with whom the fictional Gottlieb is supposed to have studied) and others, the cause of anthrax was known and its effects could be easily demonstrated:

> The assistant held the guinea pig close; Gottlieb pinched up the skin of the belly and punctured it with a quick down thrust of the hypodermic needle. The pig gave a little jerk, a little squeak, and the co-eds shuddered . . . He said quietly, 'This poor animal will now soon be dead as Moses.' The class glanced at one another uneasily. 'Some of you will think that it does not matter; some of you will think, like Bernard Shaw, that I am an executioner and the more monstrous because I am cool about it; and some of you will not think at all. This difference in philosophy is what makes life interesting.'

In his reference to Shaw, Gottlieb is probably thinking of the preface to *The Doctor's Dilemma*, where Shaw – a staunch anti-vivisectionist – asks rhetorically why 'If a guinea pig may be sacrificed for the sake of the very little that can be learnt from it, shall not a man be sacrificed for the sake of the great deal that can be learnt from him?' Indeed, as Gottlieb is preparing to demonstrate the effects of anthrax, he compares the bright-eyed guinea pig to one of his bored and uninspiring students and asks himself, 'Why should I murder him to teach *Dummköpfe*? It would be better to experiment on that fat young man.'[1]

Medical research in laboratories had made kind, benevolent healers into cold, clinical killers; that, at least, was how some saw the new scientifically trained doctors. Later in the novel, its hero, Martin Arrowsmith, is working as a family doctor in the Midwest when he tries desperately to save the life of a child with diphtheria, but fails. As he stands horrified over the dead child, 'he raged with desire to do the impossible. She could not be dead. He had do something', but there is nothing he can do. 'She is dead? Dead?' demands the child's mother. 'You killed her, with that needle thing! And not even tell us, so we could call the priest!'[2]

Like the real doctors on whom he is based – men like Sinclair Lewis's father, who was a family doctor in Minnesota – Arrowsmith is largely powerless against the microbes. He knows how to locate them, grow them and identify them, but in the 1920s, administering anti-toxins or vaccines was still a hit and miss business. But in Lewis's novel, Arrowsmith makes a remarkable breakthrough: he discovers a bacteria killer, a mysterious entity that causes microbes themselves to get sick and die and uses his discovery to fight a plague. Although he is only partially successful, the novel offers hope that one day real doctors will vanquish real bacteria.

Most of the book's readers probably assumed that the enigmatic bacteria killer was as fictional as Arrowsmith himself, but Lewis was describing the real science of his time, so recent that few scientists and almost no doctors would have heard of it. The bacteria killer's real discoverer believed, like Arrowsmith, that it was the ultimate weapon in the war against the microbes. It proved to be even more extraordinary: in time it showed us exactly how genes work, what the mysterious 'hereditary material' is, how it is preserved, copied, mutated and passed on. It ultimately helped us to understand – and treat – diseases in ways Martin Arrowsmith could never have imagined, but it would take more than half a century for us to understand what this mysterious bacteria eradicator had to tell us.

Miasmas to microbes

Humankind has been wondering about illness – and cure – for as long as we have been wondering about anything. As we have seen, some ancient Greek doctors believed that, just as the world was made of four elements – earth, air, fire and water – so humans were partially composed of four associated 'humours': black bile, blood, yellow bile and phlegm. As we have seen, major imbalances of the humours could also make us sick, which is why for hundreds of years doctors would treat disease by trying to rebalance the humours.

Although the doctrine of humours had a long career in Western medicine, its proponents had difficulty explaining epidemic diseases. Why, for example, did diseases seem to sweep through cities (and it was more often cities than small towns or villages), affecting thousands of people at once? In the early sixteenth century, when Italian cities suffered outbreaks of fever the citizens attributed them to unfavourable planetary influences; which is why the Italian for 'influence', *influenza*, came to be the name we still use for the disease. However, by the early nineteenth century, astrology had largely given way to two rival theories of infectious disease. One school was convinced that miasmas, airborne poisons such as those generated by decaying animals and plants, caused illness. The Italians had long known that those foolish enough to wander into swamps and marshes often became ill, which they attributed to the bad air, or '*mal aria*', of such places. By the late nineteenth century, it was discovered that the swamp disease was spread by a parasite carried by the mosquitoes who bred there, but the illness is still called malaria, commemorating the original 'miasmatic' theory.

However, the miasmatic theory had its critics: doctors found that patients who had been nowhere near a marsh or polluted city still came down with diseases that spread rapidly through families and villages. They wondered if these illnesses were caused by some kind of particle, which infected the body and was then spread when the patient touched other people. The theory became known as 'contagionism' (from the Latin, *tingere*, 'to touch'). Some assumed the contagious particles were simply inanimate chemicals, like poisons, but that did not seem to explain how epidemics grew and spread. So other contagionists argued that the particles must be 'animalcules', minute animals, like those the early microscopists had discovered to be swarming in every drop of water. They speculated that perhaps some of these were responsible for illnesses, creating epidemics when they multiplied and spread.

The animalcule version of contagionism got a boost in 1835, when it was shown that muscardine, a disease of silkworms, was caused by a fungus. This was the first time that a living organism had been proved to cause a disease. At about the same time, Theodor Schwann (one of the founders of the cell theory) demonstrated that the crucial process of fermentation was caused by a living organism, yeast, and was not – as had been widely believed – a purely chemical reaction. Pasteur had also investigated fermentation – largely at the insistence of French wine-growers – and he discovered that several of the 'diseases' that spoiled wine and beer were also caused by tiny living things. These were visible under a microscope, and since they looked like little sticks they became known as bacteria (from the Greek for 'little stick'). In 1865 Pasteur was also asked to investigate silkworm diseases, and after five years of tests and experiments he was able to identify two distinct diseases, each caused by a different microscopic creature; these invisible killers were christened 'microbes' or germs and the 'animalcule' version of contagionism became the germ theory of disease.

Pasteur's work encouraged dozens of other scientists across Europe to start microbe-hunting, identifying the once-mysterious causes of sickness. In 1877, Koch published his work on anthrax, which clearly showed that whenever the disease was present, so was the microbe. His critics responded by claiming that the microbes were merely a side-effect of the illness, so Koch grew them on a glass dish in the laboratory and then injected them into healthy animals, which quickly sickened and died (precisely the demonstration Gottlieb performs on his unlucky guinea pig). At much the same time, Pasteur showed that if the microbes were filtered out of the blood of infected animals, the blood could no longer cause anthrax.

Between them, Pasteur and Koch convinced the medical world that microbes caused illness. The publicity that accompanied their successes led to the assumption that every disease must have its own associated microbe, giving birth to a new science – microscopic biology, or microbiology. As universities and medical schools built labs to teach their students how to kill guinea pigs with germs, efforts were made to improve laboratory equipment. For example, as the demand for microscopes boomed, their manufacturers found themselves competing in an increasingly crowded marketplace, so they invested in making their micro-scopes more and more powerful. At the same time, biologists were improving the techniques used to prepare specimens, such as staining, which made the component parts of the specimen easier to see. And as each new tool revealed more about microbes, the researchers demanded further improvements. Like Calvin Bridges in Morgan's fly room, every lab seemed to have a tinkerer or two, working to improve their tools. The round, flat-bottomed glass dishes that are still used to grow microbes were named after their inventor, Julius Richard Petri, a German bacteriologist who worked with Koch. Also in Koch's lab was a young German doctor, Walther Hesse; one hot summer afternoon he became frustrated by the gelatin on which microbes were grown, because it melted in the heat. Gelatin (derived from animal tissues) was of course also used in cooking, to make jellies set, and Hesse was intrigued to notice that his wife Angelina's jellies did not melt. When he asked her why, she explained that instead of setting them with gelatin, she used a powder called agar. Agar, which is made from seaweed, had been used in Far Eastern cooking for centuries, and while Angelina had been growing up in New York she had learned about it from a Dutch neighbour who had been born in Java. The more stable agar gel quickly became standard in Koch's lab and is still in use all over the world.

Once they were properly equipped, students – just like those in Arrowsmith's class – were taught what had become known as 'Koch's postulates'. These were the microbe hunters' creed: first, every patient with a particular disease had to be shown to carry the same microbe; then the microbe had to be isolated and grown in the lab, to produce what was called a 'pure culture', free from contamination; and finally, the pure culture had to be capable of producing the disease when it was introduced into a healthy patient (or, more usually, a guinea pig).

That, at least, was the theory. In practice, there were still many more diseases than there were microbes. Smallpox was one of the recalcitrant ailments that resisted Koch's postulates: no one could find a definite microbe that would grow in the lab, or at least, whatever it was that *could* be grown in the lab seemed not produce the disease in healthy subjects. Microbe hunting had become a mature science, confident in its methods and reluctant to accept work that failed to conform to high professional standards; in cases like smallpox, the microbe-hunters became increasingly confident that the only reason they could not find a microbe was that there wasn't one.

By 1890, even Koch had to admit that there were many diseases – including smallpox, rabies and influenza – for which there was no sign of a bacterium. He suggested that some microbes might be too small to be seen. This was a distinct possibility: it had long been realized that no light microscope – however sophisticated and expensive – could ever resolve objects smaller than the wavelength of visible light itself. By the 1890s, microscopes had reached that theoretical limit of visibility and bacteriologists speculated that some microbes would remain forever beyond their reach. Perhaps these were responsible for the mysterious seemingly microbe-less diseases?

Microbiologists had been using filters to remove microbes from blood for some decades. One early attempt used the placenta of a living guinea pig to filter anthrax bacteria, but for once the hapless guinea pig did not become standard laboratory equipment – unglazed porcelain or plaster filters were soon used instead (and by the 1880s, it had become clear that some bacteria could cross the placental barrier in any case). A whole range of filters were developed and at London's 1884 International Health Exhibition, an array of standardized commercial filters were on display, which could be used for such purposes as making the city's polluted water supply safe enough to drink.

Filters played an important role in the hunt for ever-smaller microbes. In the 1890s, several European researchers were studying a disease of tobacco plants known as tobacco mosaic disease because it caused the leaves to become mottled with light and dark spots. The disease attacked many valuable crops, so there was considerable interest in finding its cause. In 1898 a Dutch researcher, Martinus Willem Beijerinck, discovered that whatever caused the disease easily passed through a bacterial filter and seemed able to survive indefinitely, spreading from plant to plant, from generation to generation. That ruled out a poison – it would rapidly have become too diluted to have any effect. Whatever caused the infection was reproducing, somehow renewing itself in the infected plants, infecting others, and then repeating the process. Beijerinck did not think it could be any kind of microbe, since it passed through the filters, nor could it be destroyed in any of the ways microbes were usually killed. And no matter how hard he looked with his microscope, Beijerinck could detect nothing.

Beijerinck wondered if the infectious agent was a liquid of some kind and devised an experiment to test the idea. He took a Petri dish covered in agar gel and put infected plant sap on it, leaving it for ten days so that any liquids would diffuse through the agar.

Then he removed the sap and the top layer of agar, which had been in direct contact with it. He then tested the lower, untouched layer of agar: would it still transmit the infection? It did. He concluded that the infectious agent could not be a bacterium or fungus – it must be a liquid or something that dissolved in a liquid.

Pasteur had called the seemingly unidentifiable cause of rabies a virus, and Beijerinck adopted the term. The Romans had first used the word 'virus' to mean anything bitter or unpleasant. By the Middle Ages it had come to mean a poison, and eventually was used to refer to anything that caused infection. The seventeenth-century German Jesuit scholar Anathasius Kircher seems to have been the first to use the word in this sense, when he referred to whatever it was that caused the plague as the *virus pestilens*. By the eighteenth century, 'virus' was regularly used in English to mean an infectious agent; Edward Jenner, one of the founders of vaccination, used it in his classic treatise on smallpox, referring to the substance extracted from the pustules or pocks that characterized such diseases as 'cow-pox virus'. By the early nineteenth century it was used in two senses: it referred to the specific cause of a well-defined disease, but also to any unknown disease-causing agent. By the 1870s and 80s, Pasteur and others were using 'virus' as a general term to refer to all the various kinds of disease-causing microbes. Koch claimed that in discovering the microbe responsible for TB, 'we have the true virus of tuberculosis'.[3]

Yet, however ancient the word's pedigree, there was no escaping the fact that 'virus' effectively meant 'we do not know'. Beijerinck realized that his tobacco mosaic virus could not be a simple chemical like a poison, because it reproduced in the plant; it must be a living infectious fluid, which he simply referred to as 'living infectious fluid', though in Latin (*contagium vivum fluidum*), presumably because that sounded more impressive. His proposal violated the central dogma of the cell theory, that every living thing is made of cells and all cells come from cells; whatever his mysterious *fluidum* was, it did not conform to the dogma. And, despite being alive, it resisted all attempts to grow it outside the plant: it would only multiply in living, dividing cells. The very novelty of Beijerinck's ideas made them hard to accept, especially his notion of a living thing that could not reproduce itself independently – after all, the ability to reproduce was a key part of what defined life. He was contradicting both cell theory and germ theory, two of the great triumphs of nineteenth-century biology, and, not surprisingly, few were willing to listen to him.

Nevertheless, by the beginning of the twentieth century, science had reached a clear understanding of what viruses were not: they were not blocked by filters; they were not visible under the microscope; and they would not grow in the lab. All very intriguing, but what *were* they?

The bacteria eaters

On the eve of the First World War, the English microbiologist Frederick Twort noticed something rather peculiar about his agar plates. In the middle of his growing colonies of bacteria there were clear patches. He followed procedure: stained and fixed them and put them under the microscope. All the bacteria were dead, nothing but empty carcasses were visible. He investigated further. Whatever was killing the bacteria could pass through a filter and still remain lethal. Twort published an article about his discovery, in which he discussed several theories, including the possibility that he had found a microbe that attacked other microbes. But he was too cautious a scientist to choose between his hypotheses and before he could pursue the problem, the war intervened. Not long after his paper had appeared, Twort joined the British Army Medical Corps and was sent to Salonica (now Thessaloniki), in Northern Greece, where British troops were digging in for one of the long, bloody and largely pointless campaigns that characterized the war. By the time it was over, others were at work on the bacteria-killer and so, as Twort put it, 'I passed on to other work.' His original paper was largely ignored.[4]

While Twort was in the army, a French-Canadian, Félix Hubert d'Hérelle, was one of dozens of researchers who were trying to answer a deceptively simple question: given that the world was literally swarming with microbes, many of which caused diseases, why were not we all sick all the time? This raised other questions, such as why, when there was an epidemic, did not everyone succumb? One proposed explanation of resistance to disease came from the Russian embryologist Elie Metchnikoff, who observed strange cells in starfish and sponges that were produced in massive numbers when a creature was under attack from potential diseases. The special cells seemed to be able to protect the organism from infection by engulfing and consuming the invaders, so Metchnikoff called them phagocytes ('eating cells'), arguing that they were the basis of the body's natural immunity.

In 1915 d'Hérelle was working at the Pasteur Institute in Paris, studying a patient who was suffering from severe dysentery. He was busy culturing the dysentery bacteria, just as Koch advised, growing them until the test-tubes were cloudy with swarms of thriving microbes. After repeating the experiment several times, d'Hérelle was startled when one morning he found a completely clear test-tube. The bacteria were all dead. Even more surprisingly, he discovered that if a drop of the liquid from the clear tube was put into one full of thriving bacteria, they died too. He had inadvertently rediscovered the phenomenon Twort had first observed. After several years of experiments, in 1917 d'Hérelle announced that whatever killed the bacteria both passed through filters and seemed to get more lethal with successive cultures. That suggested the unknown assassin was growing and multiplying – it was clearly a virus like the one Beijerinck had observed in tobacco, and d'Hérelle christened it 'bacteriophage' ('bacteria eater'; the same word is both the singular and the plural).

D'Hérelle suggested that the bacteriophage virus was nature's defence against bacteria, an idea that had come to him as he had been searching for a microbe that would kill locusts. He had discovered that some locusts were immune to his supposedly lethal microbes; on examining them, he had found in their guts evidence of something that killed bacteria. He began searching for other examples of this phenomenon – his dysentery experiments being one approach – and was intrigued to discover that the dysentery-killing bacteriophage were much more common in the stools of people who had just been ill than they were in healthy people. That suggested that bacteriophage multiplied to fight off an infection, like Metchnikoff's phagocytes.

D'Hérelle wrote a book about his discovery, which was translated into English as *The Bacteriophage: its role in immunity* (1922). That same year his work was discussed in the *Proceedings* of the Royal Society and he and Twort were invited to address the British Medical Association's annual meeting. News of the discovery spread rapidly: Michigan's Academy of Science, Arts and Letters discussed bacteriophage the following year. Soon, teams were investigating bacteriophage in several countries: Andre Gratia and Simon Flexner in the USA, d'Hérelle in France, and in Belgium the Nobel Prize-winner, Jules Bordet, who led a research team in Brussels.

As the number of researchers working on the bacteria-killer grew, disputes flared up about what it was they were actually studying. No one could even agree on what it should be called: a few used d'Hérelle's term 'bacteriophage', thus implicitly adopting his view that it was indeed a virus (whatever that was), but others disagreed – just as they had disagreed with Beijerinck. One thing that the researchers did agree on was that the mysterious microbe-killer caused bacteria to disintegrate (or undergo 'lysis' in scientific terminology). Bordet – who insisted that the phenomenon was purely chemical, not organic – called it 'transmissible autolysis' ('self-disintegrating that can be passed on', in English). Scientists were well aware that the term they chose to use aligned them with a different camp in a small but increasingly fractious scientific world, so some tried to distance themselves from the debate by adopting deliberately neutral phrases, such as 'the Twort-d'Hérelle phenomenon'.[5]

Bordet was convinced that his 'transmissible autolysis' was – as its name suggested – in some way hereditary, but he was equally sure that bacteriophage was not living – it was purely chemical, something produced by the bacteria themselves. Biochemists had established that organisms produced substances called enzymes (from the Greek for 'in ferment'), which acted as catalysts, controlling the speed of the body's chemical processes. Perhaps, Bordet speculated, autolysis was caused by a mutation in the bacteria that led them to produce too much of a normally useful enzyme, causing a runaway chemical reaction that eventually killed the bacterium. It might even serve some useful function; perhaps the disintegration killed off undesirable mutants, so as to 'discipline the evolution of the species'.[6] However, this was a very controversial idea. Bacteria had no nucleus and no visible chromosomes, so the idea that they might be subject to the same genetic laws as other organisms seemed unlikely. Indeed, in the early 1920s it was still not clear whether bacteria could even be considered organisms in the usual sense. Yet one of Bordet's colleagues, Andre Gratia, joined the campaign against d'Hérelle's view that bacteriophage were alive. So what if the phenomenon could spread? Gratia sarcastically pointed out that fire spreads and 'reproduces' itself and yet 'fire is not living'. Bubbles appear on the inner surface of a glass of soda water, just like the clear patches that appeared on a dish of infected bacteria, 'yet gas is not a virus'.[7] Gratia took the anti-d'Hérelle line with him to the US, when he joined the Rockefeller Institute in New York.

Standard oil and snake oil

The Rockefeller Institute for Medical Research was only one part of John D. Rockefeller Sr's vast philanthropic empire. Rockefeller had made over $900 million by 1912 (about $80 trillion in today's dollars), mostly in the oil business, yet he was haunted by what he had heard in church as a boy, that it was easier for a camel to pass through the eye of a needle than it was for a rich man to enter the kingdom of heaven. His deeply religious mother had taught him the importance of charity and so Rockefeller began a systematic campaign of giving away his money. His generosity was as legendary as his fortune, and he soon faced a torrent of begging letters from charities. In 1891 he appointed a full-time financial manager, the Reverend Frederick T. Gates, who took care both of increasing and reducing Rockefeller's fortune, by managing both his investments and his philanthropy.

Rockefeller's son, John D. Jr, urged his father and Gates to put money into medical research, conscious that there were still few effective treatments available for most diseases. The younger Rockefeller may also have felt a familial guilt that his grandfather, William Avery Rockefeller, had also made his money in oil – snake oil. A self-proclaimed 'Doctor', he charged $25 a time (well over $500 in today's money) for a worthless cancer cure. Whatever their motives, the Rockefellers' philanthropy had in 1901 resulted in the Rockefeller Institute for Medical Research (now Rockefeller University): the family had given it over $50 million by the 1930s.

In 1920 a young scientist called Paul de Kruif arrived at the Rockefeller Institute, eager to become a microbe hunter. He was immediate struck by the Institute's luxurious facilities: 'What a temple of science the Rockefeller Institute was!' he recalled, especially when compared with the medical building at the University of Michigan where he had trained, which stank of rats and guinea pigs; 'at the Rockefeller you did not smell the animals. They were brought to you from a beautiful animal house in the bowels of the Institute by a servant.' The facilities were new and expensive and life was easy: 'Lab servants washed the glassware and cooked the culture medium,' de Kruif reminisced, 'and if you had a well-enough trained technician, he could even do your experiments for you.'[8]

De Kruif knew everyone in the small but flourishing world of bacteriophage. He had visited the Pasteur Institute in Paris during the war, where he had heard of (and may even have met) d'Hérelle. Back in the US, he had studied with Frederick Novy, the first serious American bacteriophage researcher, and the year de Kruif arrived at the Rockefeller Institute, he had met Bordet and shared a lab with Gratia.

However, De Kruif's work at the Rockefeller Institute was not concerned with the enigmatic bacteria-killer, but an apparently unrelated phenomenon: bacterial impurity. As we have seen, Koch's postulates demanded that a bacteriologist produce a pure colony of the microbes suspected of causing a disease; it was an essential first step, since if there were more than one type of bacteria present, it would remain unclear which one actually caused the illness. The procedures for doing this were well known; de Kruif would have learned them from Novy in a class like the one described in *Arrowsmith*. However, sometimes what appeared to be a pure colony would separate into two types, only one of which proved to be vulnerable to the anti-toxin that ought to have killed it. De Kruif called this puzzling behaviour 'dissociation' and suggested that the bacteria might be undergoing a de Vriesian mutation, forming a new species. His idea was greeted with considerable scepticism by most biologists: the popularity of the Mutation Theory was fading, largely because Morgan's fly researchers were focusing attention on to the Mendelian behaviour of chromosomes.

De Kruif was just one of the many who were trying to understand why bacteria were so changeable. Not only could they radically change their form when watched under a microscope, they seemed able to live off one substance when in one shape, but lose their appetite for it when they changed. Bacteria seemed to be highly adaptable, but were these genuine genetic changes, or was some other mechanism at work? When the early microbe hunters had discovered bacteria they often referred to them as a 'ferment', more like a frothy mix of chemicals than a population of individual organisms. And in the twentieth century, bacteriologists still tended to talk about the properties and behaviour of bacterial 'cultures', as a whole, rather than about the properties of individual, variable bacteria. Even de Kruif, who thought he might be watching the effects of individuals mutating, by calling the phenomenon 'dissociation' was implicitly referring to the behaviour of the culture as a whole, not to the individuals that composed it.

De Kruif had begun his career as a medical student, but as he became increasingly aware of doctors' helplessness in the face of most diseases, he switched to pure science. The Rockefeller Institute was the obvious place to be; Frederick Gates had assured John D. Rockefeller Sr that, given enough money, science would soon conquer the microbes, major diseases would be cured within years and Rockefeller would find himself a place both in the history books and in paradise. But the promised cures did not arrive and de Kruif began to wonder why. He became increasingly cynical about the medical profession, which he and his fellow researchers were supposed to be serving; he began to suspect that most doctors were only interested in making money.

De Kruif began to think about a new career as a writer. There were more magazines and newspapers in the United States than ever before; by 1910 the country had about 2,600 newspapers, all struggling for circulation and advertising and in desperate need of stories to fill their pages. De Kruif wondered whether he might put his scientific training to use, explaining the latest break-throughs and discoveries to the general public while unmasking frauds and failures. One evening, at a literary party in New York, he was riding his favourite hobby horse, complaining about the self-interested greed of America's supposedly selfless healers, when he met the historian Harold Stearns. Stearns was amused that someone who worked at the temple of medical science – the Rockefeller Institute – could be so cynical about the contrast between the apparent worthlessness of most medicines and the prestige and wealth of those who dispensed them. He invited de Kruif to contribute a chapter on the state of American medicine to a book he was editing; de Kruif, increasingly tired of research, accepted – on condition that he remain anonymous. When the chapter appeared, the editor of the *Century Magazine* invited de Kruif to expand it into a book, to be serialized first in the magazine.

De Kruif began writing, but when the first instalment of 'Our Medical Men' appeared in the *Century*, it was not his criticisms of doctors that got him into trouble; his byline cost him his job. Despite his insistence on anonymity, he had mentioned his $500 advance to many of his colleagues at the Rockefeller, so when the article appeared, signed 'K—, MD', his boss, Simon Flexner, had a pretty good idea who was responsible. Flexner summoned de Kruif, to express his outrage that a PhD had apparently passed himself off as an MD, a medical doctor. In fact, de Kruif was mostly innocent: a sub-editor had made the change without his knowledge, but de Kruif had failed to spot the error because he had not troubled to read the proofs the magazine had sent him. He resigned from the Rockefeller before Flexner could fire him and set about inventing an entirely new specialization – he became one of the world's first full-time science writers.

In 1922 de Kruif was researching a piece on fraudulent medications. In a doctor's office he encountered 'a young red-headed man, very tall and slightly stooped, nervous, his face spotty red . . . An unearthly character, not to be forgotten once seen'. This was the novelist Sinclair Lewis, whom de Kruif described as 'the then most famous author in the wide world'.[9] Lewis had just published his second major novel, *Babbitt*, which like its precursor, *Main Street*, had been both a commercial and critical success. He was becoming famous around the world; some British critics acclaimed him as America's Dickens.

Lewis and de Kruif hit it off immediately, not least because of their shared fondness for bootlegged liquor. Lewis was looking for a subject for his new novel and in de Kruif's impassioned attacks on the unscientific state of American medicine, he felt he had found it. The two men shared a desire to see a genuinely scientific medicine come to the aid of honest but largely helpless family doctors (like Lewis's father), who would then be able to drive the cynical snake-oil salesmen out of medicine. Just a few months after their first meeting, Lewis and de Kruif set sail on the SS *Guiana*, bound for the Caribbean, where they planned to set their novel of science and medicine; De Kruif was going to provide a solid scientific basis for Lewis's writing skills and practised imagination.[10]

De Kruif had been a fan of H.G. Wells's novels ever since his student days and was impressed by the way Wells's fiction had always been rooted in real science, some of which Wells had learned directly from Darwin's disciple, Thomas Huxley. As de Kruif and Lewis discussed possible real scientific topics that might be comparably dramatic, de Kruif suggested that their fictional hero, Martin Arrowsmith, might fight a Caribbean plague using bacteriophage. Few in the medical world knew of d'Hérelle's work and its potential therapeutic uses, and practically no one in the wider world would have heard of it, yet here was a real science that, if d'Hérelle was right, could vanquish the microbes that made human life so precarious.

According to de Kruif's (not altogether reliable) memoirs, the two men were soon on fire with this idea; they worked hard every day, drank equally hard every night, and explored each Caribbean island they visited, unaccountably wearing pith helmets (apparently so that they would look British). The collaborators had brought with them a trunk-load of medical books, maps and letters of introduction to various island doctors and administrators. From Lewis's surviving notes, it is clear that de Kruif gave him a fairly intensive course in microbiology during their voyage; his notebooks are full of technical details of lab protocols and techniques, together with sketches of scientific equipment. De Kruif acknowledged that Lewis 'never tried to make do with phoney movie science . . . He kept teaching me to let myself go to dramatize real science'.[11]

Lewis put de Kruif to work writing a brief scientific biography of each character in the novel. Arrowsmith's mentor, Max Gottlieb, was a composite of Jacques Loeb (who also worked at the Rockefeller Institute) and de Kruif's old teacher Frederick Novy. For Arrowsmith himself, Lewis drew on several of de Kruif's colleagues and indeed on de Kruif himself. By the end of the voyage, they had the skeleton of the book and all the scientific background Lewis would need. When the novel appeared, in 1925, it was a sensational triumph; Lewis won the Nobel Prize for Literature in 1930, largely on the strength of it.

Arrowsmith was widely reviewed and praised. *Science* reviewed it in its section on 'Scientific Books'. The magazine was delighted that it was not only 'a novel of the first rank', but had 'a scientist for its main character'. *Science* even suggested that the novel's publication was 'an added bit of evidence of a certain shift in our civilization shown by the growing interest of the layman in scientific matters', a shift that was taking place because the scientific 'High Priests' had finally 'taken off their false whiskers and have given Mr. Average Citizen a peep at the ceremonies going on inside the Temples'. The journal congratulated Lewis, not merely for showing 'no small amount of courage' in making a scientist his hero, but for describing his researches, 'clearly and intelligently, without yielding to the temptation to write down to the technical knowledge of a novel reading public'.

De Kruif's contribution to the novel was acknowledged by Lewis (although not to the extent that de Kruif wanted), and *Science*'s reviewer observed that 'much of the verisimilitude of the action and characters' was undoubtedly the result of de Kruif's efforts. After praising aspects of the novel ranging from the convincing idea of laboratory life to Lewis's satirical sense of humour, the review urged that 'Every medical student who feels vague rumblings of scientific curiosity or the urge for pure research, should read it.'[12] This advice seems to have been widely followed: *Arrowsmith* was one of the first novels to feature a scientist as a hero and many idealistic young people were inspired by it – and by de Kruif's next book, *The Microbe Hunters* (1926) – to follow medical or scientific careers.

One result of *Arrowsmith*'s success was that bacteriophage became a popular topic for journalists. Before the novel's appearance, only two short pieces relating to the phenomenon had appeared in English in the non-specialist press. But soon, bacteriophage were written about so regularly that the *Lancet* was tetchily complaining about 'The musical comedy spirit which reduces 'bacteriophage' to its final syllable'.[13] The magazine's complaint was in vain. Within a year the term 'phage' was so common that it had been used in the *Encyclopaedia Britannica*; it has remained in use ever since.

What is life?

The popular American magazine *Science Monthly* discussed the question 'Do Bacteria have Disease?' a year after *Arrowsmith* appeared. The article noted that, despite all the interest in phage, there was still considerable debate about what viruses actually were. Scientists had tried to measure them by passing them through smaller and smaller filters until they were finally caught. Based on these tests, they were estimated to be about 1,000 times smaller than a bacterium. That was far too small to be observed through even the most powerful microscope, but there was a second reason for concern: could something that small possibly be living? If the measurements were right, viruses such as bacteriophage were not much bigger than a single protein molecule and, as *Science Monthly* put it, that 'seems to leave us in a dilemma'. Since it was assumed that proteins were the fundamental units of life, 'is it possible to have living organisms smaller than the protein molecule!'[14]

The possibility that proteins were the basis of life had made them the focus of intense study among biochemists, who were not only sceptical that something as tiny as phage could be alive, but were also unwilling to see such an interesting and potentially important phenomenon fall into the domain of their rivals in bacteriology. Biochemists were convinced that phage were protein molecules, and since the study of proteins was part of biochemistry, phage therefore 'belonged' to the biochemists.

Leading the protein research at the Rockefeller Institute was John H. Northrop, who had proved that enzymes were proteins (he and two colleagues would win the Nobel Prize in 1946 for this work). He also investigated the way enzymes catalysed reactions, building on the earlier work of J.B.S. Haldane and others. Northrop's team discovered that the speed at which chemical reactions proceeded within the body was controlled by the concentration of the enzyme – the greater the enzyme concentration, the faster the reaction. With this discovery in mind, Northrop turned his attention to phage, which had been found to consist almost entirely of protein. If, as Bordet and others claimed, the phage phenomenon was simply a chemical reaction, researchers would expect it also to depend on simple measurable quantities, such as the concentration of phage. Opponents of the living virus theory argued that the bacteria burst as a result of purely chemical changes within the cell: osmotic pressure drew in water until the bacterium disintegrated. If they were right, the phage were something entirely lifeless produced by the bacterium itself.

To test these ideas, Northrop's co-worker, Alfred Krueger, developed a method for estimating the amount of bacteria and phage in a culture, treating them as if they were enzyme and substrate (the substance on which an enzyme acts). His experiments revealed that the concentration of phage was indeed the crucial factor in whether or not the bacteria disintegrated – and that suggested that phage did indeed act like enzymes and so fell within the territory of the biochemists rather than that of the bacteriologists. It was also significant that viruses like phage and tobacco mosaic virus (which was also being worked on at Rockefeller) could be crystallized, just like any other protein. That seemed a most unlikely property for a living creature, but it might explain why viruses had some 'life-like' properties since, of course, crystals grow without being alive. Perhaps viruses were some kind of self-catalysing enzyme: they began a reaction that in turn produced more of the virus; since the virus itself catalysed the reaction, as the concentration of virus rose, the reaction accelerated, producing yet more virus. This idea explained the multiplication of the virus and the apparent runaway nature of the reaction, ending in the bacterium's disintegration.

Part of Northrop's argument that viruses were purely chemical was the principle of Occam's Razor. This is (perhaps surprisingly for what has become a scientific precept) named after a medieval Franciscan friar, William of Occam. He proposed that '*Pluralitas non est ponenda sine neccesitate*' ('multiple entities should not be assumed unnecessarily'); in other words, if a simple explanation will do, there is no reason to look for a more complicated one. Scientists have often suggested using Occam's Razor to decide between rival theories: if a simple theory explains the facts, it should be preferred to a more complicated one. The principle does not, of course, assume that nature *is* in fact simple, merely that it makes sense to avoid further complexities unless it becomes unavoidable – usually as a result of some experiment or observation that cannot be explained by the simplest theory. In Northrop's view, the biologists were making matters unnecessarily complicated. They were far too prone to assume that there was something mysterious, almost mystical, about 'life', which made them reluctant to fully accept the implications of the fact that living things were built out of simple chemicals and their life processes were based on straightforward chemical reactions. This was not, of course, a new argument – it has probably been going on in various forms ever since we first evolved the ability to ask the deceptively simple question: what is life?

For those with an interest in this issue, viruses such as bacteriophage posed provocative questions, but scientists were also interested in the practical problems they presented. When Emory L. Ellis completed his PhD in biochemistry at the California Institute of Technology (Caltech) in 1934, he began a post-doctoral project on viruses, trying to investigate if and how they were involved in causing cancer. He decided to work on phage, which might seem an odd choice, since there was no evidence that they cause cancer and he was not particularly interested in bacteria. What Ellis wanted to understand was the role viruses play in cancers – first in animals, but eventually in people. However, testing viruses on animals 'required a large animal colony, with all its attendant problems and expense'.[15] So Ellis abandoned the mice he had been working with and switched to phage. Ellis and his wife, Marion, visited Pasadena's sewage treatment plant to gather samples of untreated effluent. They found the sewage full of different kinds of phage, many of which seemed specialized to attack particular types of bacteria.

Ellis chose to work on a phage that preyed on *Escheria coli* (usually known simply as *E. coli*). This is a very common bacterium (every one of us has millions in our guts) and one of Morgan's students (the Morgan Drosophila team had now moved from Columbia to Caltech) happened to be working on it, so he had plenty to spare. Ellis was careful to choose a phage that was not too lethal – so that the clear areas it produced when it had killed the bacteria (known as plaques) were fairly small – he was able to get about fifty of them on a single Petri dish. (D'Hérelle had first hit on the idea of counting plaques as a way of calculating how many phage were present.)

Not only were phage small and cheap, they were fast. Testing viruses on animals involved waiting for them to get sick, which could take days or weeks, whereas it only took a few hours to complete a phage experiment (Ellis would later help reduce this to just two hours). 'Clearly, then,' he decided, 'bacteriophage was by far the best material from these points of view.'[16] He hoped that his small, fast, cheap phage would enable him to understand the basic biology of viruses, as a first step towards understanding the hypothetical role of viruses in cancer. The phage were to be used as a model organism, just as the USDA had kept guinea pigs as model farm animals, allowing costly large-scale experiments to be done in miniature; the phage were to be used rather like a scale model of a plane that is tested in a wind-tunnel before an expensive prototype is built.

As he was busy working with his phage one morning, Ellis was interrupted by a visitor, Max Delbrück, a German physicist who was looking for a paradox that would reveal new laws of physics. Yet he had not wandered into the wrong lab; he was hoping Ellis's phage might be able to help him.

In 1935 Max Delbrück had been a nineteen-year-old student in Berlin when he had sat alongside Albert Einstein and Max Planck to hear Werner Heisenberg present quantum theory for the first time. He later admitted that he had understood very little of what Heisenberg said that day, but he had grasped enough to realize that he was present at the birth of something extraordinary – an entirely new understanding of the atom, of matter itself. The following year, Delbrück joined Heisenberg to study at the University of Göttingen. He had planned to be an astronomer, but was soon completely absorbed by the new quantum physics.

In the late nineteenth century, physicists had discovered much about the atoms, which the ancient Greeks had first hypothesized as the basic components from which everything was made. It was clear that these once-hypothetical entities definitely existed, but it was also becoming apparent that the atoms physicists had identified were not in fact the fundamental entities the Greeks had imagined. As new phenomena such as X-rays and other kinds of radiation were explored, it became clear that atoms consisted of even smaller particles, such as electrons and protons. And as these new, sub-atomic particles were investigated it became clear that they were not simply very tiny versions of the billiard balls and planets whose behaviour was described by the laws of classical physics. In this new realm of the unimaginably small, those laws ceased to operate and new, strange ones took over. It became apparent, for example, that energy, instead of flowing smoothly and continuously, came in discrete packets, called quanta. Only some quantum states were stable and electrons seemed to jump between them, jerking – inexplicably and discontinuously – from one to another. Another puzzle was that ordinary light took on a paradoxical quality: sometimes it behaved in ways that could only be explained by assuming that it was a wave, while at other times its behaviour made it clear that it was made of particles. The deeper the physicists investigated, the more mysteries they unearthed. Heisenberg, Niels Bohr and Erwin Schrödinger eventually resolved these paradoxes by wrapping them in mathematics. It was impossible to picture these implausible phenomena that were sometimes waves and sometimes particles, and appeared either to be nowhere or to be in more than one place, until experiment actually tried to pin them down. But the maths worked and it worked brilliantly.

Inspired by the excitement of quantum physics, Delbrück went to Copenhagen for a year to study with Bohr, the brilliant Danish physicist who had won the Nobel Prize in 1922 for his work on atomic structure. Delbrück recalled that Bohr 'incessantly worked and reworked his ideas on the deeper meaning of quantum mechanics'. At the heart of these was the principle of complementarity, which states that it is impossible to describe all the aspects of any situation in atomic physics in a way that produces a single coherent picture. Each kind of experiment can only furnish one kind of information and the different experiments are mutually exclusive, so cannot be used to establish a complete picture. This was what Heisenberg summed up as his famous uncertainty principle – there seemed to be absolute limits on what we could know. Some interpreted Heisenberg's maths to mean that sub-atomic particles simply lacked fixed, knowable properties; a view that came as a shock to some physicists. Einstein rejected this approach on principle and remained convinced to the end of his life that quantum physics would eventually be replaced by an even newer physics that would combine quantum ideas with some of the knowable, stable reality of classical physics. But Bohr was entirely comfortable with the new indeterminacy, convinced that it was a major step forward.[17]

Of all Bohr's ideas, the one that most inspired Delbrück was the thought that complementarity might apply to all of science, including biology. Delbrück remembered Bohr repeatedly asking 'whether this new dialectic would not be important also in other aspects of science'? Perhaps biologists' failure to understand the nature of life resulted from a similar kind of mutual exclusion: 'you could look at a living organism either as a living organism or as a jumble of molecules', but not both. Some kinds of experiments revealed 'where the molecules are', but quite different ones were needed to 'tell you how the animal behaves'.[18] Part of Bohr's argument was that the atomic structure of an organism could not be investigated without killing it first, but perhaps living matter had genuinely unique properties – such as the power of replication – that were lost when it died. In which case, the traditional methods of biology, such as dissection, were doomed; if Bohr was right, biology needed to follow the same path as physics – to go smaller, to search for the real, fundamental 'sub-atomic particles' of life. When these were examined, perhaps paradoxes like those of the quantum world would emerge, and resolving them might reveal new scientific laws that would explain the mysterious properties of life. In effect, biology might hold the key to the next step forward in physics.

Having absorbed Bohr's thought-provoking vision, Delbrück returned to Berlin, where he met Nikolai Vladimirovich Timoféeff-Ressovsky, one of the Russian biologists who had become a Drosophila geneticist as a result of encountering Hermann Muller on the latter's first visit to the USSR. Like Timoféeff-Ressovsky and his collaborators, Delbrück soon became strongly influenced by Muller's ideas. For a physicist, perhaps the most interesting aspect of Muller's work was that in 1926 he had used X-rays to produce artificial mutations in Drosophila (work for which he would eventually win the Nobel Prize). This proved extremely useful to biologists, who could now generate mutations more or less to order, instead of waiting for them to arise accidentally. They could not control what kinds of mutations were caused – they were still random – but they appeared so quickly that it became much easier to find those that were interesting or useful. The X-ray experiments helped accelerate the pace of chromosome mapping but, even more crucially, they persuaded many scientists that genes really did exist; if they could be affected by X-rays, they had to be real, physical entities with definite properties. This was another important step in replacing Mendel's *Anlagen* with definite physical entities, open to full investigation.

For someone with Delbrück's background, Muller's X-ray work opened up an intriguing possibility: perhaps genes were in some ways like atoms. Both were stable – genes could be passed on from generation to generation – but genes (like atoms) could be destabilized, mutated by a burst of energy. Perhaps the heritable mutations produced by X-rays were caused by a gene being 'flipped' into a new stable state, analogous to an electron being flipped from one stable quantum state to another? If so, genes might prove to be the true elements of life, the real biological atoms, so investigating them could perhaps reveal the new paradoxes – and thus the new laws of physics – that Bohr and Delbrück were hoping for.

Delbrück, Timoféeff-Ressovsky and Karl Günter Zimmer collaborated on a joint paper, which became known as the Green Paper (simply because that was the colour of the cover on the copies they sent out). It suggested why genes were usually unchanging but also why they could be changed by radiation: genes were stable chemical molecules, but the energy of radiation was enough to rearrange their constituent atoms into new forms. The Green Paper connected physics and biology in the most direct way.

During the 1930s, many scientists were keen to connect physics and biology, for a variety of reasons. Biology had always been rather a poor relation to more rigorous and prestigious sciences such as chemistry and, in particular, physics. One reason why biologists were interested in establishing new connections between the so-called 'hard' sciences and their softer cousins, the life sciences, was that they hoped to enjoy some of the prestige – and funding – of their physicist colleagues. They were supported in this ambition by some of the physicists and chemists themselves, who – perhaps for the first time – wanted to associate themselves with the life sciences precisely because they were sciences of *life*. As we have seen, biochemists emerged from the First World War as the discoverers of vitamins, savers of lives; chemists as the discoverers of the poison gas that had maimed and killed so many. However unfair the characterization, the hard sciences had begun to be seen as sciences of death. By the 1930s, it seemed to some that an unholy alliance of physicists and chemists had unleashed new technologies of killing on the world: in 1937, as fascist aeroplanes bombed defenceless civilians in the ancient Basque city of Guernica, many wondered what new horrors science might have in store for them. Meanwhile, biologists were unravelling the basic principles of life, such as respiration and digestion – at worst, this knowledge was harmless, at best, it could save lives, free humanity from disease, even conquer death itself.

Biology benefited from the unease that had begun to surround the physical sciences. In the 1930s, thanks in large part to the

Rockefeller Foundation's money, biologists were tackling a new problem, investigating the shape of large, complex molecules such as proteins in order to see how they worked. Protein structure began to be seen by many as 'the problem of life', the key to understanding life itself. Traditional biochemistry did not seem to be making much headway in solving it, so gradually in the 1930s a new discipline began to take over. It used new tools, such as X-ray crystallography, and in 1938 it acquired a new name, 'molecular biology'. The name was coined by Warren Weaver, director of Rockefeller's Natural Sciences Division. He defined the field as the 'biology of molecules' or as 'sub-cellular biology', shifting from the cell itself as the object of study to a more fundamental level of analysis. Weaver made an explicit analogy with the sub-atomic world of the quantum physicists; to make headway, biology had to go deeper by going smaller.

Weaver was an engineer with no biological background; his sense that biology was the 'science of the future' was shaped by biologists writing in the press, describing their ambitions to control life and conquer disease. However, there seemed little evidence to support these grand claims, and Weaver's view was that biology still seemed to be 'lacking laws and beyond rational analysis'.[19] The Rockefeller Foundation's response would be a healthy injection of cash and physics – and especially of new technology. That, Weaver believed, would enable biologists to establish the laws of their field, make rapid progress and save a civilization threatened by fascism, communism and a global economic depression.

It is debatable whether Weaver and those who shared his views were really shaping biology's new direction, or simply jumping on to a bandwagon that was already gathering pace. Probably both, but the new money certainly allowed – and encouraged – biologists to adopt the exciting – but expensive – new technologies which the physicists had created and apply them to biological problems. In addition to generating mutations, X-rays could also be used to reveal chemical structures, using the process of X-ray crystallography. This technique relied on the fact that when a substance crystallizes, all its molecules are arranged into a regular, evenly spaced structure. That means that, in some important respects, the crystal is like a single giant molecule. When a beam of X-rays is shone through it, the beam is scattered and the scattering pattern can be detected using photographic film. This may be easier to understand by imagining shining a torch into a box. Inside the box is a chandelier, but the chandelier is out of sight; all that is visible is the pattern that the light makes when it is refracted on to the wall beyond the box. Using the laws of optics, the design of the chandelier can be deduced from the pattern on the wall. The same principle is used in X-ray crystallography – the pattern the X-rays make when they are scattered by the latticework of molecules within the crystal can be used to deduce the shape of an individual molecule.

X-ray crystallography was first used to investigate the structure of simple, inorganic compounds, such as diamond, but, using the rules thus developed, biochemists realized that it was possible to apply the technique to a range of much larger, more complex organic molecules, such as proteins. One of the results of such investigations was the realization that the three-dimensional shape of these molecules had a direct connection to the way the molecule behaved, the way it was able to perform its chemical – and ultimately its biological – function.

In 1935, the same year that the Green Paper appeared, one of Northrop's colleagues at the Rockefeller Institute, Wendell Stanley, managed to produce crystals of the tobacco mosaic virus (TMV) and announced that a virus was simply a protein molecule. His announcement caused a sensation: Delbrück was one of hundreds of biologists excited by an experiment which seemed to bring the properties of proteins, supposedly life's fundamental building blocks, firmly into the terrain of physics. In 1937 Delbrück was invited to apply for a Rockefeller Institute fellowship in molecular biology and went to Caltech to work with Morgan and his relocated fly boys.

As work on tobacco mosaic virus and similar plant viruses proceeded, it became clear that they belonged to a small group of proteins whose chemical structure included traces of phosphorus. Fifty years earlier, chemical analysis had shown that chromosomes were also made from this kind of protein, which – because chromosomes were found in the nucleus of a cell – had been named 'nuclein'. But in the early twentieth century the name gave way to 'nucleoprotein'; the fact that it could be crystallized allowed teams of researchers to use X-rays to start probing its structure, in the expectation that the precise shape of the molecule would reveal how it worked.

Muller was one of many scientists intrigued by the discovery that viruses were made of the same stuff as chromosomes. He had been interested in viruses – and phage in particular – for some time. Back in 1922 he had wondered whether bacteriophage 'were really genes, fundamentally like our chromosome genes'? If they were, that 'would give us an entirely new angle from which to attack the gene problem'. And it now appeared that phage might be something very like naked genes, genes that had somehow got outside cells and were able to survive on their own. Muller acknowledged that 'it would be rash to call them genes, and yet at present we must confess that there is no distinction known between genes and them'. That opened up the possibility that 'we may be able to grind genes in a mortar and cook them in a beaker after all'.[20]

The crystallization of tobacco mosaic virus seemed to confirm Muller's view: if viruses were not genes, they were so similar that they could provide the ideal tool for investigating the mechanism of inheritance. When Delbrück went to America, he was already interested in phage and viruses, wondering if phage might be naked genes, and if genes were the elements of life. When he arrived at Caltech, he found Morgan's group waiting to welcome him: the fly boys were very interested in collaborating with physicists – largely thanks to Muller's X-ray work – but few of them could understand the maths of Delbrück's Green Paper. One of his first tasks upon arrival in Pasadena was to explain it to them.

The Phage Group

While the fly boys got to grips with his maths, Delbrück tried to master Drosophila genetics, but remembered spending his first few months at Caltech struggling – and failing – to understand the fly's genetic complexities. By this time, fly studies had become a mature field, with its own terminology and literature: engaging with it entailed reading a vast number of academic books and papers, as well as mastering all the practical aspects of the fly business. Delbrück later admitted that he 'did not make much progress in reading these forbidding-looking papers . . . I just did not get any grasp of it'.[21] He decided that the flies were too much for him and in any case, to a physicist, they seemed both too large and too complex. Physicists had often found that the first step in solving a problem was to simplify it, strip it down to the most basic possible components; working on Drosophila was like trying to deduce the essential principles of quantum physics from vast, problematic molecules like proteins. Delbrück wanted a biological equivalent to the hydrogen atom – one proton, one electron, no complications.

Before leaving Europe for the United States, Delbrück had already been vaguely aware that viruses might be interesting to work on. En route to Caltech in 1937, he had visited Wendell Stanley at the Rockefeller Institute labs, but was disappointed to find that even tobacco mosaic virus seemed too complex for the kind of simple experiments he had had in mind. When the Caltech flies got too much for him, Delbrück took a holiday and went camping. On his return, he discovered that he had missed a seminar on phage, given by Emory Ellis. 'I was unhappy that I had missed it,' Delbrück remembered, 'and went down to ask him afterwards what it was all about. I had vaguely heard about viruses and bacteriophages . . . I had sort of the vaguest of notions that viruses might be an interesting experimental object.'[22] This was what had led him to Ellis's lab.

Delbrück found Ellis eager to show him his phage experiments and he was impressed by what he saw. Despite starting out with no knowledge of microbiology or viruses, and using some very primitive equipment, Ellis had mastered the basics of phage farming: growing *E. coli*; letting the phage attack them; and measuring the results. Delbrück was 'absolutely overwhelmed that there were such very simple procedures with which you could visualize individual virus particles'. At last, here was some biology that was as clean and simple as physics, which produced clear-cut, mathematical results. Phage seemed to Delbrück to have the potential to be biology's hydrogen atoms – the simplest possible example of life's ability to reproduce. And he decided to work with them until they provided the paradox he was after. 'This seemed to me just beyond my wildest dreams,' Delbrück remembered; finally he could do 'simple experiments on something like atoms in biology'.[23] Delbrück and Ellis agreed to work on phage together.

Delbrück was convinced that phage were not simply a chemical phenomenon, but were genuinely alive; they reproduced and grew like other organisms. He had to believe this if they were to be usable as models for more complex living organisms, but the idea that viruses were alive remained a minority view in the 1930s. The work of the Rockefeller Institute's protein chemists had persuaded most biologists that viruses, like phage, were some kind of enzyme, not an organism at all.

Northrop was sure that even if *some* viruses were alive, phage and tobacco mosaic virus were purely chemical – they were just too small and simple. Stanley had originally agreed with him, but – like Beijerinck before him – had decided that tobacco mosaic virus must be more than a chemical because it was impossible to produce except in living cells, a requirement that simple chemicals did not share. That was also part of what persuaded Delbrück that phage were living viruses; they only multiplied in the presence of living cells. They were also highly specific to their host – the phage that attacked *E. coli* did not attack other bacteria – and that pattern had been found in other animal and plant viruses. Also, phage were roughly the same size as other viruses and – most importantly of all – they were made of the same stuff, nucleoprotein.

As the arguments about what phage were continued, Delbrück concentrated on finding collaborators. He had originally decided to treat the bacterial cell as a black box, not to be opened because that would disturb the living system. His idea was simply to treat the number of phage that infected the bacterium as an 'input' and the number that were released when the cell disintegrated as an 'output'. He reasoned that such a simple approach would enable him to produce a mathematical equation that precisely described phage reproduction. Phage were, as he put it, 'a fine playground for serious children to ask ambitious questions'.[24] He had originally assumed that the project would take only a few months and he would be able to announce the 'secret of life' before his Rockefeller fellowship expired. Like many physicists, he soon realized that he had underestimated the complexity of biological problems: it took him two years simply to devise a reliable way of counting phage.

Long before he had discovered the secret of life, Delbrück's Rockefeller money ran out. He had to take a job as a physicist because there still was not much interest in the kind of hybrid 'biophysics' that he had begun with phage. Soon after, Delbrück met the Italian biologist Salvador Luria, who had heard of Muller's X-ray mutation work in Rome. Luria had also come across the Green Paper and Delbrück's physics-based concept of the gene. Luria soon had to leave the University of Rome; he came from a Jewish family and was forced out of his job by Italy's fascist government in 1938. He went to Paris for a while, but as German troops advanced on the city in 1940, Luria fled on a bicycle, eventually managing to leave Europe, via Spain and Portugal, for New York.

Delbrück realized immediately that this Italian refugee was the colleague he was looking for: Luria was a microbiologist; he had post-doctoral experience of physics, and he was already working on phage. Delbrück and Luria had both discovered that some bacteria were immune to phage attack, so they decided to tackle the central question of whether or not bacteria really had genes by finding out if phage-resistance was in fact a genetic mutation.

Many bacteriologists believed microbes had no genes (in part because bacteria have no nuclei), and that they must evolve in some kind of Lamarckian fashion, since they could apparently, for example, acquire phage resistance so rapidly once they were attacked.

One evening, Luria was considering this question as he watched his colleagues playing the slot machines at a club. The sporadic clatter of winning coins cascading from the machines set him thinking about randomness and probability. If the bacteria underwent genuine mutations, these must arise at random and as a result, the percentage of phage-resistant bacteria that grew on any given Petri dish would also be random. However, if the bacteria developed resistance in response to the attacking phage, the percentage of the resistant type should always be proportional to the number of phage. Luria did the experiments and Delbrück did the maths, and the numbers of resistant bacteria were indeed random. In 1943, they produced a jointly authored paper that convinced their colleagues that bacteria do indeed have genes, which behave in much the way as fly genes. (As we shall see in a later chapter, it was eventually realized that bacteria belong to an entirely separate kingdom of living things that lack a nucleus but do nevertheless have genes like those in other organisms.)

During their work, Delbrück and Luria had come across papers by Alfred Hershey that interested them. Luria met Hershey when he gave a paper at Washington University, in St Louis, Missouri, where Hershey worked on bacteriology. Hershey was also impressed by the potential of phage genetics and so what became known as the Phage Group was born. Delbrück, Luria and Hershey would add a new strand to molecular biology by providing a new tool – bacteriophage – with which to tackle some of its most important questions.

In 1944 molecular biology received an unexpected contribution from a physicist, in the shape of a little book with the ambitious title: *What is Life?* Its author was Erwin Schrödinger, one of the founders of quantum physics, who had come across Delbrück's Green Paper and been inspired by it. Schrödinger believed that fully working out Delbrück's concept of the gene would indeed 'involve hitherto unknown "other laws of physics" which, however, once they have been revealed, will form just as integral a part of this science' as the existing laws. Schrödinger suggested that genes might consist of some kind of irregular (or 'aperiodic') crystal made up of several molecules that had the same number of atoms, but were arranged in different ways, that would give them different chemical properties (such molecules are known as isomers). He commented that 'the number of atoms in such a structure need not be very large, to produce an almost unlimited number of possible arrangements. For illustration, think of the Morse code. The two different signs of dot and dash in well ordered groups of not more than four allow of more than thirty different specifications.'[25] Schrödinger's choice of metaphor – that genes might be like codes – was to profoundly shape the way genetics developed after the Second World War.

Molecular biology also received an unexpected boost from the tragedy of the atomic bombing of Japan in 1945. Many argued that the bombs saved more people than they killed by bringing the war to a speedy end, but some young physicists were more persuaded by Robert Oppenheimer's assessment that 'the physicists have known sin, and this is a knowledge which they cannot lose'.[26] As direct military investment in physics of every kind increased to unprecedented levels, some of those who read Schrödinger's book were inspired to abandon what increasingly seemed to be the death-dealing science of bomb-making for the new field of molecular biology.

As the Phage Group's work progressed, it soon became clear that even these three exceptionally brilliant men were not going to discover the secret of life in a few months. So Delbrück set out to recruit disillusioned physicists. He organized annual phage meetings and, once he became a biology professor on his return to Caltech, took on graduate students who went out to spread the word about phage. The year after *What is Life?* appeared, Delbrück began teaching a summer course at Cold Spring Harbor Laboratory to instruct newcomers in the basics of phage.

The courses continued on an annual basis for twenty-six years and – as we will see – many of the twentieth century's most influential biologists graduated from Delbrück's Phage Course.

Although Delbrück later recalled that 'the phage group was not much of a group', its members communicated with each other constantly and that was vital to its success: the open spirit of sharing information and results was one of many borrowings from physics. Delbrück acknowledged that this aspect of the group's approach was 'copied straight from Copenhagen and the circle around Bohr', where 'the first principle had to be openness. That you tell each other what you are doing and thinking.'[27] Like Bohr's lab, the Phage Group enjoyed an atmosphere of amiable impertinence, at the centre of which was Delbrück, always delighted to talk to everyone and anyone. The many temporary workers who passed through Caltech and the annual summer phage course took Delbrück's methods and techniques with them, but they also tended to adopt his doctrine of openness and pass it on to their own colleagues and students. In 1944 a newsletter was established, the *Phage Information Service*, which, like the *Drosophila* newsletter it was modelled on, soon became a vital tool for the pioneers of phage genetics. Rapid communication of results was seen as vital, and Delbrück would declare 'pipette-free days' when everyone had to leave their benches and write up their results for publication. Once published and shared, openness led to rapid progress: a proliferation of interesting new results resulted in lots of publications, which in turn recruited numerous graduate students, and they, in turn, increased the attention being paid to phage and raised the profile of the phage researchers. The Phage Group's success meant that its philosophy and methods would be copied by many of the geneticists who followed them.

Another significant way in which physics influenced the new genetics was the drive for standardization. In the early days of phage each research group had its own strains, which had usually been procured from sewage farms or similar sites. To Delbrück, this was comparable to every physics lab using its own set of weights and measures, or its own definitions of acceleration and force. Such disorder dissipated many of the benefits of co-operation, since it made it harder to compare one lab's results with another's. So he used his rapidly growing influence in the new field to enforce what became known as the 'Phage Treaty': any researcher who wished to be part of the developing phage network had to agree to work on one or more of a specific set of seven 'well-behaved' phages (the T series) that infected one specific strain of *E. coli*. Once every researcher was working on the same phage and using the standard techniques they had learned on the Cold Spring Harbor phage courses, it became possible to treat all the different Phage Groups scattered around the world as if they were one, big cooperating team. It was almost as if the phage themselves bound the groups together and made the cooperation possible.

As this scattered team worked on their viruses, they gradually became convinced that both bacteria and viruses do indeed have genes and that they behave in a standard Mendelian fashion, much like those of Drosophila. However, the central question remained unanswered: what were genes and how did they actually work? It gradually became clear that phage did not in fact feed on or consume bacteria; what the virus did was to take control of the bacteria's own cellular machinery, the equipment necessary for reproduction – that was why they could not be grown on a Petri dish, but needed living bacterial cells to duplicate themselves. Somehow phage hijacked the bacterium's repro-ductive apparatus so that the cell no longer reproduced itself, but produced phage instead. Delbrück and others could now explain the time lag between the phage attack and the death of the bacterium: that was the period during which a new generation of phage was developing – once there were too many for the bacterium to contain, it ruptured and died, releasing the new phage. This pattern strongly suggested that each phage contained some kind of template for making more phage, but what was the template composed of and where was it?

One important clue came from one of the expensive tools microbiology had acquired from physics, the electron microscope.

In the early 1930s, novel technologies, especially the cathode ray tube (the heart of the traditional television set), had made it possible to build microscopes that 'illuminated' objects with beams of electrons instead of beams of light. Because the wavelength of an electron beam was much shorter, the new microscopes allowed scientists to observe much smaller objects – several hundred times smaller – than had been possible with even the best light microscopes. The first electron microscopes were built in Germany in the mid-1930s, and when news of them reached America, some scientists thought the miraculous machine might be a Nazi hoax. But it was not. Soon, American companies like RCA were manufacturing them and, in an attempt to stimulate the market, started to investigate new applications. In 1940, RCA offered a $3,000 grant for exploring biological applications; it was won by a researcher called Thomas Anderson, one of the first American electron microscopists.

In 1941 Luria approached Anderson to see if he could take electron microscope photos of phage to see how big they really were. Anderson thought it possible, though Luria had first to apply for security clearance (the Rockefeller Institute's lab was involved in classified defence work at the time). Their first attempts failed, because the solutions of phage were not sufficiently concentrated, but by March 1942 Luria had managed to increase the concentration enough for successful pictures to be made; they revealed phage to be shaped something like tadpoles with a distinct head and tail. The fact that phage had such a relatively complex anatomy added support to the argument that they were really alive. The other intriguing aspect of the early pictures was that they all showed the phage with their heads directed towards the bacterium, as if they were swimming towards it, like sperm towards an egg. (It took Anderson eleven years to demonstrate conclusively what he had long suspected, that this arrangement was in fact simply an accidental result of the way the specimens were prepared.) The sperm-like appearance suggested that phage infection might be something akin to fertilization. Even more intriguingly, the electron microscope photos seemed to show that the phage did not actually penetrate the bacterium, instead they remained immediately outside it – and yet somehow they were still able to transfer their reproductive templates into the bacterium.

Chemical analysis had confirmed that phage consisted of nothing more than a protein coat surrounding a core of a nucleoprotein called deoxyribonucleic acid, or DNA. Presumably one or the other provided the template for producing new phage, but which? In the 1940s, several researchers had proposed that the still-enigmatic template was composed of DNA, but this claim was greeted with considerable scepticism; DNA seemed such a small, simple chemical compared with the relatively vast protein molecules. Despite his preference for a simple system to experiment on, Delbrück went so far as to reject DNA as a 'stupid' molecule, much too simple to provide the basis for building a complete new organism.[28] However, in 1952, Hershey and his colleague Martha Chase made use of another technology from physics, radioactive labelling, to reveal what the templates were made of. They took advantage of the fact that the proteins which made up the phage's coat contained sulphur but no phosphorus, while the DNA contained phosphorus but no sulphur. They labelled some phage with radioactive isotopes of each chemical and then infected bacteria with their labelled phage. Their analysis showed that the protein coat remained outside the bacterium, while all the DNA in fact entered the cell. The bacterium's protein-synthesizing equipment was hijacked by the phage's DNA to make new phage, each containing a copy of the original phage's DNA. Thanks to phage, the picture was now complete: genes were DNA.

As is well known, the year after Hershey and Chase's work, two researchers in Cambridge – James Watson (a graduate of the phage course) and Francis Crick – worked out the chemical struc-ture of DNA. Their announcement of the famous double helix was made possible by the work of many other people: Rosalind Franklin and Maurice Wilkins at King's College London had done the X-ray crystallography which had suggested the helical form; Erwin Chargaff had performed the chemical analysis that showed that there were always equal amounts of the four key constituents that made up DNA: equal amounts of the bases adenine (A) and thymine (T), on the one hand, and of guanine (G) and cytosine (C) on the other. These exact matches proved crucial to understanding how the two strands of the double helix were attached to each other, and how it was possible for DNA to accurately copy itself; adenine always pairs with thymine, and guanine with cytosine, so that each strand of the helix is a complementary copy of the other. Phage and the Phage Group had provided the crucial evidence that phage and bacteria actually possessed genes – thanks to Hershey and Delbrück's experiments on phage resistance; they then demonstrated that those genes were composed of DNA. None of that diminishes Watson and Crick's achievement, but it is a useful reminder that – like any brilliant theoreticians – they could not have done it on their own.

Medical phage

The discovery of DNA's role in heredity made modern genetics possible, but it had some unexpected, even paradoxical effects. One might assume that Delbrück would have been delighted that 'his' phage programme had led to Watson and Crick's triumph, but in fact DNA was to prove the ultimate disappointment for him. Once the structure of DNA had been worked out, it became clear that it copied itself through a simple chemical process: it demanded no new laws of physics. Delbrück lost interest in phage, handing over the programme to his younger colleagues, and took up new biological challenges, still searching for the paradox that would generate radical new theories, but never finding it.

Another irony is that Felix d'Hérelle never played any part in the phage genetics revolution. He spent five years at Yale University, trying but failing to interest American doctors in the possibility of phage therapy, using phage to treat illness by attacking the bacteria that caused the disease. In 1933 he accepted an invitation to join a Soviet phage research institute in Tiflis

(Tbilisi). The economic depression in the West was forcing Yale to cut its funding for d'Hérelle's work (he had even made up the shortfall in his department's budget out of his own pocket). Meanwhile his Soviet counterparts received substantial state support, not least because the USSR was still suffering regular outbreaks of epidemic diseases such as cholera. In Soviet Union, he found himself, for the first time in his life, treated as a scientific star and surrounded by attentive, well-trained staff and servants (he even had a chauffeur), and all the modern facilities he could imagine. But – fortunately, as it turned out – he also maintained a private lab in Paris and spent every summer there. Although he was not a communist, d'Hérelle found the intellectual climate attractive: his neo-Lamarckian views on bacterial inheritance fitted in well with the prevailing Soviet dogma of Lysenkoism.

D'Hérelle was also disillusioned by the callous helplessness of the democracies in the face of the Depression and hoped that the USSR might be more active and organized in working to improve the lives of its citizens. He argued that it was crucial in studying disease to observe the 'natural host', humans:

> Because all illnesses studied by significant authors were 'artificial' illnesses (neither the rabbit nor the guinea pig are affected by cholera or typhus in the natural environment) they have bearing only when talking about the artificial illness and not at all practical for application to real, natural illnesses which occur in humans . . .[29]

He hoped that the resources of the USSR would finally allow progress to be made, especially as he saw it as a state run on rational principles which was thus unfettered by what he perceived as the prejudice against his ideas which had held back phage therapy elsewhere. He was to be disappointed. In 1937, the director of the Tiflis institute was arrested and shot; although he was largely uninterested in politics, he had somehow made an enemy of Lavrenty Beria, the head of Stalin's secret police. D'Hérelle was in Paris when he heard the news; he never returned to the USSR. However, the use of phage therapy remained widespread in the USSR and in other Eastern Bloc nations; it remained so after the war and many successes were claimed for it. D'Hérelle survived the war and the German occupation, and in 1947 received the belated recognition of being invited to lecture on phage at the Pasteur Institute, where he was presented with its medal (despite some opposition from within the institute). He never became interested in molecular biology and clung to his neo-Lamarckian beliefs until his death in 1949.

One might also have thought that the unravelling of how phage reproduced would finally have settled the argument about whether or not they were alive, but – perhaps surprisingly – Northrop and some of the protein chemists were still not persuaded. Northrop stuck to his guns, which could look like inflexible stubbornness, but is really a good example of how some scientific answers depend on the questions asked. Approaching phage as a chemist, using a chemist's tools and a chemist's understanding, they look chemical; but if you tackle them from a biologist's point of view, they seem biological. In a sense, both Northrop and Delbrück were right, and they were both wrong; some aspects of the way phage reproduce turn out to be very like the ways cells normally make proteins, which are in turn very like simple catalysis. Whether or not viruses can really be considered to be living is still an open question: it all depends on what is meant by 'living'.

The successes of the Phage Group inspired many geneticists to look for other small, simple, fast-breeding organisms that could accelerate their research. Meanwhile, working in a field alongside the phage workers at Cold Spring Harbor was a geneticist who wanted to slow things down, to take a large, complex, slow-breeding organism and adapt her work to its pace; her name was Barbara McClintock and she built her reputation by working on one the Americas oldest, most important crops, corn.

CORN

Zea mays

Chapter 9
Zea mays:
Incorrigible corn

Before Columbus, no one in Europe had seen a tomato or a chilli, and no Italian kitchen smelled of polenta. Maize (the golden-yellow corn of which polenta is made), tomatoes and chillies are all indigenous to the Americas. On 5 November 1492, within weeks of Columbus's landing on Cuba, two of his sailors brought their leader samples of 'a sort of grain they call maize, which was well tasted, baked or dried and made into flour'. As Columbus recorded, the local people called it *mahiz*, but he, like so many explorers, renamed it after something familiar that it somewhat resembled and called it *panizo*, the Italian word for millet.

In the Bible, Ruth says to Naomi, 'Let me now go to the field, and glean ears of corn' (Ruth, 2:2), and – inspired by this verse – Keats had Ruth standing 'in tears amid the alien corn' in his *Ode to a Nightingale*. But it was wheat that Ruth picked: the group of translators who produced the King James Bible used the familiar word 'corn', which simply meant 'grain'. Corn derives from the verbs for 'to grind' in some of Europe's oldest languages – it is from the Latin version, *granum*, that we get 'grain', a word that meant the same as 'corn' to the King James translators; they simply referred to anything that was ground. The word even included grains of salt, which is why salted beef is called corned beef.

Maize is American, and for many centuries after Columbus came across it, most of what the rest of the world knew of the plant had been learned from the indigenous peoples of the Americas, who valued and revered it.

Zea mays: Incorrigible corn

In the century after Columbus, the Spanish friar Bernardino de Sahagún wrote of the central role of corn in Aztec cookery and life, describing the tortillas and tamales filled with meat or beans that are still familiar in Mexican food today. Sahagún records an Aztec speaking of maize: '[it] is precious, our flesh, our bones. It is wonderful, marvellous, coveted, desirable – a coveted thing. I honour. I desire it. I venerate it, esteem it. I consider it with respect. I prize it.' He added, just in case the Spaniard had missed his point, 'it is our sustenance'.[1]

Columbus and the conquistadors who followed him took maize back to Europe, from where it went forth and multiplied. It became so common that the sixteenth-century herbalist John Gerard, not realizing that its origin was uniquely American, shoehorned the plant into one of the ancient categories familiar to him, and referred to it as 'Corne of Asia'. On the other hand, his contemporary Carolus Clusius, who ran the Imperial Botanic Garden at Vienna, was among the first to recognize that America contained a new flora, which included what he christened 'mayz'.

No indigenous American plant, not even tobacco, has matched maize's success in spreading itself across the planet; according to the UN's Food and Agriculture Organization, wheat is the only crop that covers more of the earth's surface, but by weight, even the world's wheat harvest is overshadowed by the more than 700 million metric tons of corn that was harvested in 2004. By the time the English were settling in North America, corn was well known in Europe; it was certainly familiar enough for the starving Pilgrim Fathers to know it when they saw it – and steal it. The *Mayflower*'s inhabitants found their first season at Plymouth Colony very hard until they discovered a heap of sand that had clearly been made by the Indians. The pilgrims dug it up with their bare hands and 'found a little old basket, full of fair Indian corn; and digged further, and found a fine great new basket, full of very fair corn of this year, with some six and thirty goodly ears of corn, some yellow, and some red, and others mixed with blue, which was a very goodly sight'.[2] And, as every American child learns at their first Thanksgiving, the Pilgrim's guide Squanto (also known as

Tisquantum), a Native American of the Wampanoag tribe, helped the Plymouth Colony to survive by teaching them how to grow and plant the corn they had 'found'.

Corn was crucial to every culture in the Americas. For the Iroquois it was, along with beans and squash, one of the 'Three Sisters' – 'those on whom our life depends'. White invaders recognized the indigenous peoples' reliance on maize and realized that burning their cornfields could be easier than fighting the people who farmed them. During the American Revolutionary War, one of George Washington's commanders, Major General John Sullivan, was sent on a bloody and futile campaign against the Iroquois and told his troops that it was important to burn the cornfields in order to prove to their enemies that 'there is malice enough in our hearts to destroy everything that contributes to their support'.[3]

Corn built the great civilizations of the Americas, but it also enabled Europeans to conquer them. Most of the indigenous cultivation, storage and cooking techniques were adopted by the settlers, including interplanting with the other 'sisters', beans and squash. Modern nutritionists have realized that the three complement each other perfectly: beans provide the proteins and vitamins that corn lacks, and squash provides vitamin A as well as fat from its seeds (people forced to subsist on corn alone develop the deficiency disease pellagra). The plants also grow well together: the squash vines choke the weeds and the beans twine their way up the cornstalks into the sunlight, but without shading the corn overmuch. By imitating local practice, the settlers survived, prospered and multiplied, pushing their way further and further into the continent, eventually reaching across it from ocean to ocean – all thanks to corn.

Corn became central to the Americans' sense of themselves. In 1766, during the agitation over the Stamp Duty that preceded the Revolutionary War, Benjamin Franklin wrote to a London newspaper to dispute an Englishman's claim that Americans would not be able to sustain their boycott of tea, since they would have nothing for breakfast. Franklin retorted that: 'Indian corn, take it for all in all, is one of the most agreeable and wholesome grains in the world . . . and that johny or hoecake, hot from the fire, is better than a Yorkshire muffin.' The Englishman's ignorance of corn would, Franklin felt sure, only 'strengthen us in every resolution of advantage, to *our* country, at least, if not *yours*'.[4]

Eighty years later, in the decades following the Civil War, corn's fortunes suffered as the railroads snaked their way across the land, eventually connecting the whole continent. Because the railways made long-distance grain shipping practicable, they opened up vast areas of prairie for farming, but initially it was wheat that was colonial America's national staple. During the 1870s US wheat production boomed, partly because crop failures in Europe had boosted world prices, making wheat hugely profitable. By 1884 US wheat production was five times what it had been in 1830, but by then, European farming had recovered and wheat prices were falling rapidly. The United States Department of Agriculture (USDA), which had been established to serve and protect the American farmer, set about encouraging farmers to diversify, and it promoted new and improved crop varieties of anything that was not wheat.

Corn was the big winner from the USDA's diversification programme. Between 1866 and 1900 US corn acreage tripled, while production quadrupled – by the end of the century, twice as much corn as wheat was being produced. 400 years after Columbus, the Corn Belt Exposition at Mitchell, South Dakota, featured a palace built of corn, which is still standing and attracts thousands of tourists each year. In 1893, the last great expo of the nineteenth century – the Columbian Exposition in Chicago – offered visitors 'the earth for fifty cents', while also providing a glimpse of what America had become and where it was going. Alongside the usual interminable displays of machinery and inventions and the palace of fine arts – stuffed with over 8,000 paintings – were zoos and balloons, a 'world's congress of 40 beauties', and Buffalo Bill's Wild West Show. Exhausted and hungry visitors must have been relieved to discover the 'Indian corn kitchen', run by a delegation of local Illinois women, where every dish was made from corn and, if visitors were not yet tired of maize by the end of their meals, they could even buy a souvenir recipe book – devoted entirely to corn.

Most corn was fed to animals, to produce meat, and a lot more was distilled into whiskey, or turned into popcorn, usually covered in salt or sugar. Some Americans disapproved of these uses: the meat, the sugar, the salt and – most of all – the distilled forms of corn. For some advocates of healthy living, these misuses of corn were causes of physical, spiritual and moral decline. In 1877 a book entitled *Plain Facts for Old and Young* urged readers to give up alcohol and tobacco, tea and coffee, as well as 'candies, spices, cinnamon, cloves, peppermint, and all strong essences'. According to the author, sugary or salty popcorn and corn-based bourbon were equally responsible for 'the solitary vice', 'the most dangerous of all sexual abuses', whose 'frequent repetition fastens it upon the victim with a fascination almost irresistible!' Consumers of both strong drink and supposedly innocent candy would find that they 'powerfully excite the genital organs', leading the sweet-toothed into temptation, so that 'with his own hand he blights all his prospects for both this world and the next. Even after being solemnly warned, he will often continue this worse than beastly practice, deliberately forfeiting his right to health and happiness for a moment's mad sensuality.'[5]

Plain Facts listed the devastating effects of masturbation on health, weakening body and mind so as to leave both vulnerable to disease. The book also outlined the causes of the practice; familiar ones such as 'pernicious literature' were joined by 'exciting and irritating food, gluttony'.[6] It was obvious to the author that 'tea and coffee have led thousands to perdition in this way'; the effects of tobacco were so strong that he doubted whether there could be any boy who smoked 'who is not also addicted to this vile practice'; but the condemnation of meat, spices, sugar and salt might be more surprising to modern readers.

However, readers of *Plain Facts* need not despair if, through some misfortune, they were suffering from the terrible addiction. For boys, the best cure was circumcision, which 'should be performed by a surgeon without administering an anaesthetic, as the brief pain attending the operation will have a salutary effect upon the mind, especially if it be connected with the idea of punishment'.[7] If that sounds bad, spare some pity for the girls: 'In females, the author has found the application of pure carbolic acid to the clitoris an excellent means of allaying the abnormal excitement.'[8]

It must have come as some relief to readers anxious to save themselves from physical and spiritual perdition, to turn from these 'cures' to the vital subject of diet. They learned that 'a man that lives on pork, fine-flour bread, rich pies and cakes, and condiments, drinks tea and coffee, and uses tobacco, might as well try to fly as to be chaste in thought'.[9] The would-be recovering masturbator was urged instead to 'eat fruits, grains, milk, and vegetables. There is a rich variety of these kinds of food, and they are wholesome and unstimulating.'[10]

The author of this stirring tract was John Harvey Kellogg, inventor of the cornflake, a wholesome and unstimulating breakfast cereal, which he hoped would save millions from the dire consequences of self-pollution. However, at the time it was written, cornflakes did not yet exist, so Kellogg recommended to his readers that 'Graham flour, oatmeal, and ripe fruit are the indispensables of a dietary for those who are suffering from sexual excesses'.[11] Graham flour was named after Sylvester Graham, an evangelist and temperance campaigner from whom Kellogg derived many of his ideas. Graham had been a popular and charismatic spokesperson for the Pennsylvania State Society for the Suppression of the Use of Ardent Spirits, urging that abstinence from alcohol and sex, a healthy diet (no tea, coffee, meat or tobacco), fresh air and exercise increased the body's natural resistance to disease. Central to his philosophy was the belief that the consumption of meat promotes carnal desires, which weaken the body. His alternative was Graham bread – homemade bread, from home-grown, stone-ground, whole-wheat flour. Abstinence from just about everything apart from brown bread was in effect Graham's philosophy.

Despite his presumed self-control, Graham died prematurely,

but his ideas were taken up by various health food enthusiasts and sanatorium directors, including James Caleb Jackson. Jackson found Graham bread too perishable for his sanatorium, so he developed a more practical alternative, Granula, the world's first cold breakfast cereal. It consisted of a twice-baked wholemeal biscuit ground into crumbs, which had to be soaked in milk or water to make it even vaguely palatable. Graham flour and Graham crackers were similar innovations, all of which would have horrified Graham, who believed in people growing and preparing their own food; no packaged product could supply either the exercise or the connection with the land and its seasons that true health required. But of course few Americans had time to bake their own bread, much less grow their own grain, so the prepackaged versions caught on. Among their supporters was Ellen Harmon White, a prophetess of the Seventh Day Adventist church.

White's prophetic revelations included sermons against masturbation, and she regularly quoted Graham and Jackson on the benefits of a healthy diet and abstinence. Eventually, God instructed her to open her own sanatorium in Battle Creek, Michigan, where the healthy food would be twice as effective because it was eaten in an appropriate spiritual atmosphere. However, the Lord did not reveal to her how to make the sanatorium pay until John Harvey Kellogg, the son of a local Adventist family, took over in 1876.

After studying both alternative and conventional medicine on the East Coast, Kellogg returned to Battle Creek and took over the sanatorium. He invented a healthy breakfast cereal based on dried and toasted wheat, oats and corn, with no added salt or sugar, and called it Granula, but Jackson sued him, so Kellogg changed the name to Granola. A few years later, further breakfast food experiments resulted in the accidental invention of the cornflake, first launched at the Seventh Day Adventists' General Conference in 1895. The flakes were initially made from whole kernels and did not prove very popular, but Kellogg's younger brother, William Keith, found that using only the 'grit' or heart of the corn and flavouring it with malt created a much tastier cereal. The Kellogg brothers were eventually expelled from the Adventist church by Ellen White for refusing to accept spiritual and financial guidance from the Church elders. In 1906, they set up the Battle Creek Cornflake Company, which later became the Kellogg company.

One of the Battle Creek sanatorium's patients, Charles William Post, had observed the Kelloggs' original experiments and launched his own brand of cornflakes, originally called 'Elijah's Manna'. Responding to complaints that the name was sacrilegious, he changed it to Post Toasties. Faced with this competition, young William Kellogg decided to start using the family name to promote their cornflakes: his signature was printed on every box alongside the slogan 'The Original Bears This Signature'. For John Harvey, such branding and naked commercialization was bad enough, but the brothers really fell out when William started adding sugar to the flakes. Despite falling out with the Adventist Church, John remained preoccupied with health, and specifically with the stimulating ill-effects of inappropriate diet. Among his later books on the subject was the gloriously titled *Man the Masterpiece, or Plain Truths Plainly Told About Boyhood, Youth and Manhood* (1906), a heartfelt tract against the evils of masturbation, and in favour of his sugar-free, cornflake-based cure.

In contrast to his older brother, William Kellogg seems to have been more concerned with self-enrichment than self-abuse. The two brothers fell out over both ideals and money and sued each other over the ownership of the idea of cornflakes. William – like C.W. Post – died a millionaire. Ironically for John Harvey, the processed food industry has become one of the major consumers of corn; it is used – whether in the form of sweeteners, starch, oil, meal or syrup – in almost every junk food, from potato chips to soups, from mayonnaise to peanut butter.

Despite his disappointment over the contamination of his invention, John Harvey Kellogg lived to be ninety-one years old. As his obituary in the *New York Times* observed, he 'was perhaps the best example of the truth of his own dogmas'.[12] He had also helped make corn even more central to the American diet. By the beginning of the twentieth century, Americans were consuming corn in dozens of different forms, from breakfast to bedtime. Much of the meat they ate was corn-fed; corn products were found in everything from plastics to embalming fluid, from gunpowder to baby powder. During the First World War, corn cellulose was even packed between the inner and outer hulls of warships: because it expands enormously when it gets wet, it acted as a sealant when a hull was breached.

Corn had become America's most important crop and, for many Midwesterners in particular, its national symbol. In 1885 the popular poet Edith Thomas proposed that maize should be formally adopted as the USA's national floral emblem, since 'a single full-grown plant of Indian corn, though but a fleeting, annual growth, possesses presence and dignity no less than does the oak itself'.[13]

However dignified corn might be, it still had to be husked or shucked, removing the outer leaves from the ear ready for eating or storage. This was tedious work, and from early colonial days the custom developed of husking bees, which turned shucking from a time-consuming chore into a communal celebration. Friends and neighbours came from miles around to be divided into two teams which competed for an appropriate incentive, such as a jug of corn liquor, buried under the pile of unshucked corn to await the winning team. Picnics, games and contests accompanied the bee and as the corn whiskey circulated, the shucking songs could become increasingly bawdy, spurred on by the vaguely phallic shape of the corn cobs. When people still lived on isolated farms, husking bees became significant social events, enabling farmers and their families to get together. News was exchanged, friendships made and renewed; and of course many a romance began, or was broken off. And corn leant Cupid a hand; now and again as a husk was removed, a red ear of grain would appear amid the white or yellow rows. The red kernels, sometimes known as pokeberry ears, entitled their finder to demand a kiss from whomever they had their eye on. The red ears were on occasion surreptitiously recycled and reused.

Edith Thomas quoted Longfellow's celebration of this custom (in his 'Hiawatha'):

> And whene'er some lucky maiden
> Found a red ear in the husking,
> Found a maize-ear red as blood is,
> 'Nushka!' cried they altogether
> 'Nushka!' you shall have a sweetheart!

Thomas asked, 'Has the botanist an explanation of this anomaly?'[14] In 1885, the answer was still 'no', but attempts to find an explanation would eventually transform every cornfield in America.

A variable feast

By the second half of the nineteenth century, the millions of acres of corn, such a source of national pride to some, had become something of an embarrassment to others. As American agriculture surged ahead, raising yields and improving varieties, corn remained much the same. The basic varieties – such as dent, flint, sweet, waxy and pop (named for the particular qualities of their kernels) had all been known and cultivated before colonists arrived in America. 400 years of farming by an allegedly superior race had produced virtually no improvement. Corn was still Indian corn.

Corn shows were held to encourage farmers to improve their maize; the best ears were judged by size and colour, but most of all by uniformity, the similarity of the kernels to one another. The shows were also social events and over time corn-showing became a craze that nearly rivalled the Dutch tulipmania of the seventeenth century; a grand champion ear of corn could fetch $150. The competition became so intense that a little skulduggery was not unheard of: metal rods could be driven into corn-cobs to increase their weight and some show judges began to X-ray the ears to reveal them. Others doctored their corn, carefully removing defective kernels – including pokeberry ears – and gluing new ones in their place to make the rows more regular and even.

Organizations like the Illinois Corn Breeders' Association drew up lists of desirable features for show corn and even printed official score cards to guide the judges. Comparable shows had already produced better dogs and pigeons; selecting the best of one generation from which to breed, be it dogs or corn kernels, was, as we have seen, the standard way to improve the quality of a breed. It was the power of precisely this kind of selective breeding – which Darwin called artificial selection – that was the basis for the idea of *natural* selection.

Corn shows should have produced better corn and richer farmers, but somehow they did not. Because champion ears were so precious they were seldom planted, but when they were, yields were not particularly impressive. Some growers suggested that chasing prizes for fancy corn was a waste of time – unless the actual yield itself was judged, the judging criteria might even be lowering yields.

Some of the government's scientific breeders even began to campaign against the shows, arguing that it was time for science to investigate what would really improve corn. William James Beal, Professor of Botany at Michigan Agricultural College, was one of the first to criticize selection as a means of improving corn. His opposition was not anti-Darwinian – quite the contrary: he had learned his botany at Harvard from Asa Gray, a close friend of Darwin's and one of his earliest and most passionate American supporters. It was through Gray that Beal learned of Darwin's experiments with fertilization in plants. As we have seen, Darwin had spent years trying to ascertain whether cross-fertilization made plants more vigorous and had concluded that for the most part, it did, especially if the varieties were not too closely related. Among the plants Darwin had worked on was maize, and even before Darwin's book on the *Effects of Cross- and Self-Fertilisation in the Vegetable Kingdom* had appeared, Beal began his own experiments in what he called 'controlled parentage'. He planted flint and dent corn varieties together, but he 'de-tasselled' one – removing the male flowers before the pollen ripened – to ensure it could only be pollinated by the other. This simple technique remained the basis of maize hybridization for many decades.

Beal's experiments did indeed result in more productive varieties. He became a vocal critic of the selection based purely on the appearance of the ears that was encouraged by corn shows, urging farmers instead to de-tassel any plants with unfavourable characters to prevent them being passed on. His desire to see inferior corn 'castrated' was gently mocked by the science writer Paul de Kruif in his follow-up to *The Microbe-Hunters*, a book called *Hunger Fighters* (1929). He describes Beal, aptly, as 'the first fanatic for corn eugenics', and in de Kruif's view, 'nearly as foolish as modern folks who without humour advocate picking out human fathers by science instead of letting nature do it'. As Beal went round to farms, trying to persuade farmers to sterilize any inferior corn stalks, the good folks of Michigan listened respectfully, but could not help wondering what 'this tall loon of a professor was raving about'. How could Beal have imagined that busy farmers had time to spend their days pulling the tassels off every stalk that looked a little less than perfect?[15]

De-tasselling corn was simply too time-consuming for most farmers or commercial corn-breeders, but the US government was pleased to pay scientists to do it, so the idea of breeding pure lines of corn was investigated at government research stations and in agricultural colleges. And as increasing numbers of researchers studied maize, they realized that the anomalous red ears were worth much more than a kiss: they were in fact the key to under-standing and ultimately to improving corn.

As the Pilgrim Fathers had noted, back in the 1620s, the kernels of what they called Indian corn varied in colour, 'some yellow, and some red, and others mixed with blue'. These mixtures had attracted breeders to experiment with corn, long before anyone knew what was responsible for the coloured kernels. Thomas Andrew Knight, who as we have seen did early experiments on plant hybridization, had tried using corn precisely because the different coloured kernels on a cob were a clue to the plant's pedigree. Many plant-breeders were intrigued by the tall American plant; Mendel himself did experiments with corn. And when Darwin's friend and botanical correspondent John Scott, who had collaborated on the passionflower experiments, asked the older man to suggest a species he might investigate, Darwin suggested maize, as 'such experiments would be pre-eminently important'.[16]

Because of its long history of use in breeding experiments, even in Europe – where it was not of great agricultural significance – maize was to play an important part in the early days of Mendelism. When Hugo de Vries first became interested in applying statistics to biology in the 1870s, he began by checking that normal distribution really did apply to plants as well as to animals. He counted the number of rows on cobs of corn and plotted the results on a graph; sure enough he discovered that a bell-shaped curve did indeed appear. He continued to grow maize even after he became interested in Oenothera. While he was investigating corn fertilization in 1899, de Vries crossed white and yellow corn and counted the kernels in the next generation: the result was 3,176 yellow and 1,082 white maize seeds – the significance of the 3:1 Mendelian ratio would leap out at a twentieth-century geneticist, but de Vries did not consider the possibility that there might be a statistical law behind his numbers, it was only after he had read a copy of Mendel's original pea paper that he registered the crucial underlying pattern. Another of Mendel's 'rediscoverers', Carl Correns, was also experimenting with corn, trying to establish the rule behind the mysterious multicoloured cobs.

Yet despite two centuries of accumulated expertise, maize still proved recalcitrant, especially for American plant breeders and geneticists, whose interests in the plant were more pragmatic than those of Europeans. How were American farmers to feed a rapidly growing population? New immigrants were pouring into the country; they were widely welcomed as vital additions to

America's workforce – and as consumers of the goods that American factories were producing in ever-increasing quantities. But what were they to eat? Would every inch of the country have to be ploughed up to feed America's growing population? Increasing productivity – producing more corn from the same number of acres – was the obvious answer; it would also involve substantial profit for American farmers: in 1908, the US Secretary of Agriculture observed that 'the value of this crop [corn] almost surpasses belief. It is $1,615,000,000', enough 'to pay for the Panama Canal and fifty battle ships'.[17] Small wonder the USDA's network of agricultural experiment stations were set to work raising yields to feed America's population, to reduce dependence on imports and to open up new export markets.

But the very factor that made maize in interesting subject for experiments, its unpredictable variability, frustrated those with more practical goals in mind. One reason is apparent in one of the first English descriptions of the plant, written by the seventeenth-century English botanist Henry Lyte, who described maize in his *New Herbal* (1619). Lyte noted that it was 'a marvellous strange plant, nothing resembling any other kind of grain; for it bringeth forth his seed clean contrary from the place whereas the Flowers grow, which is against the nature and kinds of all other plants'. While at the same time, 'at the highest of the stalks, grow idle and barren ears, which bring forth nothing but the flowers'.[18] What intrigued Lyte was that maize produced 'flowers' (the tassels) at the top of the plant, while the corn-cobs (bearing the edible seed), rather than growing in the vicinity of the flowers, as in most plants, appear much further down the plant. Maize has separate male and female flowers and, as botanists had long realized, that represented a crucial difference between maize and other grains: most of the common European cereal crops, such as wheat, oats and barley, grow male anthers (which produce pollen) and female ova (which develop into seeds once fertilized) on the same flower.

The maize tassels – at the top of the plant – are the male flowers, which produce pollen but no seed. The corn-cobs, which form lower down the plant – away from the tassels – each bear a tuft of slender threads; shop-bought corn that is still in its husk (the next best thing to home-grown) usually still has its silks poking out from the top of the cob. These long, thin structures connect the stigma (where the pollen lands to fertilize the flower) with the plant's ovary; in maize, each silk is covered in fine hairs that trap the pollen. Corn's ears are its female flowers; once fertilized the tiny ova swell and grow into the kernels. Corn produces large quantities of extremely light pollen – because the slightest breeze blows it away, it very rarely falls on its own silks, so corn is invariably cross-pollinated, unlike most cereals, which fertilize themselves.

Crops like wheat are well-behaved, monogamous plants; self-fertilization is, as Woody Allen once almost observed, sex with someone you really love – it certainly involves displaying about as much fidelity as a sexual species is capable of. By comparison, corn is wildly promiscuous – each corn-cob can have several male parents. That is what causes the pokeberry ears – a grain of pollen from one of the naturally red-kernelled varieties has wafted in and landed on the silks. A single corn-cob can be a rich genetic mixture.

Corn's pollen is so light that it has been known to travel more than five miles on a puff of air. The eighteenth-century farmer Cotton Mather (the notorious Salem witch-hunter) observed in 1716 that his neighbour's field was planted with yellow corn, with one row of red and blue corn near the edge. The yellow corn produced some red and blue kernels, and there were more of these on the side of the husk on which the prevailing wind blew (because more pollen blew in from that direction). It is almost impossible to keep stray pollen out of a cornfield – there may be no red corn for miles around, but pokeberry ears will still pop up occasionally.

However, while corn's footloose flowers make it hard to create or maintain a pure-breeding line, the separate male and female flowers have one great advantage for the breeder; they make it easy to cross-breed different strains, making Beal's 'corn eugenics' possible. Cereals like wheat have the tiny, tuft-like flowers that are typical of grasses – cross-breeding them involves carefully opening up each ear of wheat and 'castrating' it: removing the anthers before they can ripen and produce pollen. Even Mendel's troublesome hawkweeds are easy to work with by comparison, so cross-breeding strains of wheat is utterly impractical in a farmer's fields. Corn is different.

Corn eugenics

The eccentricities of corn also intrigued George Harrison Shull, who – as we have seen – had worked with Daniel MacDougal trying to artificially encourage Oenothera to mutate. Shull had grown up on a farm in Ohio. He had in fact grown up on a series of farms because his parents were too poor to buy land and were continually forced from one rented farm to another, pursued by irate bankers and bad debts. Their son never managed to complete a full year of school, because his father was always keeping him home to help tend the corn. Yet Shull did not develop an abiding resentment of maize, he managed to get himself into Antioch College, and then paid his way while there by working for the college. When he could not get a position as custodian of the laboratories or libraries, he was willing to rise before dawn to light the college's furnaces, perhaps reflecting as he did so on his seemingly prestigious job title, 'engineer in charge of the steam-heating plant and water works'.[19] After graduating from Antioch, Shull worked for the US National Herbarium (part of the Smithsonian Institution in Washington, DC) before being transferred to the USDA's Bureau of Plant Industry. He started graduate work at the University of Chicago, working with Charles Davenport, and after gaining his PhD in 1904 he went to Cold Spring Harbor, which is where he did the Oenothera work with MacDougal.

Despite his time at the USDA, Shull was not really interested in the practicalities of plant-breeding. He was 1,000 miles from the corn belt and he 'had no notion whatever that the prosperous heart of the American land would ever need anything but the energy of its sweating men'.[20] Like Mendel and so many others before him, Shull worked on corn because it was attractive as an experimental subject.

In *Hunger Fighters*, De Kruif describes Shull as 'a scientist purely, with his head in the theoretical clouds, mooning over such abstruse problems as the Galtonian Regression in Pure Lines – whatever that means . . .'[21] What that meant was that Shull had learned to apply statistical methods to maize breeding by studying with Davenport, who was a biometrician. Biometry was an obvious approach to take when dealing with a large field of plants each producing several ears, each of with can have several male parents; statistical analysis is the only way to handle the resulting enormous quantities of data. Because of his statistical training, Shull was aware of Galton's ideas, especially of regression to the mean, the tendency of offspring of even exceptional parents to be closer to the average than either parent, which, as we have seen, was one basis of de Vries's objections to natural selection. And while Shull was completing his PhD he read about some experiments that appeared to prove that the kind of selection Darwin advocated – the selection of small, everyday variations which was used to improve crops – simply could not create a new species in the way that Darwin had envisaged.

The experiments in question had been performed on beans by the Danish botanist Wilhelm Johannsen. He had taken a field of bean plants and saved only the largest and smallest beans from the crop. He then planted them in two separate fields and collected the beans that were produced. With each successive crop, Johannsen crossed the plants that produced the largest beans with each other, doing the same with the smallest ones. As he had expected, in the first few seasons the large-bean line kept pro-ducing the biggest beans, and the beans got steadily bigger each year; meanwhile the beans from the small-bean line got steadily smaller. But then both lines seemed to simply run out of steam. The beans stayed the same, generation after generation. This was worrying for a Darwinist, since Darwin's theory seemed to require that new variations should keep occurring, so that the big-bean line should continue to produce ever-bigger beans (and the same for the small-bean line), which would allow continued selection over many generations to eventually turn the two varieties into entirely distinct species. But Johannsen's experiments seemed to show that the variation was simply exhausted after a few generations; if selection merely preserved novelty, but could not generate, how did new kinds of plants or animals arise?

Johannsen's 'pure lines' were, of course, part of the evidence that convinced many people that natural selection could not work, at least not on the kind of small, everyday variation found in a species. De Vries was among Johannsen's admirers and argued that Johannsen's pure lines were exactly the same as his own 'elementary species'. Shull reviewed Johannsen's claims and concluded that 'if sustained by further research' they would certainly constitute an important new principle.[22]

Johannsen also introduced the term 'gene' and gave genetics two very useful terms: 'genotype', which refers to the full set of genes an organism carries; and 'phenotype' which refers to its external features, everything from size and colour to behaviour. As we saw from Mendel's original pea experiments, a pea plant with yellow peas (the yellow phenotype) might have either two copies of the yellow version of the gene (the yellow allele), or one yellow and one green; from the phenotype alone, it was impossible to say which. While a plant with green peas must have two copies of the green allele, because green is recessive to yellow.

Johannsen's work created considerable interest: it influenced William Castle and his students in deciding on their hooded rat experiments. Shull decided to explore whether Johannsen's results could be repeated in other plants. He chose maize because each variety had a characteristic number of rows of kernels on each cob; Shull thought that crossing a variety that had, for example, ten rows with one that had eight and then counting the rows in the hybrids would enable him to see if and how the two varieties blended. The first step was to inbreed each variety, to ensure he had a pure-breeding type with the same fixed number of rows on every plant. He applied Beal's methods, which had become a standard approach for producing experimental hybrids. The male tassels and female silks were covered with cloth or paper bags to prevent accidental pollina-tion. Once the silks were receptive, the botanist could collect the pollen and apply it to the plant's own ears – where it would normally rarely fall. One maize researcher recommended that when applying the pollen, it was best to 'raise an umbrella and hold it in such a way as to keep all flying pollen from the ear, remove the bag, and apply the pollen until the silks are almost hidden'.[23]

Shull laboured away, making – to quote de Kruif – his 'high-brow paper-bag marriages of corn'. As he inbred the different strains to produce pure lines with unvarying numbers of rows, Shull found that each line did indeed become more and more uniform. Just as with the size of Johannsen's beans, maize's variability gradually disappeared. After a few generations of paper-bag weddings, every cob looked exactly like every other cob in the inbred population. That was the good news. The bad news was that the cobs were tiny, the 'runtish offspring' of an unnatural union, the 'ill-begotten children of [an] incestuous marriage'.[24] The kernels were equally stunted and the yield from each inbred field was extremely poor, but none of this mattered to Shull. He was not planning to sell his corn crop, he only wanted a uniform starting point for his experiments. Once his corn was all the same, however undersized, Shull assumed that he had isolated pure lines comparable to Johannsen's.

With his experimental materials ready, Shull proceeded to his main experiment: crossing strains with different numbers of rows: would crossing eight-row corn with ten-row result in a nine-rowed hybrid? Surprisingly, it did not – all the offspring were eight- or ten-rowed, like their parents – but the real surprise was how *many* offspring were produced. Instead of stunted runts, his fields were suddenly full of tall, vigorous plants, laden with huge healthy corn-cobs. The hybrids were as uniform as their parents, but the yield per acre was many times higher. Just as Darwin had found, hybrids were more vigorous than their parents, but Shull found maize to be an even clearer example of the phenomenon than Darwin's; he christened it heterosis, or hybrid vigour – the same vigour Sewall Wright had observed in his guinea pigs.

When he published the results, Shull noted corn's resistance to the usual breeding methods, because the inbreeding, which was normally used to maintain the stability of an improved line, 'results in deterioration'. What his generations of incestuous corn marriages had revealed was that 'an ordinary cornfield is a series of very complex hybrids'. What the seed merchants sold as a single variety was in fact a mixture that had been 'produced by the combination of numerous elementary species'.[25] His use of de Vries's term 'elementary species' was no accident: Shull was convinced that just as Johannsen had broken his bean varieties down into their basic building blocks, he had done the same for corn. As de Kruif put it, characteristically, 'In his mind, at least, [Shull] has the right to dance a fandango. He has uncovered the composition of a field of maize. George Shull has taken maize – apart.'[26]

Shull realized immediately that his dissection of maize had significant practical implications; here, he argued, was finally the key to the improvement of maize. It was a two-step process: first inbreed, to fix the trait you want; then cross-breed, to combine desirable traits and reap the benefits of hybrid vigour.

But corn was not to prove quite so tractable: breeders found they only hit the hybrid jackpot in the first generation; if farmers saved some of their seed to plant the following year, as they had always done, the hybrid vigour soon evaporated. After a couple of generations, the descendants of the tall, strong hybrids were no more productive than their ancestors had been before all the laborious business of inbreeding. To maintain the hybrid benefit, farmers would need to maintain a couple of fields of inbred strains and then re-hybridize every year to generate new hybrid seed for their crops. That meant a lot of work and land spent growing unproductive corn.

Much like Beal had done before him, Shull travelled extensively through the US Corn Belt promoting his technique. Once again, the corn farmers listened politely and laughed at him once he had moved on. What farmer had the time or energy to mess around with paper bags and umbrellas in the middle of the summer, and who could afford to cultivate fields of useless dwarf corn, year in and year out? Shull convinced the farmers of the attractions of hybrid corn, but the gains in productivity were simply not dramatic enough to persuade them to adopt his techniques.

However, Shull had found a more receptive listener when he had presented his early results to the newly established American Breeders' Association in 1907; a young plant-breeder from Illinois called Edward Murray East.

East was the son of an engineer and came from a family with a strong interest in science. After high school, he eventually went to the University of Illinois to study chemistry. He was a student when the 'rediscovery' of Mendel's laws was announced; by the time he had finished his PhD, in 1907, he had been won over to the new science of genetics. To pursue his interest in plants, East took a job at Illinois's agricultural experimental station, which was – as might be expected in a Corn Belt state – devoted to research designed to improve maize. The station's breeders were trying to produce a maize variety that would make better animal feed, by selecting strains with more protein and less oil. East performed the chemical analyses on the new strains and observed – just as Shull was doing at the same time – that the inbreeding that was used to keep the strains pure was also reducing yields. He suggested investigating the effects of inbreeding more fully, but his boss objected that 'we know what inbreeding does and I do not propose to spend people's money to learn how to reduce corn yields'.[27]

Illinois tax-payers would probably have agreed with East's boss that investigating how to reduce corn yields was no business of government researchers, but East had other ideas. He knew many farmers and was quite familiar with the practicalities of corn-growing; in one of his official reports he mentioned the need to avoid importing 'any of the rodents when bringing in the crop', while observing that if a few did get into the silo, 'a good cat is an aid not be despised'.[28] But East was as familiar with Mendel's factors as with farm cats, so he left his home state and moved to the Connecticut Agricultural Experiment Station, intent on further experiments with inbreeding.

East arrived in Connecticut with a definite agenda: he was going to apply Mendel to maize, to replace generations of rule-of-thumb corn-breeding lore with clear scientific evidence that would persuade American farmers that corn eugenics could be both practical and profitable. He attacked the corn shows and their scorecards, arguing that there was no correlation between handsome ears and good yields; many of the prize-winning ears were, he argued, 'valueless as seed corn, although they were large in size and beautiful in appearance'– they simply would not grow.[29]

However, before applying Mendel's laws to corn, the genes involved needed to be identified. The difficulty of improving corn was only partly to do with creating pure-breeding lines; the second problem, as East soon realized, was that a farmers 'yield' was not a straightforward single gene issue like pea colour. Overall yield was clearly a product of many different genes, each of which controlled different aspects of the plant and interacted with each other. Some regulated stem height, others cob size, some produced more sugar, while others resulted in different kernel colours. Selecting for all these at once was clearly impossible, but Johannsen's pure lines suggested a solution. Select for one factor at a time. Isolate it by inbreeding and then combine it with other useful factors to produce plants that had several useful qualities, such as disease resistance or high productivity, and which had the added bonus of hybrid vigour.

Both East's claims and his evidence were impressive, but once again his ideas proved impractical. He proposed that farmers keep track of the pedigree of every single plant on their farms, to make sure that they were only growing the offspring of the very best performing plants. Once again, this was all much too complicated and time-consuming for most farmers.

So when East heard Shull's talk at the American Breeders' Association, he was well-prepared to understand what he heard. In fact, he was kicking himself: he had done virtually identical experiments, but had not gone on to perform the kind of sophisticated, scientific analysis that Shull presented. East would later 'wonder why I have been so stupid as not to see the fact myself' – the fact that crossing produced hybrid vigour which overcame the problem of low yields caused by inbreeding. But he swallowed his pride, spoke to Shull at the meeting and managed to borrow a copy of the latter's still unpublished paper, which he then drew on in his report of his own experiments. East was careful to acknowledge his debt to Shull, who had, he stated, come up with 'the correct interpretation of this vexed question'. But, East continued, while Shull's concept of hybrid vigour was 'clearly and reasonably developed', it 'was supported by few data', which is why it had not made much of an impact. East suggested that his own experiments would supply the evidence Shull needed.[30]

Shull was a little annoyed that he had not been able to finish working out his ideas before East published, but he grudgingly acknowledged that 'the matter seemed to me to be of too great importance in view of the value of our maize crop to selfishly keep it to myself any longer'. The friction between the two researchers arose in part from the fact that East knew more about the realities of farming and had a tendency to accuse Shull of being impractical. He wrote to Shull: 'I wish you could have a little experience trying to get the farmers to take up anything in the least complex, and I know you would agree with me that only the very simplest things can be done by the corn grower.'[31]

The problem of how to make practical use of the benefits of hybrid corn took some time to solve. Shull returned to his laboratory and his evening primroses and never worked on maize again, but East and his students wrestled with the problem that the inbred strains were worthless in themselves, which made it hard to justify the time, effort and land involved in growing them. They also faced a second difficulty, which was that inbred strains produced fewer kernels, and of course fewer kernels meant fewer plants, so it was hard to produce enough hybrid corn seed for commercial growing. In 1915, one of East's students, Donald F. Jones, finally solved this problem. His solution was to cross the maize twice: he started out with four inbred lines, each of which was shrunken and sometimes barely fertile, but which had been selected for one specific trait, such as drought resistance, that would be valuable to the farmer. Each inbred line was crossed with one of the others, producing two different batches of hybrid corn. Each of these hybrid strains combined two of the desired features and also exhibited hybrid vigour, producing lots and lots of seeds. And lots of seeds meant lots of plants. So, when the two hybridized strains were crossed with each other they produced seed corn in much greater quantities, making it much cheaper – yet it still proved almost as vigorous (and sometimes even more so) than the parental strains; depending on the traits that had been selected for, the resulting hybrids were sweeter, more disease- and drought-resistant, and produced much higher yields. Not long after Jones went public with his method, hybrid corn production began to take off in the United States: by 1933 it was in large-scale commercial production, and by 1950, 75 per cent of all US corn was hybrid.

Maize misbehaving

As hybrid corn spread rapidly across the United States and then around the world, it seemed as though, after centuries of proving un-cooperative, corn had finally been rendered tractable. But it turned out that corn's variability was still not fully understood – it had a final confusing trick up its genetic sleeve and it would take one of the century's most brilliant and unorthodox geneticists to fully understand why maize misbehaves.

Growing maize had been women's work. Native American women did most of the work of farming while men hunted; early European settlers sometimes referred to the plant as 'squaw corn'. This pattern continued during pioneer days, with at least half the work of raising corn being done by women and children. Yet by the time a young woman called Barbara McClintock left her Brooklyn high school in 1919 and told her parents that she wanted to go to college to study science, maize had become men's work – there were certainly few women working at the research stations, teaching at the agricultural colleges or researching in universities. McClintock's mother had always supported her daughter, accepting her wish to wear boys' clothes and play their games. A woman neighbour who had scolded young McClintock for being unladylike received an irate telephone call from her mother, ticking her off for interfering. Yet, when the daughter wanted to go to college, her mother baulked; she argued that there were no jobs for college-educated women, and probably worried that her over-educated daughter might have difficulty finding a husband. Fortunately for McClintock (and, as it turned out, for twentieth-century genetics), her father took a different view and she went to the College of Agriculture at Cornell University.

Cornell suited McClintock. The university welcomed women students (at least, to a greater extent than most others of the time) and she was also lucky to find that Rollins Adams Emerson, who ran the genetics course, had grown up on a farm in the Midwest. He knew that women were more than capable of the manual work involved in corn-growing; his teenage daughter often helped out during the pollination season. Although McClintock was the only woman on the genetics course Emerson, unusually for the time, treated her no differently from his other students.

In 1900, Emerson had been a young assistant professor of horticulture at the University of Nebraska when he had got caught up in the excitement of Mendel's 'rediscovery', just as East had; and, like East, Emerson also began to study heredity, originally in beans. He only became involved with maize by accident when he gave his students what ought to have been a simple exercise, re-creating Correns's maize experiments, which involved crossing two corn strains – one that had starchy kernels and one that was sugary. None of the class came up with the 3:1 ratio of dominant starchy to recessive sugary that they were supposed to. Emerson assumed his students simply could not count, but when he checked their results, he had to acknowledge that – to his embarrassment – they were right, there were far too few sugary kernels. He began to investigate, no doubt imagining that he would spend at most a few weeks rechecking the results before returning to his beans. Weeks turned into months and he still could not figure out where the problem lay, but he was becoming more and more interested in maize. By the time he solved the problem, many years later, he was hooked. It turned out that he had inadvertently given his students a variety of popcorn (the name for those maize varieties whose kernels 'pop' when heated – and are thus suitable for making popcorn) that had an unusual linkage between the gene for sugary kernels and one of those that controlled a key aspect of pollination; the link had thrown the ratios off.

The still unresolved question of whether apparently blending characteristics obeyed Mendel's laws led Emerson to cooperate with East on further experiments with row numbers, to assess the extent of blending.[32] Like East, Emerson was well aware of maize's agricultural importance, but he had even bigger ambitions for the plant. He believed that corn would allow botanists to do genetic experiments that even Morgan and his students had not proved capable of: demonstrating exactly how genes actually worked.

The fly experiments proved that genes existed and that each had a specific function, but Emerson – like his fellow geneticists – wanted to know exactly how genes worked. The colours of corn kernels, just like those of flowers, are produced by chemicals – pigments – that the plant manufactures. Emerson was convinced that understanding exactly how corn pigments worked would reveal the biochemical link between the plant's genes (its genotype) and its physical structure and appearance (its phenotype). Investigating coloured corn kernels could finally uncover a defined chemical connection between the genes and the living plant. Emerson began promoting *Zea mays* enthusiastically, sharing his commitment, his results and his seeds with anyone who was interested; as he told one colleague, he was anxious 'to keep up the genetic end of the corn game'.[33]

One of his students would later describe Emerson as 'the spiritual father of maize genetics', who – as with those who founded similar research communities – helped inspire a remarkably friendly and cooperative maize community, largely because he was so 'completely unselfish, truly an honorable man', inspiring others to follow his example.[34] Among the recipients of his generosity was Donald Jones, working with East in Connecticut. But what Emerson did most of all was to try and pass on his passion for corn to his students; and although Barbara McClintock only took a couple of courses with him, she came to fully share his dedication to maize.

Throughout her life, McClintock liked to stand out. She was one of the first women at Cornell to wear her hair in a short bob – well before it was fashionable. She rejected the anti-Semitism of some of her fellow-students and cultivated the friendship of Jews, even learning to read Yiddish. In the evenings she played banjo in a jazz combo, attracted by the improvisation and idiosyncrasy of jazz, which suited her sense of herself as an iconoclast.

McClintock became a key member of the Cornell maize group, along with George Beadle and Marcus Rhoades. McClintock, Beadle and Rhoades formed a tight-knit group – perhaps they had to, since their shared conviction that they were way ahead of most of their fellow students is unlikely to have won them many friends. Beadle later became interested in other organisms and went on to win a Nobel Prize for his work on enzymes. They all remained friends, but it was Rhoades and McClintock who remained loyal to maize throughout their lives.

Early in her student career, McClintock discovered that she was a talented microscopist. If you have never tried to use a microscope in a biology lab, this might sound trivial, but it is much harder than it looks. Learning to prepare and stain specimens, adjusting the instrument to get the best light and interpreting what you see are all skills that take time and patience. In every biology class there are some students who pick up microscopy quickly, and some who do not, just as some are better at dissection or at handling live animals. McClintock soon found that she was not simply competent with a microscope, she was exceptional. Throughout her life, she could identify microscopic structures clearly that others found almost impossible to interpret. One of her students remembered that asking McClintock to look at something under a microscope was always done with some trepidation: she would invariably criticize some aspect of the way in which the instrument had been set up, but if she did not immediately feel the need to make some small adjustment 'you felt triumphant relief'.[35]

Thanks to her exceptional microscope skills, McClintock was the first to identify the individual chromosomes of maize. In doing so, she made use of a new technique for preparing microscope slides. Anything that is to be examined under a microscope has to be very thin – thin enough for the light to shine through it. When McClintock arrived at Cornell, the standard way of preparing specimens was to cut them into wafer-thin slices, called sections. This certainly worked well, but in this case it also cut the chromosomes into lots of small pieces. Examining one particular chromosome meant laboriously tracing it across many different sections, each on a different slide. McClintock heard of a new technique and applied it to maize for the first time. It was known – with refreshing simplicity for a technical, scientific procedure – as 'the squash': you took some cells, spread them out on a slide, stained them, and then squashed them with your thumb. That made them thin enough to look at, but it also kept the chromosomes in one piece.

The squash sounds simple – and it was – but it produced a breakthrough. Before it came into use, maize chromosomes were so difficult to identify that East had given up working on corn. However, the squash – together with new stains which enhanced the details of a specimen – made working with maize cells significantly easier; McClintock became the first scientist to identify all ten maize chromosomes. That provided the basis for a collaborative effort to work out which genes lay on each chromosome, just as had been done for Drosophila. Emerson launched the collaboration at a meeting he called the 'cornfab'.

By squashing instead of sectioning, McClintock found that some chromosomes had curious lumps at their ends, which she called knobs. In the early 1930s, she and her student Harriet Creighton used these and other peculiarities to identify each individual chromosome. That way they could observe what happened when a full set of chromosomes divided to create the half-set that would be passed on in the pollen or ova. They were the first people to observe, clearly and unambiguously, that before the half sets were created, the chromosomes really did cross over. The two sets that every normal maize cell carried (one from each of its parents) physically exchanged segments of one chromosome for those of the matching partner, creating a unique new combination of their parents' genes. As we have seen, Morgan's fly boys had reasoned that this must happen from the way in which genes on the same chromosome sometimes became separated. But the individual Drosophila chromosomes were too small to identify, so the deduction could not be checked directly. McClintock and Creighton produced the evidence that connected the way genes were expressed in the plant or animal's body – making, for example red eyes or red corn kernels – with the physical behaviour of the chromosomes.

Thanks originally to Emerson's vision, McClintock and Creighton ushered in a short-lived golden age of maize genetics. For a few years, maize had one clear advantage over the irresistible Drosophila; maize chromosomes were large enough to be identified and worked with. Unfortunately for corn, in 1933 a geneticist at the University of Texas, Theophilus Painter, discovered that the fruit fly's salivary glands contained enormous chromosomes; an unusual kind of duplication makes them about 100 times larger than any of the fly's other chromosomes, even though they contain exactly the same genes in the same sequence. As a result, by the late 1930s, the fly was back on top, but because maize remained such an important crop, the corn community survived.

McClintock was never really tempted to desert her cornfields, even though she had trouble finding work after Cornell. Jobs of every kind were in short supply in the late 20s and early 30s, but her difficulties were exacerbated by the fact that plant-breeding departments would not employ women, claiming that farmers simply would not accept advice from a woman – and persuading farmers to change their methods was a key aspect of such departments' work. So, although Emerson helped Beadle and Rhoades to get jobs, he was unable to do the same for McClintock. Or perhaps he was unwilling; McClintock's quick brain and superb experimental technique, combined with her assertive attitude, had annoyed some at Cornell. She had a habit of solving tough problems rather more quickly than some of her teachers. Things came to a head in 1929 when she published the first schematic diagram of the maize chromosomes in *Science*; Lowell Randolph, whose assistant McClintock was then supposed to be, had been working towards this goal himself and was unaware that she had even decided to tackle the problem. He complained to Emerson, who began to suspect that the brilliant but unorthodox McClintock was something of a troublemaker. Rhoades gradually helped patch things up, but McClintock always retained a reputation for being 'difficult'.

McClintock got several prestigious fellowships, including a US National Research Council and a Guggenheim, which gave her considerable freedom but little security. Eventually she got a job at the University of Missouri, but did not take to teaching; she was bored by run-of-the-mill undergraduates, preferring the company of those she found exceptional. She even talked of quitting genetics altogether if she could not find a full-time research position. Emerson had clearly forgiven her by this time, because he nominated her for membership of the American National Academy of Sciences (NAS), in part to encourage her not to quit the field. Before her nomination had been decided, she was offered a post at Cold Spring Harbor, where Shull had done his maize experiments. She had been there for a couple of years when, in 1944, the news came through that – at the comparatively young age of forty-one – she had been elected to the NAS, only the third woman ever to receive the honour.

Cold Spring Harbor Laboratory was bounded by the Atlantic Ocean to the north and was otherwise surrounded by the estates of the very rich. As a result, space was in short supply and there was no room to grow maize the way most corn geneticists did;

fields full of tens of thousands of plants, watched over by teams of technicians, students and assistants. McClintock simply did not have the resources to work on this scale at Cold Spring Harbor, but almost certainly would not have wanted to. She was probably the only maize geneticist of her generation who would have been happy at Cold Spring Harbor, because she was content to work with just a couple of hundred plants, which she could tend herself. That way she could watch them grow, walking slowly between the rows of corn each day, getting to know her plants, almost as individuals. Though she liked working alone, she did sometimes miss being part of a maize community like the one she had had at Cornell. Luckily, Marcus Rhoades was only an hour away at Columbia University in New York; they remained in close contact, discussing their ideas, exchanging information, comparing results. But trips to New York or to conferences meant leaving her plants, and McClintock did not trust anyone else to look after them. Her only regular assistant was the man she employed as a human scarecrow: she gave him a shotgun and told him to make sure none of her plants were pulled up by birds – that was as much responsibility as she was willing to delegate.

McClintock did not talk to her maize plants, hug them or befriend them, but she did expect them to answer her questions. Locked up inside their cells were the maize chromosomes; she could squash and stain them, but that involved their ceasing to function, and she wanted to know what they did in living plants. She found that watching the plants grow was the best way to learn about them, but of course she also did various experiments on their seeds and then grew them to see what would happen.

Next to McClintock's maize fields were the labs where Delbrück, Luria and their colleagues were running their phage course every summer. She got to know them pretty well, and found them stimulating neighbours, but their work was nothing like hers. Phage work was already progressing rapidly, thanks to the speed at which the tiny organisms bred. The phage geneticists worked at the pace of their organisms; they were brilliant, impatient and quick. McClintock preferred to slow things down, to consider what the genes might actually be doing in whole, living plants, big slow-growing plants; she had no interest in stripping things down to a simpler, smaller system. Maize and McClintock suited each other.

Breaks and breakthroughs

What McClintock was looking for as she walked between her corn rows was patterns. Patterns on the leaves, or in the corn kernels. Spots or stripes of colour which appear, disappear and reappear, as in the multicoloured 'Indian Corn' that is a familiar part of traditional US Thanksgiving décor.

Many plants display such patterns, which are known as variegation. Variegated plants had always had a reputation for being peculiarly fragile; in the seventeenth century the diarist John Evelyn warned his fellow gardeners that they should treat their prized tulips very carefully, 'else they will soon lose their variegations'.[36] De Vries had done one of the first scientific studies of variegation and initially, such plants seemed perfect for his purposes because they constantly mutated, but gradually, frustration set in. The 'mutations' were not inherited – variegated plants gave birth to non-variegated ones, which then begat variegated ones. There seemed no pattern to the behaviour; after months of struggling to understand why the wretched plants failed to conform to Mendel's laws, de Vries christened them 'ever-sporting' varieties and gave up on them.

Maize variegation was a subject that had also intrigued Emerson. He observed that 'Variegation is distinguished from other color patterns by its incorrigible irregularity', but he nevertheless hoped that he might eventually understand it.[37] There did seem to be glimmers of some kind of regularity. Perhaps, he wondered, a Mendelian explanation could be constructed if it were assumed that a colour-inhibiting gene was in some way linked to a pigment gene and that – for some as yet unknown reason – the inhibiting factor occasionally disappeared, so that the pigment reappeared in specific cells.

Genes such as those which were presumed to control variegation became known as unstable or mutable genes. They were studied quite extensively in Drosophila as well as maize, and there was much speculation among geneticists as to what they were: was their instability just an extreme case of the normal changeability of genes, or was there something unique about such genes? Were they perhaps faulty or 'sick' genes? Rhoades investigated mutable genes in maize and established that, much as Emerson had speculated, there did indeed seem to be pigment genes that only became unstable when other genes were also present: the colours they produce or do not produce were not only the result of the colour genes themselves, they depended on the neighbouring genes, on the overall genetic environment.

It was these unstable, mutable genes that McClintock was interested in as she spent hours peering down her microscope; she had observed chromosomes break and repair themselves during cell division, and as she watched her plants grow, she took to speculating as to how their chromosomes would look when she eventually squashed and stained their cells. The broken chromosomes seemed in some way to be responsible for the variegated patterns, but how?

During the 1930s, McClintock had tried exposing corn to intense X-rays; as she had expected, the radiation stimulated mutations, as it had in so many other species. But while X-rays had proved useful in getting her work started, McClintock found them rather too random for her way of working, so she began to collect plants with broken chromosomes instead. She painstakingly deduced the connection between what happened to the chromosomes and what she observed in the growing plants. As she did so, McClintock realized that when a chromosome broke, parts of it were effectively deleted. Sometimes, genes near the break points were only damaged, but they could also be duplicated or rearranged in other ways; and each change produced interesting effects in the plant. The results were similar to those of X-ray or natural mutations: a gene ceased to function and by studying what went wrong with the plant as it grew it was possible to work out what the missing gene ought to have been doing. Because she had learned to identify each individual chromosome under a microscope, she knew exactly which sections of which chromosomes had been damaged; that made it much easier to make the connection between a particular mutation and a specific piece of chromosome.

The breaking chromosomes had another attractive property; McClintock discovered that their fragility was inherited, so over many years of patient work she was able to breed families of plants whose chromosomes continued to break at the same place on a particular chromosome, generation after generation. Some of these lines produced kernels that were variegated, spotted or patterned with different colours – in one of these plants, McClintock identified eight highly mutable genes, which may not sound like many, but it was four times as many as had been identified in the previous thirty years of maize genetics. She spent most of the rest of her career investigating the progeny of those plants whose chromosomes broke regularly. These breeding lines gave her a new tool for generating mutations, which had two key advantages over the aggressive chemicals or devastating X-rays her colleagues used. First, it was cheaper, since it did not need any expensive equipment, but more importantly, it was more precise. The usual techniques created hundreds of random mutations all over every chromosome. Before anything useful could be done with the mutations, like building up a chromosome map, they had to be laboriously sorted through, using the frequency with which they were linked to calculate exactly where each one lay on its chromosome; this is what took the Drosophilists so many years of tedious work. The need to work this way had effectively forced geneticists to use small, cheap, fast-breeding organisms like flies and phage – they needed to mass-produce mutations in order to do the time-consuming screening. But McClintock's approach was different; she knew in advance which chromosome would break, and often had a pretty good idea of where it would break. For McClintock, her fragile maize stocks became as valuable as Columbia's carefully cleaned-up flies; what had started out as a seemingly intractable set of problems had become a usable tool.[38]

As McClintock worked on her broken chromosomes, she gradually developed a radical new idea about one of the biggest mysteries in biology. All biologists realized by this time that every cell in a plant or animal's body carried a full set of chromosomes (except of course, for the sex cells – ova and pollen or sperm – with their half set). Biologists were also generally agreed on the existence of genes (now known to be specific sections of specific chromosomes), each of which determined a specific character, such as the colour of a fly's eyes or a corn kernel. And it had also become clear that genes somehow 'made' chemicals; genes in corn's kernels made pigments, while those in the stems made the fibres that allow the plant to grow tall and strong. So the unanswered question was, given that every cell – kernel or stem, leaf or root – had the *same* complete set of genes, how was it that the genes for kernel colour seemed to 'know' they were in a kernel cell and it was time to switch themselves on and make yellow or red colouring, while the same genes in a stem or root cell apparently 'knew' that they should remain switched off?

Once phage had revealed to biologists the nature of DNA, the question of how genes were controlled became the biggest question in biology. It was clear that the precise sequence of DNA in the cells provided the basic template that ensured maize produced maize and mice gave birth to mice. But only an understanding of how and why genes switched on and off would explain how one, single cell – a newly fertilized ovum – develops into a complete plant or animal. How is it that, as the cells divide and multiply, they specialize, so that root cells become roots, and shoot cells become shoots? Not only does every cell need to be told *what* to be, it needs to be told *where* to be; some mechanism ensures not just that eye cells become eyes, but that we grow them in our heads, not our feet.

This was the puzzle McClintock thought her broken chromosomes had begun to solve for her. She found that when the chromosomes broke and rejoined, not only were some bits deleted, other bits moved. And when they moved, they switched genes on and off.

To understand what McClintock had discovered, we need to remind ourselves of the image Morgan's fly boys had of genes. When they had been trying to understand crossing-over, the exchange of different versions of a gene between chromosomes, Morgan's team had imagined the genes like beads on a string. Each string was a chromosome and during the reduction division (meiosis) the strings got entangled with each other, broke and rejoined.

McClintock's breaking chromosomes were undergoing a similar process, breaking and then rejoining. However, crossing-over occurs only during meiosis (the reduction division that occurs when sex cells are being formed), while McClintock's chromosomes were breaking during the ordinary cell division (mitosis), which is part of a plant or animal's normal growth – its cells constantly divide to form new cells. Among the eight mutable genes McClintock discovered was one that caused chromosomes to break. But as she studied it, she became puzzled by its irritating property of disappearing from time to time – and then reappearing on a different chromosome. Sometimes, after the chromosomes had broken and repaired themselves, a single 'bead' ended up at a different place in the string from where it had started. It had, in effect, jumped from one point on a chromosome to another.

McClintock realized that when one of these tiny fragments of chromosome was re-inserted back into the middle of a gene, that gene stopped working (or worked in a different way). Each gene is a series of templates that needs to be in a particular sequence to produce a specific chemical. If the sequence of templates is changed, a different chemical is made – or no chemical at all. So if, for example, the gene is one for a pigment for the cells that make up a leaf, the intruding fragment may produce a colourless cell, or a group of them.

However, what McClintock found really fascinating about variegation was that as a corn leaf grew, the colour did not simply fail to appear; it appeared, disappeared and then reappeared. She could observe this process as a plant grew – that is what she had been watching for in her cornfields – and she realized what caused the mysterious patterns. The tiny intrusive pieces of chromosomes were being inserted and then removed again, so a gene that did not work in one group of cells began to work again at a later stage in the plant's growth. On the leaves of some of her maize lines with breakable chromosomes, McClintock could see a patch of green where the gene was working, then a few pale cells where the gene had stopped, then the switch was thrown again and the green came back. On, off, on, off. Spots or stripes appeared in the developing growing leaf (just as in a tulip petal). A pattern emerged. A pattern of control.

When McClintock announced her discovery in the 1940s, she referred to the mobile gene fragments as 'controlling elements'. She had discovered that the mutable gene that caused chromosome breakage required another gene to be present before it had an effect. That made her realize that these mobile genetic elements could alter the effects of the genes that lay near them; when a 'controlling element' moved to a region of the chromosome, it would either suppress or alter the action of the surrounding genes; when it hopped out again, the genes went back to normal. She envisaged the whole process being repeated throughout every cell of every organism; countless numbers of controlling elements were at work, turning off root genes in shoot cells, or shoot genes in root cells, just as they could turn off colour in leaves. That, she proposed, was the key to understanding how a single cell could develop and differentiate into an entire organism. She published details in Cold Spring Harbor's official reports and presented her conclusions at major genetics conferences through the 1950s. The typical response to her ideas, she later remembered, was 'puzzlement and, in some cases, hostility'.[39] Very few accepted her argument that this seemingly random phenomenon in maize could hold the key to the way genes were controlled in the development of every organism. When she had finished, some of her audience quietly shook their heads as they walked away. Occasionally a few even laughed at her; she was after all, an eccentric, and a woman.

Even some of her closest colleagues found her ideas impossible to accept.

What happened next is the stuff of legend, at least among some geneticists. Realizing she was ahead of her time, McClintock gave up trying to convince her colleagues and went back to her cornfields, patiently gathering more data as she waited for the world to catch up with her. Eventually, many years later, they did, but recognition came only when two men, using one of the small, fast, fashionable organisms – the bacterium *E. coli* – confirmed her results. The predominantly male scientific establishment uncomfortably recognized that McClintock had been right all along and in 1983 she was awarded the Nobel Prize. Over subsequent years, she became something of a feminist icon, and her life was the stuff of scientific legend. She lived an almost monastic existence, always dressed in the same androgynous outfit – shirts, slacks, work shoes and an old, threadbare lab coat that she endlessly repaired with iron-on tape. She had a directness, and an intellectual sharpness, that some found off-putting, but those who knew her well loved her self-effacing, often bawdy, sense of humour. She lived alone, never married, and seems to have had few if any sexual relationships, preferring to sleep alone on a cot in her office so as not to interrupt her work. By the time she died in 1992, she was world-famous, a role model for a generation of young women scientists. Her 'mobile genetic elements' had revolutionized ideas about genetics, and in 2005 the US Postal Service even issued a stamp in her honour.

Lone genius?

It seems that in Barbara McClintock, finally, genetics produced a lone genius who changed the world. A scientist who struggled to get a job, whose heretical ideas were too advanced for her contemporaries to understand, too threatening to the scientific establishment. When she was not being ignored she was being ridiculed – and all because she was a woman.

This is the McClintock myth, but it is really as much of a distortion of history as the Mendel myth. It has been created in part by careless readings of Evelyn Fox Keller's excellent biography of McClintock, *A Feeling for the Organism*, ignoring the subtleties of Keller's argument in order to create a simplistic story with a clear-cut heroine and some suitably villainous villains. And although McClintock was an extraordinary scientist who discovered something genuinely extraordinary, as with Mendel, it was not what she thought it was.

McClintock was awarded the Nobel 'for her discovery of mobile genetic elements' but, as the historian Nathaniel Comfort has shown, she had not called them that – she referred to them as *controlling* elements.[40] For her, the mobility of the bits of chromosome was not the point, it was the way they controlled the genes that was significant. When mobile elements were eventually discovered in both viruses and bacteria, during the late 1960s and 70s, they were named 'transposons'. The name was intended as a tribute to McClintock's original term 'transposable element', but that had not in fact been her term – it had been coined by another maize geneticist, Royal Alexander Brink. He had repeated McClintock's work and observed the same phenomenon, but he did not agree with McClintock over its significance. He chose the term 'transposable element' to distance himself from McClintock's ideas, not to add his support to them.[41]

The idea that McClintock was simply ignored until her work had been repeated by men is also a distortion. What the men in question, the French scientists François Jacob and Jacques Monod, discovered in their bacteria was not transposons (or 'jumping genes', as they are sometimes called), but a completely different model of genetic control. They found there were special regions of the chromosome either side of a gene which they called regulatory regions. When specific chemical molecules attach themselves to these regions, they switch the gene on and off, but neither the genes nor the regulatory regions jump or move around.

Nor is it true that people laughed at McClintock and ignored her when she first announced her discovery. In 1951 she had given a detailed two-hour presentation to a distinguished audience at Cold Spring Harbor. Among the crowd was Alfred Sturtevant, one of the original fly boys and by this time, perhaps America's most influential geneticist. As he said afterwards, 'I did not understand one word she said, but if she says it is so, it must be so!'[42] Brink was also in the audience and his response seems to have been more typical: he admired McClintock enormously and agreed with her that transposition occurred, but they disagreed – increasingly sharply over time – over its significance. Brink never used her term 'controlling elements', because he did not think that was what they were.

When McClintock spoke, geneticists listened: she was, after all, a very distinguished and highly respected geneticist – vice-president of the Genetics Society of America in 1939, president in 1945 and a fellow of the NAS. No one laughed or called her crazy, they simply could not see how what appeared to be a *random* process, the insertion and deletion of little roving fragments of DNA, could possibly be the mechanism that controlled the actions of all the genes in an organism.

The current view among most geneticists is that transposons are some kind of molecular parasite, not entirely unlike viruses; they are DNA that has managed to multiply itself among the genes. Their evolutionary role is not fully understood, but it has been suggested that they thrive because they accidentally help to generate genetic diversity. McClintock's central argument, that transposition is a mechanism for controlling genes during development, has been largely discarded – even by those who praise her work most highly (although there do appear to be a few cases where transposons operate as she envisaged). Despite the enormous respect her colleagues had for her, McClintock won her Nobel for what she had discovered, but not for her interpretation of her own discovery. However delighted she must have been to win it, it was in one significant respect a rather bitter consolation prize, a prize for being wrong.

None of this diminishes her brilliance, nor does it diminish her hard work or the prejudice she had to overcome to do it. McClintock and her work are interesting and important enough not to need mythologizing. By contrast with her contemporaries, who tended to think of genes as unchangeable static switches, McClintock realized that the interactions between genes permit them to respond to their environment in highly complex ways. Although she was mistaken in thinking that transposition was the mechanism that permitted this dynamism, recognizing the fact and importance of dynamism has perhaps been her most important legacy.

Although her interpretation of transposition was not accepted, McClintock's discovery provided the final piece in the 300-year-old puzzle of why corn has been so hard to improve. Scientists now know that the maize genome consists mostly of transposons, which have managed to spread themselves throughout maize's chromosomes without causing any ill-effects that would lead natural selection to eliminate them. As a result, these strange molecular parasites have become dominant over the functional genes; this is another crucial reason why it is much more variable than most other crop plants. Transposons also explain why a single corn plant's genome is as large as a human being's.

What Emerson called the 'incorrigible irregularity' of maize is now well understood, but corn still causes headaches for agricultural scientists. A major project is currently under way to sequence the entire maize genome – the term for the complete genetic make-up of an organism, including all the different forms of its genes – to identify every single functioning gene, partly so that the plant's complexities can finally be tamed by direct genetic manipulation. Because corn is such a valuable crop, there is plenty of funding available for this work – among others, the USDA's Agricultural Research Service is still pouring money into maize, more than a century after it first started doing so. Yet *Zea mays* is still holding out. Despite substantial funding and the latest, most powerful technology, even a simple description of the maize genome has yet to be completed and a full understanding of the functions of all its genes may still be many years off. Finding the working genes (which is only the first step towards knowing what they all do) requires sorting through all that non-functioning or 'junk' DNA. As a result, the first plant to have its complete genome sequenced was not maize – or wheat or rice or any of the world's other major crops – but a tiny weed with no commercial value whatsoever: *Arabidopsis thaliana,* or Thale cress.

Arabidopsis thaliana

MOUSE-EAR CRESS

Chapter 10
Arabidopsis thaliana: A fruit fly for the botanists

Growing amid the rows of Barbara McClintock's corn was a tiny, insignificant weed. It was probably lurking in a corner of Darwin's greenhouse, in Mendel's priory garden, and thriving in the experimental Oenothera plots that flourished briefly in the early twentieth century. It grows in many parts of the northern hemisphere, from fields and wasteground to gardens, but few gardeners are aware of it. It is a tiny, insignificant plant, a few inches high, yet it came to depose king corn as the leading plant used for genetic research. Although it resembles its relative, edible cress, Arabidopsis is not edible nor of any economic importance, but it has led to angry protesters attacking fields of crops. It is harmless – neither gardeners nor farmers bother to remove it – but it would damage the British prime minister, Tony Blair, and threatened to start a trade war between America and Europe. It has no known medicinal properties, yet there are currently more scientists working on it than any other plant. Its name is *Arabidopsis thaliana*, a worthless and superficially uninteresting weed, but in less than thirty years it has transformed plant biology and – according to most of the scientists who work on it – it is going to change the face of the planet and feed the starving; it may even help solve the problem of global warming.

Arabidopsis thaliana is also known as Thale cress, wall cress or mouse-eared cress. It was named after the first European to describe it, the sixteenth-century naturalist Johannes Thal, who found it growing in Germany's Harz mountains. It did not get much attention during the first few centuries after its first appearance in print. When William Curtis, a Quaker botanist and apothecary, published a large, lavishly illustrated flora of the wild plants found in and around London (the *Flora Londinensis*, 1777–87), he mentioned the little cress but rather scathingly described it as a plant of 'no particular virtues or uses'.[1]

Almost nobody identified any virtues or uses for Arabidopsis until 1907, when a German botanical student called Friedrich Laibach was looking for an experimental plant. He had studied under Eduard Strasburger, an enormously influential botanist who was the first to describe cell division in plants. Strasburger also invented many of the modern methods for studying cells; he was the first to apply chemical stains to plant specimens, making the details of their anatomy easier to see. In 1894 he and his colleagues produced a massive *Textbook of Botany for Universities* (*Lehrbuch der Botanik für Hochschulen*), which was translated into eight languages and became known as the 'botanist's bible'. It defined many of the key ideas that would shape twentieth-century botany, and introduced many of what would become biology's standard terms, such as haploid, diploid and gamete.

Laibach wanted to explore the possibilities of the toolkit Strasburger had given botanists and was looking for a new plant to investigate. No one knows how or why he became interested in the mouse-eared cress, but it had properties that suggested it might be suited to life in the lab. It was small, so would not need as much space – or money – as, say, Oenothera. It also grew quickly; it took only about eight weeks to raise a new crop, so botanists would be able to grow several in a year. And it produced lots of tiny seeds, easy to collect, store and plant again. However, when Laibach stained the plant's cells as Strasburger had taught him, and put them under his microscope, he was disappointed. Arabidopsis had only five chromosomes, which was useful, but unfortunately they were tiny. Laibach was not planning to do chromosome mapping, as Morgan and his students had done; he wanted to study plant nuclei in an effort to understand whether their chromosomes were stable. He took one look at the cress's tiny chromosomes and decided they were not suitable for his research work. Laibach would not work on the plant again for many years.

Bombs and big science

In 1946 Herman Muller – the Drosophila biologist who had won the Nobel Prize for his discovery of radiation-induced mutations, was asked by the *New York Times* what he thought the long-term effects of the Hiroshima and Nagasaki bombs would be. He replied that if the survivors 'could foresee the results 1,000 years from now . . . they might consider themselves more fortunate if the bomb had killed them'.[2] He knew, perhaps better than anyone, what radioactivity in the form of X-rays could do to living things; what would the much more intense radiation released by the atomic bombs do to human genes? Even if the weapon were never used again, would atmospheric testing and the manufacture of nuclear weapons eventually devastate the human gene pool?

By the end of the war, the US military and the fledgling nuclear industry were preoccupied with assessing the effects of radiation. American military intelligence sent agents to Germany to search for evidence of the Nazi atom bomb programme. Since Germany had been a world leader in nuclear physics before the war, the United States hoped to find – among other things – documents that might shed light on the effects of radiation. The agents must have been pleased when they came across a recent PhD thesis by a German biology student with the words '*Röntgen-Mutationen*' ('X-ray mutations') in the title. The document was sent back to the USA, so that translators could discover what the Nazis knew about *Mutationen*. Sadly, the thesis turned out to be about nothing more sinister than Arabidopsis; it described experiments by one of Laibach's students, designed to reveal whether the plant could be mutated with X-rays. This PhD thus came to have the unusual distinction of being the first, and probably the only, botanical publication to appear as an unclassified captured enemy document, published by the US Joint Intelligence Objectives Agency.[3]

Clearly, Laibach had not forgotten Arabidopsis as he worked on other plants. He was evidently working on it again during the war, albeit with different questions in mind: in 1943 he had published details of some hybridization experiments, which took advantage of the plant's rapid growth and the ease with which it could be crossed. He also suggested that it should prove possible to bombard the seeds with X-rays and stimulate them to mutate, in the way Drosophila did. He concluded by arguing that Arabidopsis was worth bringing into the lab, where it could become a fruit fly for the plant people, but almost no one took up his suggestion.

Botany was, understandably, not of immediate interest to the Atomic Bomb Casualty Commission (ABCC), established in 1947 to investigate the bomb's medical effects, yet in the long run the fate of Arabidopsis would be determined by the new directions that emerged in post-war science. Before the war, science had mostly been conducted on a fairly small scale – individual researchers leading small teams. By contrast, the US effort to build the bomb, the Manhattan Project, involved industrial-scale science for the first time – by the time Hiroshima was destroyed, the bomb builders were employing tens of thousands of people in laboratories and in production plants the size of small towns. Huge teams with budgets to match became the model for what would become known as Big Science; it affected the physicists first, but it would gradually change the way every science was done.

The biologists first ran into Big Science through the ABCC, which launched a genetics project. The snarl of acronyms that provided the funding were a sign of things to come: the genetic work was co-funded by the military-controlled Atomic Energy Commission (AEC), the National Academy of Sciences (NAS) and the National Research Council (NRC). Federal funding meant more money for science than any pre-war researcher could have imagined, but it also entailed more regulations and form-filling than scientists had previously encountered. When the ABCC finally reported in 1955, it had failed to reach any very definite conclusions about the long-term genetic effects of radiation. Nevertheless, the *US News and World Report* ran the headline 'Report on Hiroshima: Thousands of babies, no A-Bomb effects', and commented that 'Children of Japanese survivors of atomic bombs are normal, healthy, happy. That is the verdict based on study of 70,000 babies, including 50,000 born to parents caught in bomb blasts.'[4]

Some Americans learned to stop worrying and love the bomb, but others were nervous about the rise of an increasingly militarized, industrial science. The implications of military-led science became even more disturbing as the anti-communist feeling that characterized the new Cold War reached a crescendo after the USSR exploded its first atom bomb in 1949. Senator Joseph McCarthy and the House Committee on Un-American Activities were involved in a series of infamous witch-hunts: among the ex-communists who decided to name names was Herman Muller, who – as we have seen – had experienced the rise of Lysenko at first hand when he had briefly lived in the USSR in the 1930s. Muller decided that communism posed a more serious threat to the long-term health of humanity than even radiation did, and so refused to support calls for a test ban. Although he eventually changed his mind, in 1955 he was still arguing publicly that *any* restrictions on nuclear weapons development would only help the Soviets because they had the lead in conventional military strength – so the US must maintain its nuclear advantage.

Maintaining America's technological superiority became a Cold War obsession. In just four years, from 1949 to 1953, US defence expenditure almost quadrupled – to over $50 billion. In 1957 the Russians revealed that they were ahead in the space race by launching the world's first artificial satellite, Sputnik. No one knew what the 'artificial moon' could or would do nor what the Soviets might have planned next, so almost overnight US defence expenditure was increased again and money was made available for anything and everything that might give America a tech-nological edge. Electronic computers were emerging as a key technology in both the arms and space races, so massive govern-ment spending went into developing them.

Although biology never received anything like as much defence money as physics, many biologists were beneficiaries of military funding, not least because of continuing concerns over radiation sickness and genetic damage – the AEC provided half of all federal funding for genetics in the 1950s.

Hard rain

Despite reassuring noises from the scientific establishment, not everyone was persuaded that radiation was safe, nor that the scientists and their allies in the military were going to make the world a better place.

Bob Dylan's song *A Hard Rain's A-Gonna Fall* (1962) typified a growing fear about the state of the world, which blended anti-war feeling ('guns and sharp swords in the hands of young children'), with concern for the environment, where 'pellets of poison' would leave nothing but sad forests and dead oceans. And the overarching fear remained the Bomb, a thunder that warned, a wave that could drown the whole world. Dylan's voice formed part of the soundtrack to the passionate anti-war protests that spread as the Vietnam war intensified. University and college students were especially active in the movement, so the scale of military funding for university-based science – especially research into new weapons – became one focus for their campaigns. Radical students urged universities to break their links with the military establishment – calls that went largely unheeded, but which nevertheless led to increasing scrutiny of where the money for science was coming from, and on what it was being spent.

In the same year that Dylan released his song, the still slight but growing public suspicion of scientists was increased by the publication of Rachel Carson's best-seller, *Silent Spring*. Carson's book raised concerns about the ever-increasing use of chemical pesticides, especially DDT; she was not anti-science – she was a trained biologist – but she was concerned that commercial interests were leading to the overuse of chemicals whose long-term effects were unknown. Her unease struck a chord with the new environmentalists, among whom a small but increasingly vociferous minority were becoming convinced that science and scientists were not to be trusted.

Although Dylan and Carson had almost nothing in common,

both became threads in the complex tapestry of ideas, beliefs and prejudices that became known as the sixties counter-culture; an inchoate but impassioned movement whose advocates blended enthusiasms for long hair, long cigarettes and long-winded music with sentiments that were anti-war, anti-establishment, anti-consumerist and occasionally anti-science as well.

Meanwhile, government funding for American science continued to grow, but some of the beneficiaries of the government's largesse were not entirely happy to be part of Big Science. Some complained that instead of being creative researchers, scientists were becoming production-line workers, ceding their independence of thought to government targets and bureaucratic management – complaints that echoed those of the counter-culture, deploring the relentless treadmill of automation, which turned interesting jobs into alienating ones.

Naturally, many biologists were more than content with both the additional money and the new style of research, but even they could not help but notice that, even as the funding increased, familiar priorities re-emerged: biology still received less money than physics; and, within biology, plants received less money than animals. Despite the US military's brief, if accidental, interest in Arabidopsis, few other people paid attention to the weed and none of the Big Science money went to botany. During the 1950s a few researchers, especially Klaus Napp-Zinn and Gerhard Röbbelen in Germany, had showed that the plant was indeed a useful laboratory plant, for exactly the reasons Laibach had identified. During the 1960s it was being used by a handful of people in Europe, Australia and the USA, where the Hungarian-born George Rédei promoted the plant as enthusiastically as others ignored it.

At a time when small, fast-growing organisms like phage were increasingly being used for experiments, most plant geneticists were still working on maize, tobacco or other significant crop plants, but their slow growth and the difficulties of experimenting with them helped to make plant genetics look rather backward, especially by comparison with the cutting edge of the new molecular biology: phage and bacterial genetics. These tiny, fast-breeding organisms allowed researchers to perform new kinds of simple and elegant experiments, answering genetic questions in days rather than years.

As we have seen, Watson and Crick's announcement of the structure of DNA suggested how the physical form of the molecule explained the way it worked. The double helix structure suggested how DNA made more DNA: because the strands were complementary, each could form a template for the other. However, it also suggested how DNA makes amino acids, the raw materials from which enzymes and other proteins are assembled. A gene consists of a specific sequence of bases – adenine (A), thymine (T), cytosine (C) and guanine (G) – that form a template from which a series of amino acids are made. It took about a decade from Watson and Crick's announcement, to understand how this worked in practice, how the DNA inside the cell nucleus makes an intermediate copy of itself out of ribonucleic acid, or RNA. The RNA then travels outside the nucleus to serve as the template on which proteins are assembled. The crucial step, in 1963, was working out that each specific set of three RNA bases forms the pattern for a particular amino acid. That equivalence is why Schrödinger's original metaphor of a code has become so ubiquitous – the sets of three bases are now known as codons. Once the amino-acid-manufacturing mechanism had been understood, the way was open for work on everything from understanding inherited diseases to mapping evolutionary relationships at the molecular level and, ultimately, for genetic engineering.

This growing understanding of DNA gradually made it easier to begin working out the relationships between different parts of a chromosome and the various biological processes that sustain living organisms. All kinds of new technologies and techniques were devised to interfere with genes, to find out what they made and how they made it. The original fly room had produced physical maps of chromosomes, detailing where and on which chromosome each gene lay. The new tools of molecular biology added the details of what the genes were, not merely their chemical nature, but the precise sequence of DNA bases that made each specific amino acid and, from them, each protein.

As the style of work pioneered by the Phage Group matured, some scientists started to look around for harder problems to tackle with their powerful new tools. Among them was Seymour Benzer, a young Jewish boy from Brooklyn whose interest in science had first been inspired by Sinclair Lewis's *Arrowsmith*. A few years later, he read Schrödinger's *What is Life?*, which brought Max Delbrück's name to his attention. Soon afterwards, he met Salvador Luria at a dinner party. It seemed as if life was pushing Benzer to Cold Spring Harbor and the phage course, which he took in 1948. He did brilliant work with phage, but as he became increasingly interested in the genetic basis of human behaviour he hit a problem: viruses and bacteria do not exhibit behaviour, at least not in any form that might help us make sense of *our* behaviour. So Benzer turned back to the roots of modern biology, back to Drosophila. In the fifties, it seemed as if the fly had had its day. But Benzer's elegant experiments helped to bring the fly back into fashion, into the new world of molecular biology, where Drosophila once again became the focus of a fly community, one that would soon dwarf its pre-war predecessor.

Benzer was not alone in itching to apply biology's new molecular tools to larger, more complex organisms. The South African-born biologist Sydney Brenner had also started with phage, bringing the virus with him to Cambridge when he joined the UK Medical Research Council's Molecular Biology lab in 1957, to work alongside Crick and Watson. But a few years later, Brenner decided it was time to tackle a new problem, by then the biggest problem of all – how genes controlled development. In particular, he wanted to explore the development of the nervous system. This again posed the problem that neither bacteria nor viruses have nerves, so Brenner also needed a new, multi-cellular organism to work with. He settled on a nematode worm called *Caenorhabditis elegans*, usually known simply as *C. elegans*. Most of us would not consider a worm less than a millimetre long as either big or complex, but compared with *E. coli* (which are about 1,000 times smaller), Brenner was moving from repairing watches to working on the jet aircraft.

Whether it was by turning to old experimental organisms, such as Drosophila, or domesticating new ones, like *C. elegans*, by the mid-1970s the new-style biologists were starting to make impressive headway in deciphering the connections between the molecular world of the genes and the familiar scale of real, living organisms. But this was all still animal work (if we allow ourselves a broad enough definition of animal); plant biologists began to think about joining the molecular revolution, the world of big scientific teams and, hopefully, larger budgets and more recognition. But they had to sow before they could reap – which plant should they be planting?

From croissants to cress

In 1978 two young Canadians were sitting at a café in Paris, planning both their own futures and that of biology. Chris and Shauna Somerville were newly married and had just completed their degrees at the University of Alberta; Chris has a PhD in *E. coli* genetics and Shauna a Masters in plant breeding. According to Chris Somerville, they spent several idyllic months in Paris, 'just talking about what we were going to do', trying to decide 'what was interesting'.[5] But life was not all croissants and conversation. The Somervilles wanted to do something useful with their lives, to make some positive contribution to solving the world's problems.

A few years earlier, the global think tank the Club of Rome had produced its best-selling report *The Limits to Growth* (1972) prophesying an imminent environmental catastrophe. Its authors argued that: 'If the present growth trends in world population, industrialization, pollution, food production, and resource depletion continue unchanged, the limits to growth on this planet will be reached sometime within the next one hundred years.'[6] They predicted a dramatic crash in human population, caused by hunger, disease, and by wars over increasingly scarce resources.

These dramatic claims caused a sensation: *Limits to Growth* sold 12 million copies in thirty-seven languages. Its warnings of doom jarred with the largely optimistic mood of the 50s and 60s; the West had enjoyed two decades of unprecedented economic growth, rising living standards and low unemployment. There had been no return to pre-war economic depression and it was widely assumed that the same formula of Western-style economic growth could be applied throughout the Third World, bringing peace and plenty to all. However, the small but growing environmental movement argued that the ecological implications of growth were being ignored. One of the Club of Rome's key predictions was of widespread famine, as environmental degradation rendered more and more farmland unusable.

These claims were widely reported and became part of a growing sense that the economic boom of the 50s and 60s could not continue for ever. The oil shocks of the early 70s – caused when the Organization of Petroleum Exporting Countries (OPEC) rapidly raised the price of oil – contributed to the despondency, as petrol rationing and soaring pump prices gave the public a taste of what might be to come when the oil finally ran out. Some reacted to the global threat by turning their backs on science and technology, arguing that it was time to abandon industrialization and go back to nature, but for idealistic young researchers like the Somervilles, science offered the only practical answer. Chris Somerville recalls reading *The Limits to Growth* when he was young and recalls that they decided to work with plants because we 'were thinking about what we could do with our lives where we would have a big *social* effect'. They were, he recalls, young and idealistic enough to think that improving plants could feed the starving and save the world.

However, Shauna was dissatisfied with the traditional plant-breeding she had been taught at university, which was largely unaffected by the recent progress that had been made in under-standing the mechanism of inheritance. As Chris remembers, she used to complain that there was too little science, instead 'you grow stuff and you weigh it, you cross material and you weigh it, you just keep growing it and weighing it'. To Chris, trained in the precise, elegant genetics of *E. coli*, this certainly sounded crude, yet 'as we talked, we could envision what was going to happen in plant molecular biology'; it was still taking shape, 'but we could see what was going to happen'.

Their vision had been shaped by a gift that Chris's PhD supervisor had received from one of his American colleagues: a sample of one of the earliest restriction enzymes, a chemical tool for cutting-up DNA. Somerville remembers that they were so excited by their new toy that they started 'more or less playing with cutting DNA'. The use of such enzymes had only emerged a few years earlier and their possibilities were still being investigated, but they had the promising property of being able to cut a strand of DNA at a specific point. The enzymes could, for example, potentially be used to extract a single gene from an organism. All previous genetic work had been done on whole organisms, and thus – inevitably – on all their genes at once. A researcher might be lucky and find or create a mutation that revealed what happened when a particular gene malfunctioned, but as the corn geneticists and others had shown, back in the early decades of the century, most of the visible traits of an organism were controlled by more than one gene. It was often almost impossible to tell what the precise effect of any single gene was.

Once restriction enzymes had been developed to work like a pair of molecular scissors, to snip out a single gene, the next step was finding out what that gene did – what chemical it made and thus what role it played in the organism's biological processes. Unfortunately, a piece of DNA in a test tube does nothing. It just sits there. Genes need the cellular machinery of a living cell in order to work. Specifically, they rely on structures called ribosomes, effectively chemical factories that move along the RNA strands, fitting the right molecules on to the RNA template in the right order to produce amino acids and eventually proteins. So, having isolated a gene by taking it out of the organism, researchers needed to put it back into an living cell to find out what it did.

345

This trick was first mastered, not surprisingly, using a simple, comparatively well-understood bacterium – *E. coli*. In 1973, Stanley Cohen of Stanford University and Herbert Boyer at the University of California, San Francisco developed a process for inserting DNA – from any species – into a bacterium. Just as restriction enzymes cut the DNA up, another set of enzymes could be used to insert it in among the bacterium's own genes. Once the gene was back in a living cell, it had the cellular machinery it needed to start making whatever it made.

Cohen and Boyer cut up lots of DNA into lots of bits, inserted it all into bacteria, and then tested the bacteria – using the same basic techniques that the microbe-hunters of the nineteenth century had invented – to find out which genes had ended up in which bacteria. They were looking for genes with interesting properties. Once they had found them, the bacteria with the target gene could be separated, grown and used for more experiments. The bacteria had been engineered, genetically engineered, to make one specific protein.

Having done his PhD on *E. coli*, Chris Somerville knew all about Cohen and Boyer's process, but initially it could not be applied to plants. However, just before he and Shauna had left for Paris, a paper had been published which showed that a microbe called Agrobacterium causes plant diseases by transferring its own DNA into the affected plant, a process broadly similar to the way viruses like bacteriophage transfer their DNA. That opened up the possibility of taking a specific gene out of one plant and using Agrobacterium to put it into another. Chris Somerville remembers how excited he and Shauna were about this; at the time, they had said to themselves, 'It'll be easy, so we should not even get into that game . . . that's done'. In fact, as he confesses, they were over-excited; 'it took ten more years, really, before the average academic could do that kind of stuff'.

Nevertheless, seen from a pavement café in Paris, the future for plant molecular biology looked very rosy. Chris Somerville remembers thinking that now 'the molecular era was upon us', 'the community was going to need a good model organism because it was ridiculous working with corn, wheat or tomatoes'. They were just too large and slow for the fast-moving biology the Phage Group had set in motion. Working on *E. coli*, he had become used to planning an experiment one day and having some results to analyse by the end of the next. The Somervilles started to look for an alternative plant, and they came across an article by George Rédei, one of the US's few Arabidopsis enthusiasts, which argued that the weed's time had come. Somerville recalls that 'we read that and we were totally convinced'.

Many of the things that had made Arabidopsis a potential botanical fruit fly also made it attractive for the new style of molecular plant genetics. As Somerville puts it, 'we had an idea that to really have a successful plant . . . it had to be something you could grow in downtown Boston', where there are several major institutions, and 'you have to put your model organisms in those institutions'. There was no space for cornfields in Boston, but Arabidopsis – like Drosophila before it – was ready for the move to the city.

The Somervilles left Paris and returned to Alberta, while they waited for the visas that would allow them to continue their studies at the University of Illinois. They knew the first step in realizing their ambitions would be to get their hands on Arabidopsis. Chris Somerville remembers that 'somehow we got put in charge of the graduate student seminar, so we invited George Rédei . . . It was a major disaster . . . George was not well known to the faculty and very few people were interested in talking to him.' And so 'Shauna and I ended up having him on our hands for a couple of days' and he proved enormously helpful. 'We sat and talked with him for hours and hours', listening intently as he described everything he knew about Arabidopsis – from how to grow it, to which chemicals would cause it to mutate – sharing every tiny detail that the couple would need and providing them with seeds of many of the mutant lines he had created and collected.

The Somervilles' goal was to create a community around the plant. Like every biologist of their generation, they had seen how biologists had gathered around Drosophila and *E. coli*, how much more could be achieved when an organism brought people from different disciplines together. For Chris Somerville, the most significant model for what they were trying to do was Delbrück's original Phage Group. In 1966 Cold Spring Harbor had published a collection of essays in Delbrück's honour, entitled *Phage and Origins of Molecular Biology*. Somerville remembers that it 'had a really *big* effect on me. I read that book like my bible for some years.' He was fascinated to read about the work that had been necessary to create the new research field, but even more by the 'aesthetics of the whole thing, the logic of it'. His first degree had been in mathematics, a training that stressed the virtues of elegance and rigour, and he loved the simple logic of experiments like those of Hershey and Chase. As he puts it, 'phage genetics was an unusually elegant form of genetics'.

However, plant geneticists were not yet thinking in terms of elegant simplicity, and they were still working on their large, slow, but economically important plants. The Somervilles decided it was vital to grab their attention, to do what they called 'a demonstration project'. So they had spent many hours in Paris 'thinking about a problem where we could use the way *we* understood genetics to solve a problem that the plant community cared about'. As Chris Somerville readily admits, at that time 'I didn't know anything about plants, so I was totally open as to what that problem might be, but I did have this feeling that because our goal was to *attract* people and create a group, that we had to put a result in the heart of the existing plant community. To get their attention.'

Finally the Somervilles' visas came through, and they set off for Illinois, their heads full of Rédei's advice, their hearts full of hopes for the revolution they hoped to launch, and their bags full of Arabidopsis seeds.

Building a community

While the Somervilles were plotting their future and that of Arabidopsis, they were unaware that others were also beginning to take an interest in the plant. Among them was Elliot Meyerowitz, who read George Rédei's review article on Arabidopsis while he was a student. At the time, Meyerowitz was working on fruit flies but he remembers 'talking to my graduate adviser and to some of the graduate students I knew who worked on plants, about using *Arabidopsis* as a system for genetics, like *Drosophila*'.[7] Although he decided not to take it up immediately, one of his fellow graduate students at Yale, David Meinke, worked on it, along with his professor, Ian Sussex; they published their first Arabidopsis papers in 1978 and Meyerowitz is convinced that Meinke and Sussex played a key role in getting the field started.[8]

A few years later Meyerowitz ended up at Caltech, still working on Drosophila. However, he had also become interested in the history of genetics, especially in Hugo de Vries and the later Oenothera work, some of which had been done there. Partly as an outcome of his historical interests, Meyerowitz and one of his graduate students, Robert Pruitt (now a distinguished Arabidopsis biologist), got interested in doing more plant genetics; Meyerowitz remembered Arabidopsis, and decided the time had come to try working with it.

Like the Somervilles before them, Meyerowitz and Pruitt faced an immediate problem: acquiring plants to work with. Arabidopsis does not grow in southern California and in any case, as Meyerowitz happily admits, 'we were not the kind of biologists who could have gone in the wild and found it anyway; we would have probably gotten the wrong thing'. At that time there were no relevant internet resources and it would have taken months in the library, reading dozens of different biological journals in the hope of finding the few articles that mentioned the plant. Fortunately, Bob Pruitt had an uncle, Andris Kleinhofs, who was a plant biologist, a barley breeder at Washington State University, 'so Bob called his uncle Andy and asked him if he knew anyone who could get us some seeds of this *Arabidopsis* plant'. It turned out that Kleinhofs had some himself. Naturally the Meyerowitz team thanked Kleinhofs for the seeds in their first publication; he later complained that as a result he was deluged with requests for more

– which was particularly annoying as he had given all he had to the team at Caltech.

The number of requests that reached Kleinhofs was a sign of growing interest in Arabidopsis. That was partly because the Somervilles had done their demonstration project a couple of years earlier and created quite a stir. To get the plant community's attention, they decided to tackle a major problem and chose a complex aspect of photosynthesis, the process by which plants make their food using sunlight. The process of photosynthesis itself had been worked out in the 1950s and 1960s, but there remained a mystery over the closely related process of photorespiration. Plants photosynthesize when it is light, using the sun's energy and carbon dioxide from the air to make their food; a process that has the positive side-effect (for us animals, at least) of giving off oxygen, which of course enables us to breathe. Once it gets dark, photosynthesis stops and plants start to respire, just as we do, taking in oxygen, breaking down the sugars they have stored during the daytime and giving off carbon dioxide. However, in some circumstances, plants respire when it is still light, a process known as photorespiration, which has the effect of reducing the rate of photosynthesis and thus, ultimately, the food value of the plant. Scientists like the Somervilles were interested in understanding this process because if photorespiration could be prevented, food plants would photosynthesize more efficiently, grow faster and be more nutritious. Characteristically, the Somervilles' interest in the problem was not purely academic.

Various competing theories existed, but the Somervilles decided they ought to be able to decide between them by simply mutating lots of Arabidopsis plants and looking for a mutation that halted photorespiration. Thanks to the plant's rapid growth, it took only two months to isolate the mutation. The mutation proved to be sufficiently detrimental that the plants needed extra carbon dioxide to grow properly; when they were grown in ordinary air, they started to produce a specific chemical – phosphoglycolic acid. The production of this acid was a result that only one of the rival theories could explain so, practically overnight, a long-running and acrimonious debate was over.[9] 'That made a splash in the plant community', Chris Somerville says, because 'it literally was the end of that problem . . . and some people really liked it because it was kind of elegant.' Finding a mutation that was sensitive to the level of a specific chemical – carbon dioxide in this case – was the kind of strategy 'we were familiar with in *E. coli* genetics, but I do not think anyone had done that in plants'.

Meanwhile, the Meyerowitz lab was planning to use Arabidopsis to tackle quite different problems. Somerville was most interested in biochemistry, while Meyerowitz was a developmental biologist, with other interests; the variety of research that could be done with Arabidopsis helped persuade botanists that it could be a general-purpose tool.

In 1981, just as Meyerowitz and Pruitt were getting started, Max Delbrück died. He had long since moved on from phage to work at Caltech on a light-sensitive fungus; his hope had been to repeat the success of bacteriophage by finding the simplest possible organism that would allow him to study photosynthesis. A young woman called Leslie Leutwiler had been given funding to join his lab, but on Delbrück's death she moved from the basement to the first floor of Caltech's biology division, to join Meyerowitz's tiny group of plant scientists, in what was still supposedly a Drosophila lab.

With two bright, young researchers to help, Meyerowitz decided the time had come to try applying the new molecular tools to Arabidopsis, by trying to extract genes from the plant and copy them.[10] This process was still in its infancy in Drosophila and no one had managed it in plants. The main obstacle was that the commonly used experimental plants, such as maize, have large genomes; as we have seen, maize's genome is as large as a human being's, but most of it is made up of repetitive or 'junk' DNA, so finding a specific gene to clone is very time-consuming. Meyerowitz hoped Arabidopsis would be more tractable, but first they needed to know roughly how large its genome was. Similar tests on other organisms had been done a decade earlier, and both the equipment and Barbara Hough-Evans, the technician who had worked on them, were still at Caltech. 'So Leslie and Barbara just pulled the stuff out of the cabinet, gunned it up and did the experiment on *Arabidopsis*'; as Meyerowitz recalls, 'we found out that it had a strikingly small genome'.

The fact that Arabidopsis had a small genome does not in itself sound like earth-shattering news. In fact, it was not strictly speaking news at all: in 1976 researchers at Kew Gardens had begun to maintain a catalogue of plant genome sizes, the first published instalment of which mentioned the small Arabidopsis genome.[11] However, the paper Meyerowitz, Leutwiler and Hough-Evans published in 1984 was much more than an accurate confirmation of the earlier work.[12] In the 1976 publication, Arabidopsis had been listed as just one plant among hundreds of others. That article had said nothing about the implications of the small genome, so the mouse-eared cress was little more than a footnote among larger, apparently more important species. By contrast, the 1984 article spelled out the implications of the small genome quite clearly. Meyerowitz began to take every opportunity to promote Arabidopsis, stressing how well suited it was to new molecular biology. In addition to the aspects obvious to Laibach, such as the weed's small size and rapid growth, new tools were now revealing more information – such as the small genome size – that added to its attractions as an experimental subject.

As the Caltech team got to work on the plant, they gradually came across the handful of other people working on it. They learned of George Rédei, and Leutwiler flew out to Missouri to meet him, coming back with packets of seeds. Rédei also played a part in starting Arabidopsis work in Britain. Ian Furner was at the University of California, Berkeley, in the late seventies doing his PhD on plants, when he first heard talk of the mouse-eared cress. Arabidopsis sounded interesting and he decided to write to Rédei, who – with characteristic generosity – 'sent me every paper he had ever written and seeds, and everything else. He was so desperate to get people to work on the area.' Furner was initially unsure, but changed his mind on reading the Meyerowitz group's paper describing the small genome; he remembers it as the critical paper that persuaded people Arabidopsis was a feasible organism.[13] A few years later Furner founded an Arabidopsis lab at Cambridge University; the only other lab in Britain that had been working on the plant had ceased doing so, so for a while Furner's was the only Arabidopsis lab in the country.

In addition to Rédei, the Caltech group made contact with the Dutch biologist Maarten Koornneef, who was then a graduate student; Meyerowitz wrote to him, beginning an exchange of mutants and letters that soon blossomed into friendship. Koornneef had done his master's research on ornamental plants. A century and a half after Carl Friedrich von Gärtner had won the Dutch Academy of Science's prize for his essay on plant-breeding, the improvement of plants was a topic that remained of vital importance for a nation for whom flowers were a major export. Koornneef's professor had an interest in Arabidopsis and, among other things, used it to teach undergraduates; after a couple of years of working for a commercial seed company, Koornneef had returned to academia to work on a PhD on Arabidopsis when the Caltech team contacted him.[14]

Another friend Meyerowitz made via Arabidopsis was Chris Somerville; they met at a 1985 plant genetics conference in Colorado. Most of the conference was devoted to maize, but a contemporary photo shows the burgeoning Arabidopsis community: Chris and Shauna Somerville with Meyerowitz, Maarten Koornneef and David Meinke; a small group, excited to meet a few other people who understood the attractions of Arabidopsis and shared the vision of building a group around it.

Gradually, the Arabidopsis researchers got to know each other and the plant started to generate more publications and conferences, and a long-standing newsletter – the *Arabidopsis Information Service* (which Röbbelen had started) – got a new lease of life. Other plants had been widely promoted as standard model plants: petunias and tomatoes were both highly favoured for a while, but in the 80s, they quickly began to lose out to the seemingly unstoppable weed.

Once again, an important aspect of the Arabidopsis group's success was a lack of secrecy; everything was shared – plants, seeds, interesting mutants, practical techniques. The very fact that it was not an agricultural crop, so the information had no immediate commercial value, may have helped foster this openness. Somerville recalls that, from the outset, he and Shauna, Meyerowitz, Meinke and Koornneef all shared 'this openness idea, that we wanted to recruit people to the organism, because of the power of having a group'. For Somerville, this was all part of a conscious plan to create an Arabidopsis community, like the Phage Group or the *E. coli* community, 'that for me was the real idea, just to get a critical mass, so that we could accomplish larger things'. Meyerowitz remembers the early days as more of 'a series of basically chance events', which may simply be modesty on his part, given how hard he worked to promote his goal of developing the plant into a model system like Drosophila. But Arabidopsis proved flexible enough to be used by many different kinds of biologist with many different ideas. In fact, one of the plant's great successes was that it helped to build links between what had been two unrelated fields: the new molecular biology and classical genetics.

Meanwhile, not everyone was happy with the growing success of Arabidopsis. In 1986, Somerville published a review article suggesting what seemed obvious to him: that the cress was becoming *the* standard plant and everyone would soon be working on it.[15] He remembers that 'a distinguished maize geneticist' came to see him 'astounded at the idea that several young people, who did not know anything about maize, could think that a weed of no economic importance could displace maize as a model organism'. The maize expert dismissed the idea and urged Somerville to stop wasting his time on it.[16] Somerville believes that 'people seriously tried to suppress funding for *Arabidopsis*' and recalls that among some of those who worked on crops, the plant became so unpopular that it began to 'be called "the A-word" in the Department of Agriculture'.

Faced with indifference or outright hostility, the Arabidopsis community initially had trouble securing adequate funding. The established sources for plant research were focused on crops, and funding providers shared the difficulty of understanding why anyone would want to work on an insignificant little weed. According to Furner, there were times when Koornneef, despite his growing international reputation, 'could not even get money for doing *Arabidopsis* work'; the Dutch government would only fund work on tomatoes. Koornneef himself recalls that in 1982, when he submitted one of the first genetic maps of the plant for publication in the *Journal of Heredity*, he was asked if it could be 'reduced a lot' and told 'a much shorter article . . . would be far more acceptable', since according to one of journal's expert reviewers, '*Arabidopsis* does not appear to have excited as many researchers as one would have anticipated.' Koornneef managed to persuade the editors that the reviewer was mistaken and the complete article was eventually accepted.[17] Perhaps it is just as well that expert reviewers remain anonymous, since researchers like Furner describe the article that almost never appeared as 'the basis of modern plant biology'.

As it became clear that Arabidopsis allowed people to get results faster and cheaper, the funding increased, but as it did, the plant's opponents became more vocal. Furner remembers that the British Arabidopsis community 'became quite unpopular with a lot of people, traditional crop scientists' who complained that 'those bastards with *Arabidopsis*, they have got all the money'. As he puts it, 'everyone's got resentments against the wealthy'.

Dollars from DNA

While the Arabidopsis community was growing rapidly, biology was changing yet again, but this latest development began not in a lab, but on the stock market. On 14 October 1980, a company called Genentech (*Gen*etic *Eng*ineering *Tech*nology) Inc. was floated. Within minutes of the stock first trading in New York, 1 million shares had changed hands and the price leapt from $35 a share to $89. Without having a single product ready to sell, Genentech had raised almost $40 million in a few hours and the

355

company's founder, Herbert Boyer, made a personal profit of $60 million – for an initial investment of just $500. The biotechnology boom had begun.

Boyer's fortune had begun back in 1973 when, as we have seen, he and Stanley Cohen had discovered the process for transferring DNA between species. Originally, their goal had been to devise a tool for studying the actions of specific genes, but they soon realized that their genetically engineered bacteria were a factory that would produce a limitless supply of whatever it was that the inserted gene produced; a factory whose labour force would work for food, and cheap food at that. Obviously, this would prove very useful for researchers, but as Cohen and Boyer immediately realized, if the gene was for a protein that does something such as making the insulin diabetics need, they had a factory for making a potentially very valuable product.

In June 1973, when Boyer had told colleagues at a scientific meeting what he and Cohen had discovered, the implications were obvious; as one attendee put it, 'Well, now we can put together any DNA that we want to.' However, concerns were immediately raised over the ethical and safety implications of this new technique, and the conference asked the US National Academy of Sciences (NAS) to investigate the new technology. NAS promptly set up a committee; its first action was to write a letter to *Science* which brought a new word – biohazard – to the public's attention.[18] Cohen and Boyer were among the signatories of the letter, as was James Watson; they all supported the idea that scientists should agree not to use the new technology until its safety had been properly assessed. A moratorium on genetic engineering began immediately – an almost unprecedented event in the history of science which, inevitably, attracted much public attention. Some of the public were reassured by the scientists' evident sense of social responsibility, but others concluded that the new technology must be really dangerous if leading scientists were taking such an extreme step.

While scientists were debating the implications of the new technology, Cohen and Boyer were trying to patent it. In the late sixties, Stanford University, where Cohen worked, had set up a pilot programme to patent inventions made by its staff, both to encourage commercialization and to increase revenue, which became the Stanford University Office of Technology Licensing (OTL). To encourage academics to participate, patent income was split: one-third to the inventor, one-third to the inventor's academic department, and one-third to the university. Stanford had a long tradition of involvement in industry, and other universities also held patents, but these were mostly in obviously applied disciplines such as engineering and chemistry. Few academics outside these fields ever thought of patenting their work.

So Cohen was astonished when OTL suggested he patent the genetic engineering technology: he had not considered the possibility, even though he had recognized the technology's practical implications from the outset. Nevertheless, the idea of patenting basic research seemed wrong to him. Like many scientists, he saw science as a collaborative activity where credit was shared by the research team – that is one reason why scientific publications always acknowledge their debts to other researchers' earlier work. To single out one or two people as the 'inventors' of a technique, which is what a patent application requires, went against the grain of this well-established ethos. And there was a particular aversion to patenting medical discoveries – one potential application of this new technology; since curing disease was supposed to be more important than profit, most researchers felt such discoveries should be freely available. But Cohen was eventually persuaded that a patent would encourage commercial development of the technology, including new, potentially life-saving drugs. Nevertheless it is a measure of how uncomfortable he felt that he chose to renounce his share in the royalties, to make clear to his colleagues that he was not motivated by profits. Herbert Boyer was less concerned and let patenting experts take over and handle the whole process.

However, the patent application had hardly been filed when it stalled – the US Patent Office announced in 1976 that it would not grant patents on living organisms. Stanford filed a new

357

application that – unlike the original – sought to protect the process, but not its products, the genetically modified organisms themselves. In the same year, the US National Institutes of Health (NIH) issued their first guidelines for genetic research, which ended the voluntary moratorium on using the new technology. In the same year, Boyer announced the formation of Genentech, the world's first biotechnology company, which planned to commercialize the new DNA technology. Boyer and some of his colleagues immediately began work synthesizing the human hormone somatostatin in their university laboratory. The fact that explicitly commercial research was being carried out in a university biochemistry department caused controversy. Boyer was criticized and even personally attacked for bringing commerce into academe.

Despite these concerns, biotechnology grew rapidly. The initial moratorium was replaced by stringent NIH guidelines, which treated genetically modified organisms as if they were lethal disease-causing bacteria, requiring researchers to use safety procedures similar to those used in germ warfare research. However, in July 1978, the NIH issued a new, less restrictive set of guidelines to encourage research and further, even more relaxed guidelines were to follow. The NIH also announced that there would be no special restrictions governing intellectual property rights, such as patents, applying to biotechnology. In September 1978, Genentech announced the production of human insulin in vitro using the new DNA technology; the breakthrough was hailed in the media as the dawn of a new era.

The biotechnology companies had further good news in 1980, when the US Supreme Court ruled that living organisms could indeed be patented.[19] The decision allowed the Cohen–Boyer patent to be granted in December 1980. Within a fortnight, seventy-two companies had paid $20,000 each for an initial licence to use the technology; total licence income passed $1.4 million within two months of the patent being granted. Biotechnology became a key sector in the 1980s stock market boom, its growth greatly aided by the pro-business and antiregulation Reagan administration. The US government was particularly eager to develop new American industries to replace traditional ones, such as car- and ship-building, in which the rapidly growing 'Tiger' economies of the Far East were becoming dominant.

The new DNA technology had been assessed as safe by the National Institutes of Health. It had been patented, making it possible for anyone to license and use it. And its potential for profit had been dramatically demonstrated by Genentech's flotation. It seemed as though the dream that Hugo de Vries and others had shared – of creating new plants at will – was about to be realized, until Arabidopsis threw a spanner in the works; it refused to be modified. Belying the optimism of the Somervilles and others when the Agrobacterium technique had been announced, it proved much harder to make genetically modified plants than they had imagined.

Somerville remembers that even as biotech stocks were soaring, modifying plants was still proving very tricky; 'it took a student about a year to get *one* plant', he recalls, 'everybody was really struggling with making transgenic plants in the early- and mid-eighties'. For a while, every lab active in the field was facing the same difficulties; Somerville is convinced that if the Agrobacterium technique had not been perfected, Arabidopsis could never have sustained its momentum. Researchers, 'would have gravitated towards the easiest thing to transform, because what most people wanted to do was take a gene out and put it back in'.

Eventually, however, researchers worked out how to get Arabidopsis to accept the genes they wanted to insert.[20] The breakthrough came from an unorthodox experiment by Ken Feldman and David Marks, who were working at a biotech company in Palo Alto. They found that just soaking seeds in Agrobacterium before planting produced a few genetically modified progeny. Their technique formed the basis for the currently used methods and Arabidopsis eventually proved to be one of the easier plants to modify. It is now possible simply to dip Arabidopsis flowers into a solution containing genetically modified bacteria and produce genetically modified plants from the resulting seeds. Somerville observed that one of his recent students 'made 140,000 transgenic plants for one experiment . . . the technology today is incredibly easy'. And Meyerowitz agrees: 'it is the easiest multicellular organism in the world to transform, so that turned out to be a great motivator of future research and not an inhibitor'. For a few years it was unclear whether the technique would work, 'but it turned out that the plant was cooperative'.

Even before the initial difficulties of modifying Arabidopsis had been solved, new biotechnology companies were being established rapidly. Investors were attracted to biotech companies, convinced by the early successes with Agrobacterium that hugely profitable genetically modified crops were just around the corner. A young British woman, Caroline Dean, who had just completed a PhD in plant biology, joined one of the brand-new Californian companies in 1983, having done, as she recalls 'very little, real hardcore molecular biology'. Fortunately, the company already employed a number of molecular biologists – what they were short of was people who knew about plants. So Dean found herself paired with a colleague who was very familiar with bacterial genes, 'but did not really know one end of a plant from the other'.[21]

While Dean was teaching her co-workers about plants and learning about the latest molecular biology, she was also planting tulips in her garden, a reminder of those she planted every year in England. At first she could not get them to grow and, realizing that the Californian climate was too mild for them, she tried putting the bulbs in the fridge for six weeks before planting, to simulate a harsher winter. The tulips now grew, and Dean remembers thinking 'this is *bizarre*' and becoming intrigued by the phenomenon. In the mild climate of San Francisco it never got cold enough to convince her tulips that winter had come – and gone again – so they remained dormant. This got her interested in the genetic mechanism that controls the process called vernalization, an adaptation to life in cold climates, which helps to ensure that the flowers are not killed by frost before they can be pollinated and set seed.[22] Dean found that there was very little information about the process and, while she was wondering how to investigate the phenomenon further, she heard Chris Somerville lecture on Arabidopsis. Among the topics he mentioned were some mutants that had turned up in his lab which responded to cold treatment – just like her tulips. Arabidopsis had made another convert.

After a few years in California, Dean went back to Britain in 1988, ready to begin her new research with Arabidopsis. There was still almost no one in Britain working on the plant, but she was able to acquire seeds and a great deal of information from Maarten Koornneef and Karl Napp-Zinn, a pioneering Arabidopsis worker since the fifties, who was also just about the only person who had worked on the genetics of flowering time.

Dean had been hired, along with several others, by the John Innes research centre outside Norwich, which was in the process of creating an Arabidopsis programme. John Innes is a major institution in both horticulture and genetics, founded (as the John Innes Horticultural Institution) in 1910 and named after the successful London merchant whose legacy initially financed it. The institution was set up to make scientific discoveries that would benefit gardeners and horticulturalists and its first director was William Bateson, who – as we have seen – was the most significant British promoter of the then newly rediscovered Mendelism. When Dean arrived she found she had the benefit of a large horticultural staff, employed to support the scientific researchers. She took the dozens of packets of seeds she had gathered from Koornneef and Napp-Zinn and handed them over to the gardeners. 'It was,' she remembers, 'quite a shock to their system' when they discovered she was interested in 'growing a *weed*'. As the gardeners were planting, she began to gather the resources she would need to begin genetically transforming her plants.

By this time, Arabidopsis research was spreading rapidly in the US, where in addition to Meyerowitz and the Somervilles, several prominent molecular biologists who had made their names working on other organisms were also taking an interest in the plant. Some European politicians became concerned that, thanks to their generous funding and extensive facilities, the Americans might establish a lead in plant science and thus in biotech. So bureaucrats from the European Union (EU) approached the John Innes with an offer of funding for additional staff. Dean began to develop an Arabidopsis programme that was, she recalls, initially modelled on the *C. elegans* work being done a few miles away in Cambridge.

The politicians had their own motives, but neither Dean nor the other scientists were interested in a race or duplicating each others' work. She gathered information from US researchers and integrated it with the work being done in Britain and elsewhere, so they could share the mutations they were identifying and begin creating a physical map of the plant's chromosomes. An informal arrangement was made so that different chromosomes were worked on by different labs: two were done at John Innes, two in the US and one in France.

In the footsteps of phage

One of the chance opportunities that helped Arabidopsis on its way was that in the late 1970s, James Watson, then director of Cold Spring Harbor, decided the lab needed to offer a plant course to complement the phage course, which meant there would have to be a standard plant. During the initial consultations with interested researchers, few favoured Arabidopsis; since it was not a crop plant, it seemed unlikely that anyone would fund the research. So when the new course first began, students mainly worked with a variety of petunia, Petunia 'Mitchell' – one of the few plants that could be genetically modified at the time. However, Fred Ausubel, who helped run the course, became an early convert and it soon became an Arabidopsis course. Just like the phage course, the Cold Spring Harbor Arabidopsis course – and a similar one run in Cologne, Germany – became a major source of new researchers for the plant.

Watson himself remained interested in Arabidopsis and persuaded the National Science Foundation (NSF) to fund research on the plant. In 1989, Watson encouraged the NSF to organize a meeting on the future of Arabidopsis research; Somerville and Meyerowitz were among the distinguished geneticists who attended the meeting that gave birth to the Arabidopsis Genome Project, a project to publish details of the complete Arabidopsis genome.

Sequencing means working out the exact order in which the bases – the As, Cs, Gs and Ts – occur along the DNA of a chromosome; it is the newest and most powerful tool in the quest to identify the genes of a plant or animal, with a view to understanding what they do – and what we can do with them. The basic technique, devised in the late 70s, relied on making the bases radioactive – a technique comparable to Hershey and Chase's bacteriophage labelling. Hershey and Chase had been able to label the different chemicals that made up phage with distinct radioactive markers, and a broadly similar technique was used to label each of the different DNA bases, which were then exposed to X-ray film to produce a visible record of the sequence of bases. The exposed films then had to be read by eye and the lists of bases entered into a computer manually; the process was both slow and expensive. But all that changed in 1986 when the first automatic DNA sequencer came on the market. The new machines used a method of making the DNA bases fluorescent so that they could be scanned by lasers and the results entered directly on to the computer with minimal human intervention. The DNA being sequenced was cut into sections, using restriction enzymes (as we have seen), to create manageable pieces, but that creates a further problem of matching up all the short sections of the emerging DNA sequence to recreate the unbroken one. The process is broadly like taking numerous copies of an unknown novel, chopping the pages into fragments each consisting of a few words, and then trying to reconstruct the original story. Each fragment of DNA is searched for sequences of bases that duplicate those on other sequences and these overlapping sections are used to reconstruct the original sequence. It would be like finding fragments of the novel that read 'it is a truth', 'a truth universally', and 'universally acknowledged that'; the overlaps allow the sentence 'it is a truth universally acknowledged that' to be reconstructed. This time-consuming work is now also done entirely automatically by computers.

The new technology made sequencing the entire genome of an organism possible for the first time, and Watson put his considerable prestige behind the idea of sequencing Arabidopsis. But according to Somerville, 'after the meeting, he explained in private that he did not particularly care about *Arabidopsis*'; his eyes were already on a much bigger prize – the Human Genome Project. However, Watson had clearly decided that it would be useful to sequence a whole range of other, simpler organisms first, both in order to refine the technology and to help identify the functions of many different genes, which would ultimately help establish what human genes did.

To begin the Arabidopsis Genome Project, the NSF brought together an international steering group which included Meyerowitz, Somerville and Koornneef, as well as Caroline Dean and Richard Flavell, director of John Innes. Dean recalls that Flavell already had plenty of experience dealing with the governmental bureaucracies who would eventually put up the money, so he urged everyone to set targets for the sequencing.

An initial goal of completing the genome by 2003 was set and labs around the world started building on the resources – and on the friendships – that were already part of the international Arabidopsis community. A crucial aspect was the creation of state-funded stock centres, which distributed seeds and genetic material to any researcher who needed them, but of course, the centres relied on researchers to donate their materials in the first place. In addition to the major US efforts, a team in Japan were involved and the European Union sponsored thirty-three labs in nine different countries.[23] Botany had finally entered the world of Big Science.

Although the scale of Arabidopsis work was expanding rapidly, the plant's pioneers were determined that their community would not change in character. In 1993 Somerville and his colleagues won a $1 million grant from the NSF for sequencing work. As their computer began to produce lists of bases, they loaded all the data on to a public database called GenBank, accessible via the internet, before they had even looked at it themselves. Somerville argued that this was in the public interest; he did not see why his group should have an advantage over others. As he recalls, he simply told everybody involved 'that we were going to do it that way to set an example' and other researchers generally followed suit. Furner recalls the excitement of these early results as one genetic region after another just 'popped up on the internet'.[24] The internet made the sharing of this kind of data quicker and easier than ever.

Ironically, there were times when the biotechnology boom – which had done so much to promote interest in Arabidopsis – threatened to derail the work. Somerville blames the 1984 Bayh-Dole Act, which encouraged American universities to seek patents on research supported by Federal grants: 'that,' he says firmly, 'was a *disastrous*, incredible thing because it encouraged universities to set up these intellectual property offices that tried to control all these useless little bits of information'. These misguided attempts to patent basic science have, he suggests, had 'an incredibly negative effect on science, it is absolutely corrosive in my opinion'. Perhaps because of the lure of biotech money, on occasion some Arabidopsis research teams were less open with their results than Somerville and Meyerowitz would have liked. Somerville recalls that on the rare occasions this happened, 'a couple of us would phone them and have a chat, and they would change their policy'. Meyerowitz agrees that setting a positive example is the best way to maintain openness, so he never refuses to share anything, even when it has been requested by a researcher with a reputation for not reciprocating. He says, 'I don't really feel it is "my stuff" . . . the last thing I want to do is be hoarding pieces of DNA in my freezer instead of having people use them. It is just stupid.'

Despite occasional hiccups, the Arabidopsis community maintained its openness and grew rapidly, enabling the sequencing to progress rapidly. In 1998, the journal *Genetics* announced that '*Arabidopsis* has joined the Security Council of Model Genetic

Organisms', the favoured few with which 'all other organisms are compared'.[25] The article's author, Gerry Fink, had done important work on yeast – itself a founder member of the genetic 'Security Council' – before he started to take an interest in Arabidopsis. His high profile helped ensure that many members of the wider biological community paid attention to the plant work.

Fink noted that only a decade earlier most geneticists had ignored the cress, but by the end of the twentieth century 'there is virtually no major academic institution or agrotech company that does not have an Arabidopsis group'.[26] The fact that Arabidopsis had spread from academic labs to biotech companies was a mark of how useful the plant was becoming, and inevitably the links between academic and industrial research became closer. Several biotechnology companies, such as Monsanto, participated in the genome project, sharing their data in exactly the same way as the government-funded labs.[27] Their participation helped to ensure that the sequencing progressed more rapidly than had been hoped – the complete sequence was published in 2000 and Arabidopsis became the first plant to have its entire genome sequenced. This in turn dramatically accelerated plant research: it is becoming easier and easier to find a specific gene and – thanks to the ease with which Arabidopsis can now be transformed – discover what it does in a living plant. As Furner observes, 'the sequence has freed us from a lot of dull work'.[28] Among many other projects that the sequence has made possible is one to discover by 2010 what every one of the cress's approximately 25,000 genes does.

The genome sequence is also proving extremely useful to the plant biotechnology companies – that, after all, was why they participated in the sequencing effort. The increasingly close connections between the Arabidopsis community and the biotech companies have brought the tiny plant into the political limelight.

The 'Prime Monster'

In February 1999 Britain's *Daily Mirror* ran this headline above a caricature of prime minister Tony Blair as Frankenstein's monster: 'Fury As Blair Says: I Eat Frankenstein Food And It is Safe'. An opinion poll had revealed that the majority of Britain's population would not eat genetically modified (GM) foods – Frankenstein foods as the tabloids delighted in calling them – and that the public wanted genetically modified plants banned pending further research. Blair hoped to reassure the public that the new foods were safe by revealing that he and his family were happy to eat them every night.[29] This tactic backfired, since it inevitably reminded people of the mad cow disease (bovine spongiform encephalopathy, or BSE) outbreak, which had struck a decade earlier. At the height of the crisis the then agriculture minister, John Selwyn Gummer, appeared on national television, feeding his daughter a hamburger and assuring the viewers that British beef was safe to eat. It was not. A form of BSE (known as Variant Creutzfeldt-Jakob disease) was discovered in humans, and it proved to have been transmitted by eating beef, just as BSE had travelled from sheep to cows via high-protein cattle-foods that contained meat.

Blair's ill-advised decision to repeat Gummer's heartless stunt was a sign of how worried his government was by public hostility to genetic modifications. Britain's middle classes had for some time displayed an antipathy for processed and synthetic foods, preferring those they imagined to be 'natural', the unspoilt, wholesome foods of their grandparents. But the *Daily Mirror's* readers had traditionally been assumed to be indifferent to such fads; the fact that the chip-eating classes appeared to have joined the chattering ones in opposing GM foods was a cause for concern, especially for a prime minister who had promoted the new biotechnology so enthusiastically.

Hostility to GM foods was not merely confined to Britain: by the end of the 90s, German opinion polls showed 80 per cent were hostile to the idea, and most western Europeans shared their concern.[30] The story of how and why the British and most other western Europeans came to oppose GM food is a complex one – beyond the scope of this book – but it was certainly connected with BSE and other food scares. Mad cow disease affected British attitudes to food even more sharply than those of other Europeans: nostalgia for traditional food – already a powerful factor in shaping attitudes to new, unfamiliar ones – became even more pronounced after BSE.

And so GM foods became subject to intense suspicion. The biotech companies have successfully modified various crop plants in ways that make them easier to cultivate; for example, Monsanto produced a strain of soya beans that were resistant to the herbicide Roundup (the world's most widely used agricultural chemical – which Monsanto also makes). The beans, sold under the trademark Roundup Ready, can be treated with very high doses of the herbicide early in their growing season; non-GM beans would be harmed or killed by such doses. The herbicide kills both weeds and their seeds, thus allowing the farmer to do less spraying later in the season and reducing their overall spending on weedkiller. Monsanto and their supporters claim that because such GM crops use less herbicide overall, they are beneficial to the environment. Meanwhile, opponents of genetic modification cite independent scientific research showing that the biodiversity of fields of Roundup Ready crops (which now include maize, cotton, alfalfa and others) is much lower than amid conventional crops. That is, of course, precisely the effect that Monsanto intended – fewer weeds – but many environmentalists are concerned by the impact that this will have on the insects, birds and other wildlife that eat the weeds' seeds.

In addition to the environmental implications of specific genetically modified crops, the very word 'genetic' links GM foods in the public's minds with all kinds of unrelated biomedical technologies – from human cloning to fears that genetic testing might lead to a new eugenics. Journalists, politicians and scientists could not but acknowledge that – whether such concerns were rational or not – the public had begun to show a strong aversion to anything that might be regarded as 'interfering with nature'. It is no coincidence that Friends of the Earth has chosen to promote its anti-GM activities as a campaign for 'Real Food'. Nor is the concern restricted to activists; in recent years, every major British supermarket has begun to advertise the fact that their products do not contain GM ingredients; their market research suggests that even a suspicion of genetic modification might harm sales. At the same time, the market for – more expensive – organically grown food has been growing rapidly in Britain, as in many other wealthy countries.

By contrast, in the United States genetically modified food is widely consumed. There has been almost no consumer resistance, perhaps because most of the public are unaware they are eating it – the biotechnology companies having successfully persuaded the US government that there is no need to label GM foods. Indeed, the issue of labelling has become a major source of friction between the European Union and the US; the Americans do not label and so do not separate non-GM ingredients, such as the ubiquitous cornstarch that is in so much processed food. Many Americans regard European demands as an unacceptable obstacle to free trade, particularly since such labelling would inevitably harm sales and reduce exports. Meanwhile, a majority of European consumers insist on knowing what they are eating; trade wars over both GM maize and soya beans have threatened in recent years.

The Blair government's determination to see GM crops growing in British fields led it to announce field trials, in order to test whether such crops were safe. Environmental groups, inheritors of the 60s protest tradition, opposed the move, concerned that if the crops were grown in fields it would prove impossible to contain them – their pollen would carry the alien genes for miles, contaminating both non-GM and organically grown crops. When the government nevertheless went ahead with the trials, protesters from the environmental group Greenpeace cut down and destroyed a field of GM maize, developed by the biotech company Aventis, before it could flower. The protesters were subsequently acquitted of causing criminal damage, after a court accepted that their fears of genetic pollution meant they believed they had a lawful motivation for the attack. Encouraged by the verdict, protesters organized similar actions which eventually succeeded in halting most of the proposed trials. Meanwhile, farmers and scientists were dismayed; the farmer whose field was attacked accused Greenpeace of 'bully boy tactics to get their point across'.[31]

Although maize was the immediate focus of the protests, Arabidopsis research ultimately underlies the development of all GM crops. As Somerville says, 'All the innovation in maize comes from these big companies'; since they want to improve maize as efficiently as possible, 'they have big *Arabidopsis* programmes'. Because flowering plants are all more or less closely related, it is relatively straightforward to apply knowledge gained from the fast-growing, cheap cress to crop species.

The Arabidopsis community is baffled by the way that plant geneticists have been turned into bogeymen by the environmental movement. Meyerowitz believes that Europeans are trapped by 'a sort of a nature religion', a hostility to science as irrational as the anti-evolutionary fundamentalism that grips so many Americans. Its tenets, he claims, are 'that the things that your grandparents did are somehow . . . more pure, cleaner and more healthful'. He is convinced that there is no evidence for this view – but plenty of evidence for the opposite one: that GM food is safer and healthier – hence his description of the European anti-GM prejudice as a religion: 'it is counter-factual and faith-based and not amenable to discussion'.

Maarten Koornneef also regards public resistance as irrational, and traces it to the food scares, which have ensured that 'some people do not trust politics and scientists'; as a result, it has become 'easy to make people afraid'. Caroline Dean agrees: 'I have stopped arguing, because no one's going to listen,' but she is also puzzled and somewhat angry at the public's hostility; 'If the organic lobby had understood what GM could do for them,' she argues, 'it would have been a very different story'. She believes groups like Greenpeace have exploited anxieties over BSE and other scares in order to 'play on people's fears about food'.

Somerville is similarly bemused. He was on the scientific board of Monsanto for many years and is CEO of Mendel

Biotechnology (Meyerowitz is on the company's scientific advisory board). Somerville regards biotechnology as a way to fulfil his original dream of making the world a better place and is enthusiastic about the company, which – precisely because it is a commercial company – has the resources to tackle more ambitious projects than are possible in an academic lab. 'And of course,' he adds, 'I think we are going to make a lot of money in the end as well, which I consider an attractive thing.' However, he is proud of the fact that he has not lost his social conscience: Mendel Biotech has donated some of its discoveries to the Rockefeller Foundation, in the hope that they will prove useful in Africa.

Somerville is sceptical about at least some of the environmentalist's claims to care for the planet, regarding them as primarily anti-capitalist in their goals. In his view, real environmentalism would be using science to solve the world's problems. His current work, for example, is on plant cell walls. Unlike animal cells, which are bounded by a thin, semi-permeable membrane, plants have a wall of complex sugar-based molecules around their cells – which is why wood is dense enough to build furniture out of, and cotton strong enough to make clothes from. Yet despite these invaluable properties scientists still know very little about how cell walls are actually made. This may sound like a purely theoretical problem, but once again, Somerville is as interested in the practical applications as in the pure science. 'If you want to harvest solar energy,' he says, 'one of the really best ways is to take a highly productive plant and grow a couple of hundred million acres of it.' Then you can harvest the plant and turn its biomass into fuel. The attraction of this is that the plants soak up the carbon dioxide that is produced when the fuel is burnt, instead of releasing it into the atmosphere, where it contributes to global warming. He regards genetic engineering as vital to achieving this goal, since 'there is no such thing as an energy crop that has been in use for tens of thousands of years, so we are starting with really wild species'. Shauna Somerville, who runs her own research lab at the Carnegie Institution at Stanford, next door to her husband's, shares this vision; they are currently collaborating on creating energy plants – and Arabidopsis remains central to the new research. They hope to create a crop that can be grown in places where environmental degradation has made conventional agriculture impossible, thus providing livelihoods for people who need them.

If Somerville's vision is realized, there is every chance that his genetically engineered energy plants will be planted everywhere – except in western Europe, where resistance to GM crops shows no sign of abating. Anti-GM campaigners believe they have clear scientific evidence that genetically modified crops are bad for the environment and for human health – and they claim they would have even more if the biotech companies and their allies in government did not suppress research that might be hostile to genetic engineering.

Ian Furner has a slightly different view of GM from some of his colleagues: he recognizes the nostalgia for 'pure food' that Meyerowitz describes, and the strong emotions that food arouses in people. But ultimately, he thinks 'the thing that makes western Europe resistant to GM products is that there is very little benefit in it for the consumer'. Because European consumers buy so much processed and prepared food, cutting the cost of the ingredients by growing GM crops would make almost no difference to the prices consumers pay. So 'they are taking the perceived risk of eating this stuff for no perceived benefit'. Even if the risk is, according to every scientific test, non-existent, if there is no benefit, there is no incentive to switch. But – if Furner is right – perhaps energy plants might change the public's mind: they are not going to be eaten, and any commercialization of what is still a purely hypothetical plant is still decades away. And concern about global warming is currently much higher in Europe than in America; Europeans may feel that genetically engineered plants are a price worth paying to maintain their living standards while reducing carbon emissions. As the fossil fuels run out, the benefits of Arabidopsis research may finally be acknowledged back in Europe, where the plant itself was first identified.

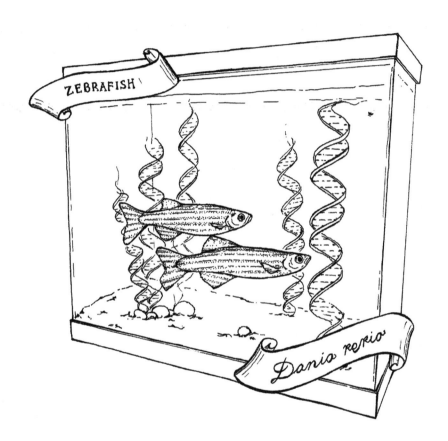

Chapter 11

Danio rerio: **Seeing through zebrafish**

In November 1814, Francis Hamilton finally became director of the Honourable East India Company's Botanic Garden in Calcutta – a job he had long aspired to. He had worked for the Company as an army surgeon for thirty years, though always hoping eventually to become a full-time botanist.

It may seem surprising that surgery might be considered a route into botany, but in Hamilton's day most medicines were still made from plants and since most doctors made their own medicines (especially in a colonial context) they needed to be sure they were dosing their patients with the right plants. It may seem equally unexpected that a commercial trading company should own a botanic garden, but the East India Company was no typical corporation. It had been founded 200 years earlier, when a group of London merchants were granted a monopoly over British trade with India. The Company gradually built up a private army to defend its increasingly lucrative Indian trade and, in 1757, stumbled almost accidentally into the role of imperial power when one of its officers, Robert Clive, defeated the Nawab of Bengal's forces at the Battle of Plassey. After a profitable period of collecting Bengal's taxes for the Mughal Emperor, the Company eventually became the effective government of India.

The Botanic Garden was added to the Company's holdings towards the end of the eighteenth century when another of its army officers, Robert Kyd, persuaded the directors to let him indulge his passion for botany and at their expense. He argued that they should follow the fashion for agricultural improvement by introducing new crops and new farming methods. His proposed garden would help identify new plants that could be developed into new crops, while at the same time it would grow spices, cotton, indigo, tobacco, coffee, sandalwood, pepper and tea – to see which could be successfully introduced into regions where they had not previously been cultivated.

Thirty years after the garden's foundation, Hamilton's directorship was short-lived; his long-anticipated reward had come late and he was already close to retirement. He returned to London in 1815 and presented the Company's Court of Directors with his entire natural history collection, the product of decades of travel and diligent collecting. He perhaps expected to be rewarded with a generous Company pension, which would allow him to write up and publish his travel notes, but he was disappointed – the Company accepted his collections with minimal thanks and less reward. An embittered Hamilton nevertheless proceeded to write *The Kingdom of Nepal and Genealogies of the Hindus* in 1819, and *An Account of the Fishes found in the river Ganges* three years later.[1]

In the latter book Hamilton described a tiny fish, a member of the minnow family, which he had collected in Eastern India. He gave it the scientific name 'danio' – derived from its Indian name 'Dhani' ('wealthy') – to which he added the prefix 'brachy' (from the Greek *brachus* – 'short'), so it became *'Brachydanio' rerio* (later renamed *Danio rerio*), but its distinctive black and silver stripes soon gave it the popular name zebrafish. For all the thousands of hours and hundreds of pages he had devoted to his enthusiasm for botany, it was this insignificant fish which was to prove his most significant scientific legacy. However, *Danio rerio*'s scientific career would have to wait until it reached America – Eugene, Oregon, to be precise – 150 years after Hamilton named it. Its adoption by Victorian aquarium fanatics and their successors eventually led the zebrafish into the lab.

Bringing the seaside home

It would be delightful, thought Mrs Bennet, if she and her daughters could persuade Mr Bennet to take them all to Brighton; 'a little sea-bathing', she believed, 'would set me up for ever'. Her sentiment would have baffled her ancestors. Just a few decades earlier the sea was widely regarded as dirty and dangerous, but by the time Jane Austen wrote *Pride and Prejudice*, salt water had come to be widely regarded as health-giving and doctors regularly advised real-life Mrs Bennets to take a discreet plunge from the back of a horse-drawn bathing machine.

Resorts like Brighton were fashionable, not least because very few could afford to visit them. However, within a few decades of Austen's death, the explosive growth of Britain's railway network eroded the seaside's exclusivity, permitting the unwashed masses to take advantage of cheap excursion fares. Once seaside holidays became cheap, Victorians of every class used the trains to reach the coast in ever-increasing numbers, but once there they confronted the problem that still faces everyone who visits the British seaside: there is nothing to do. It is too cold to swim and there is often little opportunity for sunbathing. And so Victorian England set to work to make the seaside interesting: building piers and bandstands, organizing concerts and plays, setting out deckchairs, selling ice-cream and in general inventing most of the standard entertainments for which British beachgoers are still routinely fleeced.

Victorian holidaymakers, unlike most of their modern counter-parts, often brought a passionate enthusiasm for natural history on holiday with them. Their delight in examining, classifying, studying and above all collecting samples of the natural world transformed the seaside from yards of dull, dank sand into a collection of intriguing new objects with which to satisfy the yen to collect and classify – seaweeds and seashells, crabs and fish, and the bizarre assortment of tiny creatures that live in rock pools.

As we have seen, the steam power that took Victorians to the seaside also powered their printing presses – and as the output of books and magazines soared, many were aimed at railway travellers, including hundreds of titles relating to the seaside. Among many others, the Reverend Charles Kingsley, whose aquatic interests are apparent in *The Water Babies*, also wrote a book about seaside natural history, *Glaucus: or the Wonders of the Shore*.

The founder of this popular genre was Philip Henry Gosse, now best remembered because of his son Edmund's depiction of him as a stiff-necked puritanical preacher in *Father and Son*. Victorian audiences knew him as the author of such books as *A Naturalist's Rambles on the Devonshire Coast*. Gosse *père* promoted the exploration of the seaside not only as a healthy and respectable hobby, but as a devout one; like many of his contemporaries, he believed that to study nature was also to study its Creator and learn to appreciate His wisdom and benevolence. Gosse also instructed his readers on how to bring the coast home, by building an aquarium. The year after *Naturalist's Rambles* appeared, he published *The Aquarium: An Unveiling of the Wonders of the Deep Sea*, which described how to build and maintain one, so that his readers could 'visit the caves of a miniature ocean / The gorgeous sea-flowers and worms to behold'.[2]

Gosse and other writers about the seaside helped to create a Victorian aquarium craze, which relied on the availability of cheap glass, just as Darwin's greenhouse did. All sorts of glass boxes – aquariums, 'vivariums' (for land animals, such as snakes) and Wardian cases – were mass-produced in increasingly elaborate designs and became popular items of domestic furniture. Displaying one in the parlour suggested not only good taste, but an interest in an educational, 'improving' hobby. William Alford Lloyd established his Aquarium Warehouse near London's Regent's Park, with a nearby factory to produce the aquariums. His 1858 catalogue ran to over 100 pages, advertising everything and anything the hobbyist could want, from a pint of seawater to microscopes, nets and even the aquarium's future inhabitants, for those with no time to catch their own. The warehouse displayed over 15,000 live specimens in fifty large tanks; Lloyd had a team of professional collectors to gather fresh specimens from the coasts.

American readers were also kept appraised of what became a transatlantic aquarium boom by journals and magazines. By 1858 there was enough interest in this latest British eccentricity for Henry D. Butler to publish *The Family Aquarium*, which he described in his preface as 'a complete adaptation to American peculiarities'. Butler enthused about the beauty and simplicity of the new hobby, extolling both its scientific and artistic qualities, its 'kaleidoscopic novelty; its tempting peculiarity, to thoughtful minds, as an introduction to natural history; all constitute an attraction as chaste as it is beautiful, as refined as it is irresistible'.[3] The book was a considerable success, and produced numerous imitators, along with specialist magazines and fish-keeping clubs. Americans were soon building their own aquaria and filling them with marine life as enthusiastically and piously as their British contemporaries.

In the early decades of aquarium keeping, the tanks tended to contain only local fish – the goldfish was the only exotic found in most domestic aquaria. But in the 1860s the growing hobby began to create an increasing demand for exotic and tropical fish. Importing such collector's items became a profitable business, despite the heavy losses that were usually suffered in transit. The more colourful, strange and unfamiliar the fish, the higher the price, and the aquarium market was swept by repeated fads, as today's fashionable fish became tomorrow's has-been. Life was not easy for the travelling fish: the *New York Aquarium Journal* described the fate of eighty-eight Kingyo goldfish, brought to the United States from Japan by a steamer captain. When the ship rolled, the fish were thrown against their tank's wall and died. Horrified to see his potential profit floating belly-up, the captain transferred the survivors to a smaller tank, and floated it within the larger one, like a ship's compass. Even so, only fifteen fish arrived in San Francisco alive and of these, 'eight subsequently died'.[4] A 90 per cent mortality rate was a strong incentive to improve transport conditions, and various ingenious containers were devised to bring the fish to American homes safely. Among these was a steel fish jug, with an integrated bellows to aerate the water – a bicycle pump was recommended as an alternative if more specialized containers were not available.

In addition to domestic tanks, large public aquaria grew up; huge displays centred on massive tanks that could hold much larger creatures. Almost sixty years after Jane Austen's Bennets planned their visit to Brighton, visitors could marvel at the 132,000 gallon tank in the town's new aquarium, which opened in 1872. A decade later – the year before Galton's laboratory at London's International Health Exhibition – Londoners were being entertained by an International Fisheries Exhibition, for which massive temporary displays were built. Bigger tanks needed bigger animals to fill them, and the permanent shows competed to display larger and more exotic creatures: captive whales rapidly became the ultimate attraction.

Enter the fish

Aquariums are designed to overcome the single biggest problem that prevents us observing living aquatic organisms: they can breathe under water and we cannot. However, there is a second obstacle to learning about life in the oceans, rivers and lakes; these bodies of water are substantially larger than the animals that live in them, making it difficult to locate whatever creature we might be curious about. This is also a problem for the organisms themselves – lots of space is good for avoiding competition for food, and those who would like to treat you as food, but it also makes it awkward for a species to ensure that its eggs and sperm are the same place at the same time. But finding a mate is only one part of the problem. Land-living creatures need to keep their eggs and sperm moist, so they fertilize their eggs while they are still inside their bodies and either keep them there (as in most mammals), or release them in a hard protective shell that will prevent them drying out and dying (birds and reptiles); effectively recreating the wet environments in which their distant ancestors evolved. By contrast, aquatic organisms can simply release their eggs and sperm into the water. Once one has found the other, the eggs develop and hatch in the same water in which the offspring

378

will spend their lives. However, this creates a different problem: the eggs of fish and similar creatures have very soft shells and these, like the tiny hatchlings that emerge from them, are extremely vulnerable to predators. These factors have led the majority of aquatic organisms to evolve a strategy that maximizes their chances of winning the Darwinian lottery: buying a lot of tickets.

Species like *Homo sapiens* have the opposite strategy: producing a small number of large offspring and looking after them carefully for a very long time. As every parent knows, that is expensive: it takes a lot of food, energy and time (plus an impossibly large number of brightly coloured plastic objects) to take your child from conception to college. The alternative, which every woman in labour must have considered at some point, is to produce lots of small, cheap offspring, deposit them in a pond, and let them fend for themselves. Unfortunately our biology, to say nothing of our legal and social systems, does not permit such a strategy, but most human beings would not do it even if we could – we simply could not cope psychologically with knowing that the vast majority of our offspring would never make it to adulthood.

Fish have a different approach, as do sea urchins, starfish and most amphibians, such as frogs and salamanders. Most of their eggs and sperm will be eaten before fertilization; most of the fertilized eggs will be eaten before they hatch; most of the hatchlings will be eaten before they mature; and most of the mature offspring will be eaten before they can breed. Natural selection ensures that the only way to survive this wasteful breeding strategy is to produce huge quantities of eggs and sperm – vast clouds of both. To increase their chances further, many aquatic species synchronize their spawning, hoping to overwhelm the appetites of their predators by simply producing too much for even the most rapacious. The results can render the ocean milky white, with billions of eggs and many more billions of sperm.

External fertilization, the lack of a hard protective shell, and the absence of parental care all attract marauders to the eggs of marine organisms. Among the curious predators are human zoologists, who have been gathering the eggs of aquatic organisms for centuries in order to study the process by which a fertilized egg develops into a new creature, a study known as embryology. Embryology began with aquatic species and Aristotle; he was in all likelihood the first to record the fact that fish eggs 'are small and grow quickly', which allows their development to be observed.[5] Their size was a problem for Aristotle, who of course did not have access to a microscope. Scientifically minded humans were not able to add much to what he had seen until the late eighteenth century, when microscopists began to observe developing fish eggs more closely. The study of embryology grew markedly in the nineteenth century, especially in the German-speaking world, as the techniques and methods of the lab-based worm-slicers displaced old-style natural history. Many of the biologists we have encountered – including William Bateson and Thomas Hunt Morgan – were originally embryologists. Embryology was probably the most advanced biological laboratory specialization by the beginning of the twentieth century, which is why so many of the most important biological research institutions of the century – from the *Stazione Zoologica* in Naples to the Woods Hole Marine Biological Laboratory in Massachusetts – were built by the sea, close to supplies of the research subjects on which embryology has traditionally relied.

Hamilton's little zebrafish first arrived in Europe early in the twentieth century, thanks to the demand for interesting new fish to fill the hobbyist's fish tanks. They were first recorded in German aquaria in 1905 and, unsurprisingly, German biologists seem to have been the first to make use of zebrafish, both for embryological and breeding experiments. The fish arrived in the United States soon afterwards and rapidly became popular; as early as 1917, an American tropical fish magazine was describing the little fish as the 'familiar *Danio rerio*'.[6]

Charles W. Creaser – a typical early-twentieth-century American embryologist – helped *Danio rerio* make the transition from the domestic fish tank to the laboratory. He was a professor of zoology at Wayne State University in Detroit, where he worked close to the water at the university's biological station on the shores of Michigan's Douglas Lake. Creaser did not record when or why he first noticed the zebrafish, but in 1934 he wrote an article arguing that they were ideal for embryological research.[7] He had obviously been keeping them for a while, since he included a long list of their practical advantages. First, they were cheap: at the time he was writing, 100 fish cost just $15; an initial breeding stock would keep a researcher in eggs forever. Second, they were easy to look after. Not only would they eat commercial tropical fish foods, but the average university biology department already had an even cheaper food source close at hand: Creaser's zebrafish happily ate Drosophila, 'which we have introduced as a live food for the aquarium'. He advised any would-be fish keepers to use the 'vestigial winged' Drosophila, 'the ordinary mutant genetic strain of fruit fly, raised in pint milk bottles in the manner recommended by the geneticists', because it 'has the advantage of being unable to fly'.[8] Providing fresh, easy-to-catch meals for zebrafish was probably not a purpose that Morgan and his fly boys had in mind when they first identified this mutation in Columbia's fly room, but one lab's scientific breakthrough soon became the other's convenience food.

The fish are not quite as small as the flies, but they only grow to about 1¼ inches (30mm) long, so they do not need much space; Creaser noted that 'we keep as many as 50' in an aquarium 6 × 9 × 26 inches (150 × 230 × 660 mm). Breeding was also straightforward, although 'the beginner may have some difficulty in discriminating the sexes'. Once that had been learned, however, a pair of fish will produce eggs every twelve to fourteen days, some even more frequently; although Creaser noted the need to provide some largish stones in the tank, under which the eggs could be laid, as 'a protection from cannibalistic parents'. For Creaser, the eggs were the entire purpose of fish-keeping: in less than half an hour each tiny female can produce several hundred eggs, eggs that had several especially attractive qualities for his purposes. Many fish eggs are laid glued together; they form a large sticky mass which protects individual eggs from being washed away and lost – an advantage for the fish, but rather a nuisance for biologists who want to examine individual eggs. Fortunately, zebrafish eggs are one of the exceptions.

However, the main advantages of the eggs are that they are tiny (on average, just over a millimetre across) and so 'will hatch in Petri dishes, Syracuse watchglasses, finger bowls, or other dishes satisfactory for observation with the microscope'. They also develop fast, taking only a couple of days to become recognizable fish. But, finally – and most importantly – they are completely transparent. That meant a single egg could be placed under a microscope and watched as it hatched, releasing a tiny embryonic fish that soon began to move around its minute puddle, under the biologist's intent gaze. The embryos are as transparent as the eggs, so every detail of their development could be observed. As Creaser noted, they 'are being used in our student laboratories to demonstrate the development of a living egg and the circulation of the blood.'[9] His students could watch the blood vessels developing and the fish's blood begin to move around its body.

Other American scientists were also studying the zebrafish in the 1930s. Among them was Warren Harmon Lewis, a pioneering embryologist who in 1942 took advantage of its transparent eggs to make one of the earliest time-lapse films of a developing embryo. At the same time, the first American breeding studies were being done by Hubert Baker Goodrich, at the Marine Biological Laboratory at Woods Hole, Massachusetts; Goodrich was a pioneer in fish genetics and included zebrafish in some of his earliest experiments.

Work continued through the 50s and 60s. A tiny handful of researchers maintained colonies of zebrafish, but there was no sign of the fish becoming a scientifically important organism; there were still many more zebrafish in pet shops than there were in labs.

The heat, the flies
In a 1940s lecture theatre at Cornell University, a biology class is trying to concentrate; it is eight o'clock in the morning and for

most of the undergraduates, introductory fruit fly genetics is a little more than they can take, given the time of day. But one of the students, Lotte Sielman, is sufficiently alert to register the lecturer saying excitedly that 'we have a student who has already published some research on this subject: George Streisinger. Where is he?' There was no answer; Streisinger, it turned out, was still in bed. Lotte remembered his name when she heard it again; her then boyfriend mentioned that he had a housemate, George Streisinger, who he described as a 'crazy guy', adding that he thought Lotte would 'probably like him'.[10] She did. They were married in 1949, the day before Lotte graduated.

A few years earlier, Streisinger had passed the intensely competitive entrance exam for the newly founded Bronx High School of Science, which today counts six Nobel Prize winners among its graduates. At school, George joined a reptile and amphibian club and alarmed his mother by returning from camping trips with salamanders, lizards and snakes. His enthusiasm for natural history helped him get a part-time job at the New York Aquarium, assisting one Dr Myron Gordon. Some fish-enthusiasts still feed their charges a homemade fish food called 'Gordon's Formula', which he invented.[11]

Gordon had been a tropical fish enthusiast since his teenage years; he had also done a science degree at Cornell, where among the teachers who had encouraged his interest in genetics was Rollins Emerson, who as we have seen, also did so much to inspire Barbara McClintock. In the 30s, Gordon found that despite economic depression the tropical fish hobby was booming, and he persuaded hobbyists and commercial fish breeders to help finance a series of fish-collecting trips to Mexico, where he discovered several new species.

In New York, Gordon founded a genetics laboratory to study his new finds, originally based at the Aquarium, but it soon moved to the top floor of the American Museum of Natural History, off Central Park. Gordon got three rooms the bird department was not using, with high ceilings and a glass roof that faced west; during New York's summer heat, they turned into a greenhouse and the temperature sometimes hit 32°C (90°F). However, while Gordon and his young assistant, George Streisinger, sweated over their work, they found that the fish flourished in the heat and produced eggs even more rapidly.[12]

Streisinger's passion for science also won him a summer at Cold Spring Harbor working with Theodosius Dobzhansky. Dobzhansky became a father figure and mentor to Streisinger, even taking him on a family holiday to California, and it was thanks to Dobzhansky that Streisinger had managed to publish on Drosophila prior to sleeping through his first genetics class at Cornell. For a while, flies almost displaced fish in his interests; when Lotte went to Colorado for a family vacation, Streisinger sent her some bananas and vials, so she could trap the local fruit flies for him.

George Streisinger was a naturalist by upbringing. As a child in Budapest, György, as he was then known, spent his afternoons butterfly hunting or looking after the pigeons his father bred on the roof of their apartment building. The pigeons had to be abandoned when, along with thousands of other Hungarian Jews, the Streisinger family left Budapest in 1939, as the Hungarian regime followed Nazi Germany in passing anti-Jewish laws. The eleven-year-old György became George soon after the family arrived in America, in March 1939; his future wife Lotte's family had fled Munich the previous year.

When Lotte and George finally met at Cornell, they went to chamber music concerts together, and on regular bird-watching walks. Lotte remembered that their dates often ended up in the romantic surroundings of a local swamp, where George 'would take me to observe the mating rituals of tree frogs and salamanders'. After graduation, Streisinger went to Bloomington, Indiana, to work with Salvador Luria. Meanwhile Lotte remained in Ithaca, finishing her master's degree; George hitch-hiked the 700 miles back and forth as often as he could in order to see her.

Streisinger moved on from flies to phage when he worked with Luria and naturally that took him back to Cold Spring Harbor and the phage course, where he met Max Delbrück. Like all potential students, Streisinger had to take the tough mathematical entrance exam before he could join the course; Delbrück was adamant that entrants should come from the 'quantitative side of biology, not the "stamp-collecting" side'.[13] Despite his enthusiasm for traditional natural history, Streisinger got in, and he and Lotte spent the summer in Cold Spring Harbor in 1949. Conditions there were still pretty primitive; the rather basic sewage system led to considerable debate about whether the abundant watercress that grew near the lab was safe to eat.

The phage course was informal, but rigorous – if a student's presentation seemed unconvincing, Delbrück would simply announce, 'I don't believe a word of it.' However, the fortnight ended with improvised graduation ceremonies, where everyone (including Delbrück) would dress up in comic costumes before the ceremony concluded in a traditional water fight. After his PhD with Luria, Streisinger made the obvious next move and spent three years doing post-doctoral research with Delbrück at Caltech; he and Lotte drove to Pasadena and their first daughter, Lisa, was born soon after they arrived. The family loved the atmosphere of Caltech, especially Delbrück's house, where evenings of music, spaghetti and Shakespeare were accompanied by endless talk of science. In later years, the Streisingers' home in Oregon and the parties they threw were modelled on Delbrück's.

After Caltech, Streisinger took a job back at Cold Spring Harbor, but he also managed to get funding to spend a year at Cambridge University, a Mecca for molecular biologists. Phage was still king in Cambridge: Streisinger worked alongside Seymour Benzer, another graduate of the phage course, while across town, Sidney Brenner was also at work on phage. However, as we have seen, Benzer and Brenner would soon take the elegant, stripped-down genetics of phage and apply them to larger and more complex organisms; Streisinger would eventually do the same, by applying the tools of molecular biology to his childhood enthusiasm – tropical fish.

Going west

During the late 50s, George Streisinger became a well-known phage researcher, publishing important results on key aspects of viral genetics. After his year in Cambridge, he was offered his first university post at Brandeis University near Boston; the offer was tempting, but he and Lotte were not sure whether they really belonged in the prosperous, conservative suburbs of Boston. Ever since student days, they had been involved in radical politics. Like their friends, they were always ready to drop their studies to go out and campaign, whether it was to oppose the excesses of McCarthyism, for an end to atomic testing, or against racial segregation. As the 60s dawned, America's war in Vietnam would become the defining radical cause. Both Streisingers remained activists throughout their lives. George took the job at Brandeis with reservations, but before they could move, he got a better offer: a visitor dropped in to Cold Spring Harbor to offer him a job at a new department, in a largely unknown college in the middle of nowhere.

The visitor was Aaron Novik, a physicist who had abandoned atoms for genes after working at Los Alamos on the atomic bomb, describing his move as one 'from death to life'. Like Streisinger, he had done the phage course and started working on bacterial genetics. He was looking for a place to pursue his science of life when the University of Oregon offered him the opportunity to start a new department. The offer was a direct consequence of the alarm that had greeted the launch of Sputnik. What concerned Americans was not the satellite itself, but the Soviet rocket that had put it into orbit; it had been designed for the much more sinister purpose of landing a hydrogen bomb on Washington – the satellite was merely a dramatic demonstration of the 'missile gap' that had suddenly opened up. Despite spiriting the Nazi scientist Werner von Braun, designer of the V2 rocket, out of Germany at the end of the war, the Americans knew they had nothing to compare with the Red missiles. So, the year after Sputnik, the National Defense Education Act was passed by Congress, making science and maths an educational priority. Oregon's state university was one of many across the United States to benefit from the government's panic-driven generosity.

Novik was initially unimpressed with Oregon, suspecting that he would be bored there, so he took a post in San Diego instead, but he found the bland southern Californian climate even duller than the wilds of Eugene, so he returned in 1959 to establish the university's Institute of Molecular Biology – the first university department in America to include the phrase in its name.

Novik was the son of an immigrant Jewish socialist and tailor from eastern Europe. The family mostly spoke Yiddish at home and were desperately poor during the depression; Aaron made his contribution to the family's survival by loading tomatoes on trains bound for the canneries, an experience he never forgot. Like Streisinger, Novik was a bright boy and won a scholarship to the University of Chicago to study chemistry.

After his PhD, Novik became a physical biochemist, but guilt over his work on the Manhattan project haunted him for his entire life. After the bombings of Hiroshima and Nagasaki, he could not wait to leave Los Alamos, telling one newspaper 'it smelled like death'. He – like his mentor, the Hungarian-born physicist Leo Szilard – became a pacifist and attended anti-nuclear conferences and meetings. After the war, it was Szilard who invited Novik to come to Cold Spring Harbor, to take the phage course and join him in 'an adventure in biology'.[14]

That 'adventure' had ultimately led Novik to Eugene, in the Willamette Valley, right in the heart of Oregon, where he was soon joined by Frank Stahl. Stahl had also been at Caltech and had then gone to work at the University of Missouri, but found being the only molecular biologist there too isolating. Not long after Stahl arrived in Eugene, Novik began what would become a tradition of the new Institute of Molecular Biology (IMB): he invited the newest recruit to help choose the next one and Stahl suggested Streisinger, whom he knew from his Caltech days. The Streisingers moved to Eugene in 1960 and the IMB gained its third member. Its fourth, Sidney Bernhard, would join them soon afterwards.

Like the Noviks before them, the Streisingers were a little worried that Eugene would prove too isolated, but they soon discovered the town had a lively arts scene and they took to the rural life immediately. George and his daughter Cory took up goat-keeping; at one stage they had a herd of eight Nubians. George helped set up a local goat's cheese marketing cooperative and became a qualified goat-judge, spending almost every weekend at little country fairs, using a scorecard like those issued at corn shows to assess the merits of the animals; scenes that would have been entirely familiar to Darwin, as he moved among the rabbit-breeders and pigeon-fanciers of Victorian England, gathering information for his books.

From time to time, however, the charms of even prize-winning goats would pall, and Streisinger would take the family off to the big city, to visit Chicago or New York and do a bit of shopping. He sought-out magic supply shops, where he would acquire the paraphernalia he needed to do the magic tricks with which he loved to entertain young children. When he had stocked up on decks of marked cards, silk scarves and disappearing cabinets, he would go in search of pet shops, to buy new fish for the large and handsome aquarium he had built under the stairs of the family home.

Back in Oregon, Streisinger wanted to move on from phage work, feeling the field becoming overcrowded. He wanted to do something 'big' and – like Benzer in Cambridge – was interested in applying molecular genetics to the problem of understanding behaviour. While Benzer turned to flies for his work, Streisinger decided to work on fish and, after considering various alternatives, settled on the hardy little zebrafish.

Although the fish had been around labs for decades, there was no zebrafish community, no conferences or journals, and no body of shared expertise; none of the communal resources that had helped make the fly and phage communities so successful and productive. Streisinger's main intention was to make the zebrafish into a model organism like Drosophila. He could have continued with flies, as Benzer had done, and built on the fly people's vast body of knowledge, but while flies had many advantages, they had one important drawback – they are invertebrates. Unlike most of the large animals on this planet, which includes everything from pygmy shrews to blue whales, and of course, human beings, insects have no backbones. Our spines, composed of separate bones called vertebrae, have played a key part in allowing us (and the shrews and the whales) to get much larger and more complex than insects. Vertebrates – such as fish, amphibians, reptiles, birds and mammals – have a distinctive anatomy, including a complex nervous system (one of the spine's functions is to protect its key component, the spinal cord, that connects all our peripheral nerves to our brains).

Streisinger decided that he could not hope to understand the complex behaviour of animals with this kind of nervous system unless he studied something that had a similar body plan. But obviously, a *really* big, complex animal, like the mouse, would be too great a leap in complexity from the microscopic world of phage. The fish seemed possible; simpler than a mammal, more complex than a fly, small enough to be easy to breed, and possessing all the advantages Creaser had identified thirty years earlier.

Streisinger's years of keeping fish at home had also taught him that *Danio rerio* were easy to look after; they are the classic beginner's fish, the ideal species for a first aquarium (and cheap enough to replace if early attempts fail). But he soon found that breeding enough zebrafish for experimental work was very different from keeping a dozen in tank at home. Diseases and parasites kept wiping out his fish stocks. He kept adjusting the conditions in an effort to beat the parasites, until he had the fish swimming about in a cocktail of toxic chemicals that he knew must be affecting his results. He was on the brink of abandoning the fish altogether when he acquired a new assistant, Charline Walker (later Durachanek). Charline proved to have a way with the fish, and she soon helped him clean up the tanks, installing new pipes and food sources and devising better ways to keep the water and air fresh. The fish began to flourish. Streisinger and

Walker installed their fish tanks in a Second-World-War-era Quonset hut (similar to a British Nissen hut), part of an abandoned army barracks on the edge of the campus. At times, Streisinger must have thought he was back on the top floor of the Museum of Natural History in New York – in summer the steel hut's roof became hot enough to fry eggs on, and had to be sprayed with water to prevent the fish cooking. In the winter, the Quonset hut got too cold for the tropical zebrafish, so Streisinger and Walker installed a series of small electric heaters and fans around their fish; from time-to-time they would have to stop their experiments to put out small fires that the heaters had caused.

Fish farming
It took Streisinger almost a decade to develop the zebrafish into a model organism, suitable for scientific work; a less dedicated biologist might have given up, especially as it was hard to get graduate students or post-docs to come and work on such an uncertain and long-term project. Most molecular biologists still worked on small organisms – bacteria and viruses – they were complex enough, given the tools available to biologists in the 60s, so fish looked almost impossible. So, despite Streisinger's personal reputation, there were few people willing to put money into his new project; and for several years, these fears seemed justified, because Streisinger did not seem to be producing any interesting results.

The fish work would probably have failed had it been based anywhere other than Eugene, which provided the support that kept the zebrafish afloat in its early years. Inspired by the idealism of its founders – and perhaps by the laid-back West Coast spirit that typifies Oregon – the IMB was run as what some have described as a 'scientific commune', where all resources were shared. Novik encouraged researchers who were doing more conventional projects – and thus producing publications, recruiting students and post-docs and attracting grants – to share their government grants and equipment with Streisinger. Novik may have absorbed these principles from his socialist father; he certainly practised a little practical socialism at Eugene, which allowed Streisinger to get more resources for his fish than might otherwise have been the case.

Once the practical problems of zebrafish farming had been solved, Streisinger was able to turn to the scientific ones. One of the major challenges he confronted early on was as a direct result of *Danio rerio*'s useful habit of mating all the year round; there were plenty of eggs and embryos to work on, but they were, of course, all diploid – they had inherited one set of chromosomes from their fathers and another from their mothers. That made it hard to identify rare recessive mutations, because – as Mendel had found a century earlier – the fish would need two copies of the mutated form of the gene before the consequences of the mutation could be observed and Streisinger could study its effects. Creating families of fish that he could inbreed to get the homozygous form would have been enormously labour-intensive. Once he had found a rare combination he was interested in, it would continue to be time-consuming to breed more fish carrying the same mutation. It is not possible to self-fertilize fish the way you can with plants.

Those who had followed Sidney Brenner's lead and were working with the little nematode worm, *C. elegans*, had a clear advantage – the worm is hermaphroditic, which means that it can easily be self-fertilized, as easily as one of Mendel's pea plants. Streisinger knew that some animals can reproduce either sexually or asexually, like Mendel's hawkweeds, although in animals asexual reproduction is called parthenogenesis (from the Greek, meaning 'virgin birth'). So he tried to persuade his female zebrafish to reproduce parthenogenetically. He and Walker spent several years working on this problem, trying to get the eggs to 'activate', to start developing without the usual trigger – fertilization by sperm, which of course he had to avoid in order to keep the male's genes out of the equation. They found a way of inducing parthenogenesis in the early 1970s, and over several years refined the technique to the point where they could produce to order embryos that had only one set of genes. The trick was to expose the sperm to intense ultraviolet light: that made them genetically impotent, but did not kill them, so they could still penetrate the eggs and stimulate them into development. (The fish's external fertilization made this technique much simpler, of course – doing the same thing with, say, a mouse would have been much more difficult.) Once it had been perfected, Streisinger and Walker's technique allowed recessive mutations to reveal their consequences in a single generation, directly through the female line, avoiding having to cross-breed and select for several generations. Their work allowed standard genetic techniques, like mapping, to be done much more rapidly and efficiently.

After nine years of work, Streisinger finally had the combination of practical and genetic techniques that would allow him to start making real progress with his wider project – locating the genes that controlled the fish's behaviour. However, the trick he and Walker devised to simplify their work had a much greater implication: their fish had only one parent, so instead of carrying a mixture of parental genes, they were exact genetic copies of their mothers – clones. In 1981 Streisinger had become the first person in the world to clone a vertebrate and, not surprisingly, his fish eggs appeared on the front cover of *Nature*. They made quite a splash: Streisinger was interviewed by the newspapers and often asked about the ethical and political implications of cloning. The *Chicago Tribune* even carried a cartoon showing a fisherman's wife telling her husband, 'You cloned them, you clean them.'[15] Among those who took an interest in Streisinger's work were a group of feminists of the 70s women's liberation movement; one of his colleagues recalls that they regarded Streisinger as having proved that 'the male part of the conception process [was] irrelevant'.[16]

Despite the publicity surrounding his cloning triumph, Streisinger had not published anything on zebrafish during the 1970s: at an IMB party he was satirically awarded the degree of 'Doctor of Delayed Publication' in recognition of his services to procrastination. The drought was partly a product of the long years it had taken to master the fish-keeping and genetic techniques necessary for his research, but he was also kept busy by his anti-war activities. Like many early members of the IMB, he was politically active, especially in protesting against nuclear weapons and the Vietnam war, and when the war finally ended he became active in environmental causes.

Rachel Carson's book *Silent Spring* had raised widespread concern about the indiscriminate use of pesticides. Thanks to the environmental campaigns Carson's book helped create, the US Environmental Protection Agency was established in 1972 and one of its first actions was to ban the insecticide DDT, which had been at the heart of Carson's concerns. However, environmentalists in Oregon soon discovered that the US Forest Service was using a powerful herbicide on the woodlands around their homes (forestry is a major industry throughout the Pacific Northwest), whose main component was the extremely toxic chemical dioxin. The spraying caught the attention of the anti-war movement, because dioxin was the main component of Agent Orange, the herbicide that the US military had used to try to destroy Vietnam's forests so that they would no longer provide cover for the Viet Cong. To the peace movement, this civilian use of military technology was both sinister and eerily familiar; DDT had first been used to delouse soldiers during the Second World War and the first crop-dusting planes were mostly air-force surplus. In the 70s, George Streisinger would occasionally neglect his zebrafish to get involved in the campaign against dioxin; he presented expert testimony in a successful court case against the US Forest Service, which ultimately resulted in the chemical being banned.

In 1984, the Institute of Molecular Biology was preparing to celebrate its twenty-fifth birthday. The pioneering institute in the middle of nowhere was now very much on the map, in large part thanks to Streisinger's zebrafish. Several of his colleagues at Oregon – Chuck Kimmel, Monte Westerfield and Judith Eisen – had also started working on the fish, extending its usefulness by studying the genetics and development of its nervous system.

It seemed that Streisinger would soon be proved right; the fish was beginning to be recognized as a significant experimental organism and a world-wide community of zebrafish researchers was taking shape. But on 11 August 1984, a few weeks before the IMB's birthday party, his colleagues were shocked to learn that Streisinger was dead; he had suffered a fatal heart attack during the final exam for a scuba-diving class.

The IMB's birthday turned into a celebration of Streisinger's life and achievements; a Streisinger Memorial Lecture series was inaugurated, the first one given by Seymour Benzer. Other stars of molecular biology, including Alfred Hershey and James Watson, were in the audience. Streisinger's colleagues at Eugene were determined that the zebrafish community should survive him; they had already begun to discover for themselves how scientifically useful the fish could be, but their promotion of *Danio rerio* was also influenced by a desire to see a successful, global research community become a fitting memorial for Streisinger.

Oregon rapidly became for the zebrafish what Cold Spring Harbor had been for phage; everyone with an interest in the fish spent time there. Streisinger's colleagues set up an informal course, modelled on the phage course, to teach would-be researchers the basics of looking after the fish, as well as the essentials of zebrafish genetics and their embryonic anatomy. Westerfield helped write *The Zebrafish Book*, a compendium of the Oregon labs' shared expertise, which – while obviously not a best-seller in the conventional sense – became the bible for anyone who wanted to start a fish lab. Chuck Kimmel began to map out the zebrafish's wider experimental potential, realizing that it could answer even more biological questions than Streisinger had anticipated.

Like so many biologists before him, Kimmel was intrigued by that biggest of all big questions: development. How does one cell – a fertilized egg – turn into a complete adult fish? He worked initially at the development of the zebrafish's brain, and also worked on trying to understand how and why each tissue developed as it did from the earliest stages of the embryo. He produced what biologists call a 'fate map', showing the destiny of each cell as it divides and differentiates. This brought the fish into the mainstream of vertebrate embryology and development studies – a vindication of Creaser's original predictions for it. As Kimmel unravelled each cell's path, it became clear that the zebrafish was going to be even more useful than people had realized: many of the genes that controlled the most basic developmental processes – such as determining where the front and back of the animal were going to be – turned out to be the same in zebrafish as in other animals. As Kimmel put it, the fish was not just a fish, but also a frog, a chicken and a mouse; what biologists learned from the fish could be applied to many other kinds of animals, including ourselves.

Kimmel's work was crucial in demonstrating the enormous possibilities of zebrafish for developmental work. Thousands of miles away in Europe, the fish was also making waves: the highly regarded Drosophilist Christiane (Janni) Nüsslein-Volhard and some of her colleagues in Tübingen, Germany, were also starting to take an interest in the fish, intrigued by possibilities of this new model organism. They started breeding huge numbers of fish and screening them for mutations, just as they had previously done with Drosophila, building up libraries of mutations that made it possible to understand the relationships between specific genes and their effects on the whole organism. The worm-breeders, as the *C. elegans* community had become known, were making progress, and the fish-researchers became convinced that they could do the same.

As the genetic work progressed across a range of organisms, it confirmed what Kimmel's developmental studies had shown – mother nature is a conservative. Evolution effectively operates on the principle that if it ain't broke, don't fix it; once natural selection has effectively sifted though the random variations in the genes to find and preserve a working method of getting something done in an organism, it tends to stick with it. If a gene makes a useful protein in a very simple, ancient organism like a bacterium, the same gene is very likely to be found manufacturing the same protein in many different organisms. For example, there is a zebrafish mutant simply known as '*no tail*'; an almost identical mutation in the analogous gene produces the same defect in mice (where the mutant is called *Brachyury*, from the Greek for 'short tail'). The mouse and fish genes where the defect appears are both descended from the same ancestral gene. In practice, this means that even very distantly related groups turn out to have a lot in common: not only does the same gene perform a similar function in fish and humans, it often does so in yeast as well.

Visiting fish world

Within the Department of Developmental Biology of Stanford University's School of Medicine are two rooms packed with fish-tanks, each containing carefully filtered water and between ten and fifteen zebrafish – thousands in total. Looking after this compact aquatic empire is Will Talbot, a young zebrafish researcher who trained under Chuck Kimmel at Oregon, and is now head of his own lab. Why is the medical school housing all these fish? Because Talbot hopes that his research will eventually shed light on the causes of – and help find a cure for – diseases like multiple sclerosis.

Five thousand miles away in Britain is an equally impressive institution – the Wellcome Trust Sanger Institute, just outside Cambridge, where crucial parts of the human genome were sequenced. Here Derek Stemple, a lapsed engineer with mathematical leanings, takes care of another series of fish-rooms; his zebrafish are – to his surprise – helping him unravel the mystery of muscular dystrophy.

Meanwhile at Cambridge University, Kate Lewis (another graduate of the Oregon fish labs) is in charge of a new lab, where her fish are already shedding light on the workings of specific kinds of nerve cells known as interneurons, which play a role in diseases ranging from schizophrenia to Creutzfeldt-Jakob Disease (the human form of mad cow disease). An hour's train-ride away in London, Steve Wilson has replaced his original experimental organism, the strange-looking Mexican salamander the axolotl, with rooms full of carefully quarantined, lab-bred zebrafish. These are gradually revealing the patterns that underlie the development of the forebrain, the part of our brains where it seems that our conscious thoughts, emotions and some of our memories are housed.

Talbot, Stemple, Lewis and Wilson are just a handful of the thousands of researchers currently working on zebrafish in over 400 laboratories in thirty countries.[17] Collectively, they are a tribute to Streisinger's original vision. And yet the zebrafish community is still a young one: when the current generation of lab heads first entered the field, it was still unclear whether the fish was really going to take off. When Will Talbot first contemplated switching from flies (which he had worked on for his graduate degree), he remembers wondering if choosing zebrafish 'was a crazy thing to consider doing'. There were attractions, one of which was that fish research was still new, so there were fewer people working on it. He realized that 'if the system were to go well, you would have more problems to pick from, there would be more to do and it might be easier to make a unique con-tribution'.[18] But that was only *if* the system went well; for every young researcher who was excited by the thought of developing a new organism, there were many more who were put off by the thought of having to reinvent all the basic tools that already existed for mice or flies.

The scientific newness of the fish was not the only attraction; they were not just small and easy to care for but also – compared with most vertebrates – simple. Steve Wilson was still working on axolotls when he read some of the first zebrafish papers in the mid-1980s. To him, with his interest in brain development, it was the simplicity of the fish's neural anatomy which appealed.[19]

One would expect scientists to have rational, practical and scientific reasons for choosing the fish, and they do, but what is more surprising is that many researchers seem also to have simply fallen in love with zebrafish. Even after years of watching them every day, Lewis thinks they are still 'the most beautiful research animal that I have worked on or seen'. Every year she watches her students experience the same enthusiasm: 'when people first work on them, one of the things that is most exciting is the fact that they *are* transparent, that you can see them easily down the microscope'. Not only are they easy to watch, they change constantly: 'when they have just been born, you can go away for five or ten minutes, come back and they have changed.' One day eggs are being fertilized in preparation for an experiment and the following morning they have become 'a little thing that looks like a fish and they are already moving'.

Talbot had a similar experience when he arrived at Oregon, never having seen a zebrafish before. Despite their minute size, he found the embryos 'quite spectacular. Truly beautiful'. Just like Warren Harmon Lewis, who first filmed a fish embryo half a century earlier, Kimmel and the others at Oregon had taken advantage of *Danio rerio*'s transparency to produce what Talbot calls beautiful movies. He describes himself as one of the many people who were first attracted to developmental biology because they 'just love looking at embryos'. As he puts it, 'I like to see these time-lapse movies of a fairly nondescript ball of cells turning into a vertebrate animal.'[20] You can watch this extraordinary process with almost any embryo, but the fish are among the quickest.

The fish's speed of development is not just a gift to impatient researchers and film-makers, it has a very practical benefit: lots of fish mean results to analyse and, fairly quickly, to publish. That, as we have seen, was one of the reasons Drosophila became so popular in the early twentieth century, and Talbot was initially worried the fish would be too slow to compete with flies, especially since he had to spend time developing the biological tools he needed for his research. But the fish soon proved they were fast enough; not as quick as flies, of course, but that also turned out to be an advantage. Practically every living thing obeys the rule 'live fast, die young' – the smaller the organism, the faster it lives and the shorter its lifespan. Zebrafish are certainly small, but they are still much bigger than flies and so they live longer. In a fly room, a lot of time is spent renewing the stocks, making sure any interesting mutants are constantly being fed and interbred with each other, to ensure a supply of live flies with whatever peculiarity is being studied. Because the fish live longer, a researcher can spend less time on maintenance and more on science, so although the fish breed more slowly, Talbot found the pace of research was, surprisingly, 'similar to the flies; once you have a mutant you want to analyse, [the research] can go very quickly . . . It was quicker than I'd imagined.' And it was not just the lab work that grew quickly.

Can zebrafish fly?

In 1990 the Oregon fish labs held a conference to assess the prospects for zebrafish research. One of the reports, while acknowledging that zebrafish might be '*Drosophila* with a spine', wanted to know 'But can they fly?'[21] Would the fish field take off, would it continue to grow and attract new researchers, or would it fade after the initial burst and prove to be another Oenothera, a passing fad that would soon be history?

The answer was soon clear: the zebrafish field began to grow at an exponential rate in the 1990s. A decade-by-decade comparison of the numbers of scientific papers which mention zebrafish is very revealing: in the 50s, there were only four; seventeen in the 60s and just thirty-seven in the 70s. During the 80s, 121 papers were published – twice as many as in the previous thirty years, but in the 90s there were almost 2,000 scientific papers that related to zebrafish. And the growth shows no signs of stopping – in the first five years of the twenty-first century, there were over 5,000 new papers.[22] As Judith Eisen notes, 'No other developmental model has risen to prominence so quickly.'[23] The question is why – what has made this fish fly so high, so quickly? It was not the only model organism around, or even the only vertebrate (mice and frogs both had a much longer history in the lab). It was not the smallest, cheapest or fastest-growing organism (bacteria, viruses and yeast were all way ahead by these criteria, and had well-established communities promoting them). It was not even the only fish – the Japanese Medaka fish, the puffer fish (*Fugu*) and the stickleback all had a toe-hold in the lab, and have hung on to it. But none of them have experienced anything remotely like the explosion of work on zebrafish. So why *Danio rerio*?

Zebrafish were lucky in attracting Streisinger's attention, not just because of his dedication to them, but because he took them to Oregon, a place whose relaxed pace suited them as much as it did him. Streisinger's doctorate of delayed publication seems to have set a precedent for the Eugene researchers. Kate Lewis recalls that when she was there, doing her post-doctoral work, new papers were published, apparently from the laboratory she worked in, and she would think, 'Who is this author? I've never met them' – that was because they had left – sometimes eight years earlier, but their work had only just found it is way into print. As she puts it, with dry understatement, the Oregon labs are 'not so worried about finishing things fast, or publishing things fast', probably because when the fish work began, 'they were the only people doing it, so they could take their time'. One of the luxuries of being a pioneer is that the Oregon labs could afford both to do complex experiments, and to take their time over them. By contrast with other labs, there was less pressure on young researchers to be finishing PhDs or post-docs. Derek Stemple agrees, but describes the Eugene atmosphere in even simpler terms, 'They are all sort of hippies, in some funny way – not literally – but they have a very relaxed approach.' Certainly the idealism of sharing funds to allow work to develop at its own pace, seems a reflection of the US West Coast's wider, 60s-tinged culture – and very unlike the cut-throat competition that characterizes some scientific fields.

Being among the first in the field is a luxury few scientists ever experience and it tends to promote generosity; the Morgan fly lab shared their flies and data with anyone who was interested, confident that their leadership of the field could not be eroded. The fish pioneers shared everything for similar reasons, but they were also doing deliberately something Morgan's fly boys had done almost accidentally – they were establishing their fish as a scientific instrument.

In the early twentieth century, biologists worked on problems and selected an organism – or several – that looked promising to work on. Not any more. Today's biologists typically tackle the puzzles that interest them using a 'model system': at its heart is an organism, but one that has been carefully developed, using techniques such as selective breeding, to be amenable to their purposes. Then the organism has to be documented: just as Wistar rats were shipped with an operator's manual – the book of rat statistics that made them so useful – today's model organisms come complete with instructions. These cover everything from how to feed, breed and care for them to describing their normal behaviour and development. These days, the information is more likely to be on a website than in a book – and it will include resources that were unimaginable a century ago. Zebrafish websites (like those for every other widely used organism) include downloadable DNA sequences, computer programmes for data analysis, and descriptions of known genes and their effects. There are stock centres which can supply specific strains of mutant fish; fill in a form on-line and ready-to-hatch fish eggs will arrive in the mail within days. (Currently these facilities are only available in the United States; European zebrafish researchers are still lobbying to establish similar resources on their side of the Atlantic.) Just thirty years ago, as we have seen, finding out who else was working on a specific organism or problem could take months in a library. Today, the names and addresses of everyone with a specific research interest can be located in seconds; and their publications are just as easily accessible – most scientific articles now appear on the internet as soon as – or even before – they appear in print.

The Oregon fish community knew that they needed to make all these resources available before their fish would fly. Being pioneers made it essential to share everything; getting other researchers involved would produce a community and in time, the resources. But Kate Lewis remembers it as even more communal than that. In the early years, 'It was not just people sharing data; people left whole areas of research alone, because they knew somebody else was working on it' – avoiding competition ensured that efforts and resources were not duplicated unnecessarily, but it also avoided the ill-feeling that competition can cause. And, for a few years, practically everyone who worked on zebrafish – like the fish themselves – came out of Eugene, or had been trained by someone who had been trained at Eugene, and as they learned they absorbed Oregon's quasi-hippy ethos as well. Almost the entire community belonged to what Lewis describes as the Eugene 'lineage', giving those who worked on the fish the feeling of being almost a 'close-knit family'.

Things began to change somewhat in the early 1990s, when the German Drosophila biologist Janni Nüsslein-Volhard in Tübingen and one of her former students, Wolfgang Driever, in Boston began work on parallel large-scale genetic screens of zebrafish, which together became known in zebrafish world as 'the big screen'. Stemple was a post-doc in Driever's lab, and while he acknowledges that Streisinger and the Oregon labs created the whole zebrafish community, he feels it was Nüsslein-Volhard who saw the power of treating the fish like the fly, by breeding them on an almost industrial scale and then creating huge numbers of random mutations. Screening is the process of sorting through the mutant fish, patiently working out which genes have mutated. Just as in Columbia's original fly room, eighty years earlier, spotting mutants reveals what genes the animal has and what they do, because the mutations reveal what happens when those genes malfunction. In Stemple's view, 'systematic screens were really the key'. Nüsslein-Volhard proposed to do what is called 'saturation mutagenesis': mutating and screening so many fish that new mutations are no longer found. That is much harder in a vertebrate than in a fly, because the genome is so much larger. When she originally proposed saturation it seemed impossible; it took someone of her international reputation to persuade her fellow researchers it was doable and to persuade funding bodies to finance it.

The scale of the big screen produced a self-sustaining expansion of the zebrafish community. As Lewis recalls, 'There were suddenly all these post-docs who had been involved in the screen who then got their own labs . . . and started producing their own post-docs and graduate students and the whole thing starts to snowball.' Steve Wilson agrees that the screens were a turning point, because they produced more mutants than anyone could possibly work on. Some feared the rapid growth in the community and the influx of researchers with no direct ties to Oregon might weaken the zebrafish community's distinctive openness, but while the atmosphere has undoubtedly changed, the supportive culture remains – the range of interesting projects emerging from the screens has meant that there has been no need to fight over the fish; there are too many mutations for that.

Like many of the post-docs who worked on the big screen, Derek Stemple took a selection of the mutants that most interested him when he left Driever's lab to start his own and has 'been studying them ever since'.[24] The mutations that caught Stemple's interest were those involved in the development of the notochord – the embryonic structure that will become part of the fish's backbone. The screen revealed seven separate genes that together control the way the notochord develops; if any one of them mutates so that it no longer works properly, the result is very short, dwarfish embryos. Stemple's team have dubbed the genes *Happy, Sleepy, Grumpy, Bashful, Sneezy, Dopey* and *Doc.*

The notochord is a structure that is common to every vertebrate, but also to a wider group of organisms called chordates, which also includes creatures such as sea-squirts (tunicates) and lancelets. Although vertebrates make up only about 5 per cent of all the animals that have lived on the earth, they are a group that includes many creatures humans are especially interested in, not least ourselves. Stemple's lab wanted to understand the development and evolution of the notochord because it is inherently interesting, but as they worked they discovered something completely unexpected. The notochord has a flexible sheath which protects it; once the sheath is developed fully, it inflates, making it rigid. If the inflation were to start before the sheath was complete, it would rupture, so during development a specific genetic switch has to be thrown when, and only when, the sheath is ready to be inflated. The signal is chemical – a protein that binds to a separate chemical site, called a receptor; the right protein attaching at the right position sends the message that the sheath is completed.

Stemple's team discovered that the genes *Sleepy*, *Bashful* and *Grumpy* played some important role in sending that signal. To try to find out exactly what it was, they tried knocking out the receptor for the signalling protein – no receiver would mean no signal. As Stemple says, 'it gave the fish muscular dystrophy. A very severe form of muscular dystrophy.' Bad news for the fish, obviously, but – on the face of it – not otherwise scientifically interesting; however, evolution's conservatism in fact makes it very interesting indeed. Further research revealed that there is a human equivalent to the *Bashful* gene, whose absence causes severe, congenital muscular dystrophy. Stemple's team had no idea they might be developing a model of human muscular dystrophy when they started their research – but that is what they ended up with.

Human muscular dystrophy is a group of related, inherited diseases one of which, Duchenne muscular dystrophy, is connected to a gene on the X chromosome, which means it primarily affects boys (since they have only one X chromosome, they only need to inherit one copy of the faulty gene to trigger the disease; girls have two X-chromosomes, so would need to inherit the rare gene from both parents). The gene in question makes a very large, complex protein called dystrophin, which is crucial to setting up the mechanical connection between a muscle cell and the rest of the body. Every time we move a muscle, that connection is placed under mechanical stress. In people without dystrophin, the muscle cells tear themselves apart and die. Boys with the faulty gene are usually unable to walk by the time they are twelve years old – by twenty they may need a respirator simply to breathe. There is currently no cure, but the zebrafish may offer sufferers some hope.

As Stemple explains, it 'turns out that zebrafish have dystrophin. And furthermore, that mutations in dystrophin in zebrafish lead to muscular dystrophy.' But while humans do not display any symptoms of the disease until they are six, by the time the fish are forty-eight hours old the loss of dystrophin is already apparent – their muscle cells have started to die.

The speed with which the fish show symptoms is only the beginning of their usefulness. Because the embryos are so tiny, it is possible to keep 1,000 of them alive in the tiny wells of what is called a standard ninety-six-well plate – a plastic dish, not much bigger than a playing card, that has ninety-six tiny depressions on its surface. Into each of these wells, the researchers can put a minute dose of a different chemical, to see what happens to the ten tiny embryos in that well. As Stemple explains, 'We are just looking for compounds that will rescue the fish from this mutation.' They have begun with a group of several hundred drugs that are known to be safe for humans, but are no longer patented – such as aspirin and paracetamol (acetaminophen); the intention being to try different combinations to see if they can find a new use for existing drugs. The logic is simple; human muscular dystrophy is, thankfully, rare, so 'What are the chances that a substantial number of human muscular dystrophy sufferers have been treated with a whole collection of drugs?' Stemple asks. 'Almost nothing.' So trying existing drugs is 'kind of a no-brainer – if you do not do that, you are being silly'.

If none of the hundreds of off-patent drugs help, there is a second set of what are known as 'druggable' compounds; those that have similar chemical properties to existing drugs. There are many thousands of compounds in this group, and of course it may take a combination of them to cure the fish. The number of combinations is so dizzyingly high that there would be no chance of ever screening them in humans, even if every muscular dystrophy sufferer volunteered to take part in trials. However, the zebrafish embryos, packed into their tiny homes, make it possible to screen tens of thousands of chemicals in months rather than centuries. The odds are still against the researchers, but in 2004 a similar screen of just 5,000 compounds found two which suppress a lethal mutation in zebrafish known as '*gridlock*', in which the main artery carrying blood from the heart does not form properly. Stemple is optimistic about his own search because in the case of *gridlock*, these two chemicals – found by simply screening at random – 'fix the defect and the fish survive', so 'I'm quite hopeful.'

Will Talbot is also interested in the fish's medical potential; he describes the fish as 'a test-tube, it is a way to get an answer, to learn what a gene does'. He is fairly sure that one of the reasons Streisinger was initially interested in working on a vertebrate was that he hoped to find a model for human diseases. However, any attempt at finding cures has to be built on basic research; the normal processes of development have to be thoroughly understood before researchers can begin to get a picture of what is going wrong when disease strikes. Talbot's lab works on the development of the fish's nervous system, particularly the formation of myelin, the white layer of protein and fats that insulates nerves; just like the plastic insulation around a telephone cable, myelin allows the electrical nerve impulses to be carried without interference. Myelin is an evolutionary innovation in vertebrates, so flies or worms are not directly useful to researchers. Understanding myelin formation has important implications, since diseases like multiple sclerosis are essentially caused by the myelin breaking down and failing to repair itself; the messages that should be flowing through the patient's nervous system slow down and become distorted, they can end up in the wrong nerve fibre, or simply not get through at all. The result is symptoms that vary from dizzy spells to slurred speech, chronic pain and difficulty in walking.

Just as in Stemple's lab, the fish allow the possibility of massive screens that can identify both the genes that are involved in making myelin and the mutations that cause those genes to malfunction. While the medical possibilities are significant, Talbot is anxious not to exaggerate these; if he or Stemple find anything interesting it could be many years before it results in a usable treatment – and current legislation would require that it be tested on mammals before it can be tried on humans.

Meanwhile, Talbot is excited by the prospect of understanding the fundamental biology of development. As he says, there is a general question that continues to come up over and over again in biology: 'I have one cell – a fertilized egg – it has a genome within it. Hundreds of different cell types come out of that cell and they all have basically the same genome and yet somehow they are different' – despite their all containing the same full set of genes. So the questions he is asking – which, as we have seen in earlier chapters, have been asked many times before – are, 'What turns the genes on in the appropriate cells, in the appropriate positions and times? How do they become different in the first place?'

Gradually, in labs like Talbot's all over the world, the answers to these questions are taking shape. They are universal questions, ones that apply to every living thing on this planet – many of the Arabidopsis workers have similar interests. Development can be – and is – studied in every organism, so why use zebrafish? Talbot finds that whenever he discusses his work with non-scientists they rapidly grasp that 'if you can study it in fish, all things being equal, you'd rather do that than in a mouse or a dog'. Choosing the simplest organism that will help answer your questions has a practical and an ethical benefit in his view, 'if you can still do your science and not use as many mammals, that is cheaper and just altogether better'.

Other researchers share his view. One scientist I interviewed, who preferred not to be named, told me he was driven by pure scientific curiosity, so would 'find it quite difficult to work on mice', because of the need to kill them to get their embryos. By contrast, 'I do not have any guilt about working on fish. They lay eggs and . . . if I just leave them alone, they eat them' – just as Creaser had found, seventy years earlier. For this researcher, the cannibalism is a plus, since 'if I take those eggs and do some experiments on them, I do not have any worries about what they are feeling'. He accepts that if he became fascinated with a biological problem that could only be studied in a mammal, he would have to learn to use mice, but that would be 'a little bit more ethically tricky for me'.

As we have seen, the ethical implications of using animals in the laboratory have been the subject of often fierce debate for well over a century, and some fish researchers find discussing their ethical qualms difficult. The anonymous researcher stressed that he was anxious not to pass judgement on colleagues who work with different organisms, acknowledging that 'everybody draws their lines in different places. I'm happy working on fish . . . I'm not sure I'd be happy working on mice. I definitely would not be happy working on primates.' He is certainly not alone in feeling this way, but because many research scientists feel themselves under attack by the animal rights lobby, few are willing to discuss their feelings on these issues, for fear of being seen as disloyal to colleagues whose labs are being firebombed.

Ecce Danio

The medical uses – and relative lack of ethical concerns – of research on fish represent one more factor that helps explain their rise. But what has made the zebrafish successful is the system as a whole: the tangled web that includes the fish, the researchers, their laboratories and instruments, lab technicians and pet-shop owners, and the shared knowledge of the fish that they have all accumulated over many decades. All these disparate elements are held together as much by friendships as by funding applications; generosity and genetics have gone hand-in-hand.

In many ways, the zebrafish world is just like that of Arabidopsis researchers, or the fly workers, the worm-breeders or any of the dozens of other scientific communities that have built up around research on a single organism. Except that the fish community feels that it owes its existence to one man in particular: George Streisinger. As Talbot comments, his phage work had established his credentials and 'his reputation was such that he could easily go for a while on a risky new project and continue to garner the support he needed'; he could stick with the fish during the long years it took to produce any tangible results. However, the importance of Streisinger's contribution meant that 'the system really could have died with George; it must have been touch-and-go at that point'. Chuck Kimmel and Streisinger's other colleagues in Oregon 'stepped in and kept the science going'. In Talbot's view, 'everyone who works on the fish now owes everything not just to George, but just as much to Chuck, I would say'. Kate Lewis agrees; Kimmel, Eisen and Westerfield are still at Eugene, still working on fish, and 'if they write an article on the history of the field, if they stand-up at a zebrafish conference and do some sort of address, they will always pay tribute to George'. Kate Lewis also stressed how important Charline Walker was in providing continuity after Streisinger's death: 'she was the person who trained new people in Eugene . . . Her knowledge of and love for the fish were extremely important.'

Much has been done since Streisinger's day – Nüsslein-Volhard and Driever's screening work made an enormous difference to the growth of the field and did a great deal to convince any remaining doubters that there was nothing eccentric about choosing to work on the fish. And, of course, there is still much to be done. Steve Wilson observes that other vertebrates, especially the mouse, are still much better funded than zebrafish and key resources are still lacking that will need funding; as we have seen, there are, for example, still no major European stock centres, so European researchers cannot yet order mutants off the internet.

Hamilton's tiny fish are currently vital to some of the most sophisticated, high-tech biology being done anywhere in the world; their genome is currently being sequenced and once it is completed, even more possibilities will be opened. Yet for Lewis there is something appealingly old-fashioned about zebrafish: 'you can see individual cells moving in the embryo', branching out to build the fish's nervous system or coalescing together to make its jaw. As she sits at her bench, watching, Lewis feels that the fish have 'allowed us to go back to a much earlier stage in biology', with nothing more than a pair of eyes and a microscope, 'suddenly, we can learn a lot again, just by looking'.

Perhaps more than anything else, the simple pleasure of looking – and learning to understand what you see – has driven much of the history of biology. While I was visiting the fish labs, I got to look down a microscope and see the tiny, day-old embryos moving around. With the slightest touch of the focus knob, you can see right through them, watch their tiny hearts beat and follow the blood as it moves through their perfect glassy bodies.

Watching the fish, I felt I understood the fascination of biology more clearly that I have ever done before. I understood how Charles Darwin could spend whole days in his greenhouse watching the 'wonderfully crafty & sagacious' tendrils of his passionflowers, twining and growing. How Hugo de Vries could be so passionate about his primroses, and why Sewall Wright could not give up his guinea pigs. They needed to know and before they could know, they needed to see.

OncoMouse™

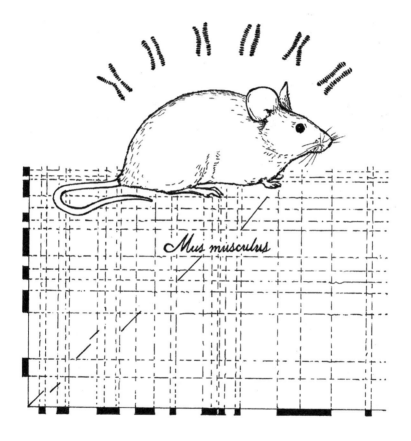

Mus musculus

Chapter 12

OncoMouse®:
Engineering organisms

In 1985, researchers at Harvard University produced a genetically modified mouse. That in itself would not have made news: putting genes into mammals had proved difficult, but modified mice have been around since the 70s. Yet this mouse made headlines; what had been inserted into their genes was not mouse, but human DNA – a gene that is implicated in causing breast cancer. Three years later the university was awarded a US patent for what was dubbed OncoMouse® – the first patented, transgenic animal.

For the scientists involved, the mouse was the logical culmination of many decades of work. Beginning with William Castle's work at Harvard, mice had followed a scientific career very similar to their relatives, the rats and guinea pigs. Nineteenth-century 'mouse-fanciers' bred various strains with distinctive coat colours and in the early twentieth century these had been used to test Mendelian inheritance. While Sewall Wright had decided to focus on guinea pigs, another of Castle's students, Clarence Cook Little, decided to adopt the mouse, specifically to investigate how genes are involved in human diseases.

In 1929 Little founded the Jackson Laboratory, in Bar Harbor in Maine, and within a few years his researchers and mice had together established the link between cancer and viruses: the connection Emory Ellis was exploring at Caltech when he met Max Delbrück. The inbred, genetically pure mouse strains used at the Jackson Lab were soon in great demand among researchers of all kinds, and the lab had to start charging for them to cover their costs. The JAX mouse (its name derived from the lab's telegram address), like the Wistar rat, became a familiar sight in many laboratories; almost a piece of standard equipment comparable to the Bunsen burner or the Petri dish. JAX is now copyrighted and it has been estimated that more than 95 per cent of the world's laboratory mice are descended from JAX mice and that at least seventeen Nobel Prizes have ultimately originated in the Jackson lab.[1]

The researchers who developed OncoMouse could reasonably claim that if Castle, Wright and Little had been able to use genetic-engineering technology, they would have done. They would have seized the chance to produce organisms with the precise combinations of genes they were interested in without having to go through the tedious business of breeding and selecting over generations. It is often argued that genetic engineering is ultimately no different from selective breeding – just dramatically quicker. Nor was selecting strains of mice that were predisposed to cancer a new development: it was the existence of such strains that had made them so useful in earlier research. Cancer runs in some mouse families, and breeding these luckless animals allowed new treatments to be tested much more rapidly than would have been possible if scientists had waited for naturally occurring cancers to develop. The cancer-prone mice also led to the discovery of what are called oncogenes, genes that produce a predisposition to cancer and have equivalents in humans (it was an oncogene that had been inserted into OncoMouse, hence its name). Nor was OncoMouse the first animal to be commercialized, as the Wistar rats and JAX mice show. And the possibility of patenting living things was, as we have seen, also not a new question: the US Supreme Court had permitted the first such patent eight years before OncoMouse received protection.[2]

So why the fuss about OncoMouse? One source of controversy was within the scientific community itself: the patent brought a renewed intensity to the debate about the way patents threatened the long-standing scientific practice of sharing research materials and knowledge freely. But the wider public were more often disturbed by the implications of inserting a human gene into a non-human animal; for many, the boundary between us and other animals is too important to be crossed so casually.

Any experiments on mice – including genetic ones – arouse the same objections to vivisection and cruelty that have been voiced ever since animals first entered laboratories. All use of animals in laboratories is controversial, as indeed it probably should be: given that our taxes fund much of this research and many of us will benefit in some way from its results, the public is arguably obliged to consider what is being done in its name and on its behalf. Some argue that all animal experiments are morally wrong, a position that has the merit of being unambiguous; but most of us are willing to tolerate some suffering if the intended benefits are clear and we consider them important enough. To take two purely hypothetical cases: if it were possible to cure AIDS by killing one single mouse, most of us would not hesitate to sacrifice it; but we might baulk at being asked to sanction the torture and killing of a million mice to cure dandruff. In reality, each of us draws our own uneasy line somewhere between those two extremes.

Ideally, the smallest possible number of animals should be used – none at all should be the goal, but that does not yet seem possible. Those who claim that all animal experimentation is unnecessary are probably mistaken; if researchers could stop using mice, they would already have done so, because although mice are much cheaper to keep than larger animals, they are still expensive. Most research requires thousands of them to be fed, cared for, kept free of diseases and housed in such a way as to prevent them injuring themselves or each other. (These are, incidentally, conditions that are in many respects much more pleasant than those endured by wild mice.) When there are cheaper ways to do the research, most scientists adopt them enthusiastically – that has been a key factor in the rise of the zebrafish. But, as we have seen, fish cannot be used to investigate unique features of mammals, any more than flies can be used to investigate the unique features of vertebrates; some mammals are still needed. And current legal requirements for drug-testing mean that, even if the fish were to reveal potential treatments for human diseases, such treatments would still have to be tested on mice before human trials could begin. Although considerable research is currently being done on alternatives to using animals, more is undoubtedly needed. The regulations that presently require using animals need to be reviewed regularly as alternatives become available, to ensure that we are not inflicting unnecessary suffering through sheer force of habit, but for the moment animal testing is widely considered essential. Given the amount of suffering humanity has been spared by generations of mice, rats and guinea pigs, it would seem appropriate to show our gratitude by eliminating their use in laboratories as rapidly as possible.

The right organism for the job?
One objection to using animals for medical research is that they do not provide a good enough model of human diseases, implying that any results will be useless and the animals will have suffered in vain. When Felix d'Hérelle went to the USSR to pursue his work on phage therapy he was, as we have seen, partly motivated by his belief that Western medicine was studying 'artificial' diseases, since neither guinea pigs nor rabbits normally suffer from cholera or typhus. His hope was that that the Soviet Union, which he believed to be more rational and scientific than the West, would allow him to experiment on people.

However repellent d'Hérelle's reasoning, his objection does highlight another reason why mice, guinea pigs or, indeed, any other organism might not be appropriate for research into a particular problem, quite apart from the ethical questions. As we have seen, the discovery of vitamin C relied on using guinea pigs because they, unlike the pigeons that had been used initially, are unable to manufacture vitamin C and so have to eat it – as do humans.

Obviously scientists are aware of these pitfalls and take great care to try to ensure they have found the right organism to work on, but as we have seen, a researcher can often develop a deep investment in 'their' particular creature, not least because they have usually spent many years learning how to work with it. Although most of the scientists I have interviewed said they would happily switch organisms if that were the most effective way to pursue a problem of interest, one or two admitted that it might make more sense to switch problems, especially for older scientists who may have more research years behind them than ahead of them. There comes a point when there simply are not enough years left to master a new organism, or to read everything that has been published about it.

In addition to the risk of being 'trapped' by an organism, there is a subtler pitfall awaiting the unwary scientist. As a research organism becomes established, it becomes an obvious choice for tackling a new problem, which can lead to organisms being used to tackle the wrong problem. Scientists do not like to recall such misadventures, most of which are quickly corrected in any case, but one such story is worth telling, since its consequences are very much with us.

Angus John Bateman was a botanist at the John Innes Institute, near Norwich. In 1948 he made a rare venture into zoology and produced a short research paper that became a classic – cited both in the pages of numerous scientific journals and in *Playboy*.

Bateman wanted to search for evidence that would explain one of Darwin's claims about the differences between males and females. Alongside natural selection, Darwin had put forward a second theory – sexual selection – to explain features of organisms that seemed of no practical benefit to their possessors, or even seemed to be a positive disadvantage. The extravagantly brilliant peacock's tail is an obvious case: huge amounts of biological energy go into growing such an ornament, which can surely only catch the eyes of predators *and* make it harder to escape from them. It therefore seemed impossible to Darwin that natural selection could have been responsible for such magnificent plumage. Equally baffling were the staggering antlers of the misnamed – and extinct – Irish elk, which was neither exclusively

Irish nor an elk, but did possess the largest antlers of any deer – up to twelve feet (3.6 metres) across. Not only could such extravagant weaponry serve no practical purpose, it had even been suggested that it had condemned the poor creatures to extinction: Victorian naturalists wondered if these animals had become involved in a kind of biological arms race, their antlers getting bigger and bigger until they could no longer lift their heads and had sunk into Ireland's bogs, never to be seen again.[3] Although this particular theory was mistaken, such biologically costly armaments posed a genuine challenge to natural selection.

Sexual selection claimed to solve both these problems. Biological success – in Darwin's terms – did not mean that an individual animal had to survive, only that it must leave the most offspring. So imagine if among the distant ancestors of the peahen a few had possessed an initially purely random preference for bright tail feathers and so tended to mate with males who had such feathers. As a result, those males would father a higher percentage of the next generation than their dowdier competitors, producing a generation in which more birds had inherited either their father's feathers or their mother's attraction to such feathers. That would ensure that the brightly coloured feathers would become an increasing advantage over successive generations. If a predilection for gaudy tails became strong enough among successive generations of peahens, the advantages to the males of possessing bright tail feathers would outweigh even the most severe drawbacks. The result could be a process of runaway sexual selection as the preference became more decided and the feathers got grander, generation after generation. The same thing could happen to male weaponry, which might become as much a sexually attractive ornament as a practical means of offence or defence. However, it was more likely to be used for combat, real or symbolic, between males as they competed to inseminate the most females: any advantage in such competitions would out-weigh any difficulties a male might face as he dragged his heavy antlers across the Irish bogs. As Darwin put it, 'the males have acquired their present structure, not from being better fitted to survive in the struggle for existence, but from having gained an advantage over other males, and from having transmitted this advantage to their male offspring'.[4]

Bateman noted that although the idea of sexual selection had received some support from biologists since Darwin had first mooted it, it had not been universally accepted, primarily because few experiments had been done to verify it. This was an oversight he decided to rectify. Darwin had claimed that 'it is certain that with almost all animals there is a struggle between the males for the possession of the female. This fact is so notorious that it would be superfluous to give instances.' He acknowledged that females also played a role, because they 'are most excited by, or prefer pairing with, the more ornamented males, or those which are the best songsters, or play the best antics'.[5] It was, therefore, an almost universal truth: the fact 'that the males of all mammals eagerly pursue the females is notorious to every one. So it is with birds; but many male birds do not so much pursue the female, as display their plumage, perform strange antics, and pour forth their song, in her presence.'

Reading Darwin's thoughts on sexual selection, it is hard not to notice that he regarded the animal kingdom as behaving in much the same way that Victorian ladies and gentlemen were supposed to behave; he noted that a female 'with the rarest exceptions, is less eager than the male' and so she 'requires to be courted'. Like a demure Victorian maiden, 'she is coy, and may often be seen endeavouring for a long time to escape from the male. Every one who has attended to the habits of animals will be able to call to mind instances of this kind.' As, of course, could everyone who had watched a country dance, or a society ball. Yet the ladies undoubtedly also played their part in the marriage game, and Darwin concluded that 'the exertion of some choice on the part of the female seems almost as general a law as the eagerness of the male'.[6]

Examples from the animal kingdom appeared to reinforce Darwin's assumptions about the natural roles of male and female. Half a century later, Charles Creaser explained to beginners how they could tell their male and female zebrafish apart: 'Behaviour is by the far the best index of a true breeding pair. The male pursues the female with a very persistent drive which the female when ripe tolerates. Occasionally a female chases other fishes but never with a really persistent drive.'[7] However accurately the behaviour was being described, Creaser's choice of the word 'tolerates' for the female's behaviour suggests that the ideal of a properly modest female lay at the back of his mind.

Bateman's experiments – on *Drosophila melanogaster* – were designed to find a biological basis for these allegedly universal truths. It might seem odd to choose a fly as a model for complex human behaviour, but it is not unusual; the US National Institutes of Health still fund Drosophila work precisely because they consider these flies as models for aspects of human biology and behaviour.[8]

Bateman chose the flies for the same reasons as other fly workers: they are small, cheap and easy to get hold of. What was especially useful to him was that the work of Morgan's team and their successors meant that there were many different strains of fly available, each with clearly defined visible mutations, whose patterns of inheritance were well understood. These 'marked' strains made it easy to know which flies had successfully passed on their genes, which was to be Bateman's original contribution to the long-running debates over sexual selection. Instead of counting matings to see if the apparently dominant males really did buy the most tickets in the Darwinian lottery (which would, in any case, not have been an easy task with a bottle-full of flies), Bateman decided to use flies with different genetic markers to prove exactly how successful each male and female had been.

Among Bateman's results, one simple fact stood out: 4 per cent of the females failed to mate, while 21 per cent of the males left no offspring. This difference proved that, as he put it, some males 'mate excessively', and if some males were getting less than their fair share, others must be getting more. He defined the fertility of flies in terms of the actual number of offspring they left; while male success varied widely, that of females did not; they all produced roughly the same number of offspring. Bateman

418

concluded from the difference between the success rates of the males and females that 'there is greater competition for mates between males than between females'.[9] Those males who competed most effectively left more of their competitive genes in the next generation, gradually producing males so eager that few of the females failed to be impregnated, which is why the females' contribution to the next generation did not vary as widely.

Bateman's question was, why? Why should male fecundity vary so much more than female, why was not either competition between females or choice by females equally important? His explanation was that in flies, as in most animals, 'the fertility of the female is limited by egg production which causes a severe strain on their nutrition', while for males, 'fertility is seldom likely to be limited by sperm production but rather by the number of inseminations or the number of females available to him'. In other words, eggs are larger than sperm, making them more costly to produce; because females need to spend more of their energy on making their gametes, they can never produce as many as the males do, churning out tiny, low-cost, sperm. As a result, 'the fertility of an individual female will be much more limited than the fertility of the male'.[10]

Bateman concluded that this explained why 'there is nearly always a combination of an undiscriminating eagerness in the males and a discriminating passivity in the females'. And, just in case anyone missed the implication of his work, he added, that even in 'monogamous species (*e.g.* man) this sex difference might be expected to persist as a relic'.[11]

Bateman's paper became a classic. He had apparently found the biological basis for what many believed to be true: when it came to sex, men were undiscriminating and always eager, while women were naturally virtuous – coy and relatively uninterested in sex. In 1979, the science writer Desmond Morris (author of the best-selling *The Naked Ape*) wrote an influential article in *Playboy* entitled 'Darwin and the double standard', which used Bateman's work to prove 'scientifically' that a man's desire for multiple sexual partners was an unalterable fact of human nature.[12] The equality that the women's movement demanded might be an attractive idea in theory, but it seemed that their biology was against them. Men were effectively hard-wired to mate with as many females as possible, while women maximized their evolutionary chances by being monogamous and, if possible, persuading men to overcome their natural instincts and instead stay home and help raise the children.

The message *Playboy*'s readers were encouraged to take from Bateman's research was that there is a natural, unavoidable conflict between male and female reproductive strategies. It is a claim that is still circulating, in more or less subtle forms, especially in many popular biology and evolutionary psychology books. This is somewhat surprising, since there is very little evidence that 'Bateman's principle', as it became known, applies to humans, while there is ample evidence to show that it does *not* even apply to all species of fruit flies.

The first suggestions that Bateman might have generalized too far came in the 1970s, as more women researchers became involved in the study of our closest evolutionary relatives, the great apes. Like any newcomers to a field, they looked for new problems to investigate. Some women primatologists decided that since their male colleagues had concentrated on the often dramatic conflicts between male apes (which had been a major focus of research in the 1960s), it might be productive to investigate female behaviour more closely. It rapidly became clear that while the males were fighting, the females were frequently busy having sex. Female apes regularly seek copulation even when they are not in heat – and not just with the supposedly dominant males in their troop. The more dominant the male, the more active he is in defending his territory and seniority, which leaves less time to prevent 'his' females from seeking copulation elsewhere. This was unexpected: thanks to Bateman's principle, there was no theoretical basis for predicting female promiscuity. However, once such female behaviour had been recognized, other researchers have looked for examples and found them in a great variety of species, from lions to fish. It turns out that there are a great many species in which the males, far from living a playboy lifestyle, have to devote much of their time to attempting to ensure the female's fidelity. In many species, male animals are more likely to be close to home, keeping an eye on their existing mate and offspring, than they are to be out seeking further mates.[13]

It also became clear that the species of Drosophila Bateman used is not even typical of its genus; in many other species the females are as promiscuous as any primate, and have various strategies for manipulating the males, including the ability to store sperm until they have mated several times, so that different males are inadvertently competing to fertilize the eggs.[14]

The range of animal mating strategies should be enough to make us wonder if fruit flies are necessarily the best model for trying to interpret human behaviour. Bateman's mistake is a useful reminder of how difficult it can be to make useful generalizations about human behaviour by studying the behaviour of non-human animals. At the very least, the variety and complexity of mating strategies – even among the great apes – makes it unlikely that we can use their behaviour as a basis for assessing which (if any) human sexual behaviours we can reasonably describe as 'natural'.

The kingdom of the blind

Bateman's fly experiments are also a useful reminder of how careful scientists have to be when they try to use a familiar organism to tackle a new research problem; as Hugo de Vries found with Oenothera, what is true for one organism does not always apply to others. As most scientists realize, the idio-syncrasies of each plant or animal species have to be known and acknowledged before any general scientific conclusions can be formulated.

However, the case of the 'passive' female fruit flies has even more interesting implications. It is possible to assume that Bateman's principle arose from a simple mistake, one whose effects on our view of ourselves have arguably been rather unfortunate, but a mistake nonetheless. The fact that we are now able to identify it *as* a mistake should perhaps help us to avoid similar ones in future. Moreover, identifying and correcting an error like Bateman's is also modest evidence that science actually works: individual scientists may get things wrong; indeed, the entire scientific community may be mistaken for a while; but, over time, reality catches up and the scientific method helps to ensure that incorrect beliefs are eventually corrected. And, to some extent, that is true, but it is not the whole story.

Science is the kingdom of the blind: there are no sighted – or even one-eyed – people, because we have no way of looking directly at reality to assess what it is like. This might seem like a perverse claim, but establishing the accuracy of even the simplest scientific statement is fraught with difficulty. Is it, for example, true to say that 'it gets dark at night'? Obviously, we believe the answer to be 'yes', but how can we prove it? We could step outside and look, but we do not all have perfect eyesight. Even if we do, there are lights from streetlamps and houses; we could qualify the original claim by saying that it is dark when there are no artificial lights, but what about starlight? We qualify the statement by simply claiming that it is dark*er* at night than in the day: but is that true if we compare night under a full moon with a day spent in a dust-storm during a total eclipse? The potential complexities are virtually endless; establishing agreement about how to verify even a very simple statement can be nigh-on impossible.

However, even if we can agree that it is indeed dark at night, science is rarely interested in such simple issues. The scientific question is more likely to be *why* is it dark at night? Try to answer *that* question and it soon becomes clear that almost any simple observation can be explained by more than one theory. Each scientist's decision about which of the as yet untested theories they find most plausible will affect the kinds of evidence they will then look for and the kinds of experiments they will do. An early-twentieth-century Mutationist would not have recorded evidence of standard Mendelian ratios, any more than a Mendelian would have been looking for a 'mutation period'; as we saw with phage, one scientist's crucial evidence is just an anomaly to be explained away when seen from a different theoretical perspective.

It is also important to realize that – until they found them – no one had been looking for genes; they were working on mutations, gemmules, *Anlagen* and so forth. When the geneticists came to write the history of their triumph, they tended to imply that their less enlightened predecessors were – had they but known it – really looking for genes all the time, but that cannot be literally true. The philosopher of science Thomas Kuhn liked to emphasize that scientists can never know where they are going; as he put it, 'scientific development must be seen as a process driven from behind, not pulled from ahead – as evolution from, rather than evolution towards'.[15] While the debates between the biometricians and the Mendelians were in full swing, no one knew who would win: had the world's biologists been questioned in 1900 about who would prove to be the most important scientific figure of the twentieth century, it is quite possible that a majority would have named Hugo de Vries rather than Darwin or Mendel. Perhaps only historical perspective enables us to judge, but we can never be sure a battle is over; we have little idea how future generations of scientists will view our current ideas about genes. However, history shows that what today's scientist 'knows' about genes is quite different from what every scientist 'knew' just fifty years ago, so it is at least possible that in fifty years' time the scientific consensus will have shifted just as radically. This is a key reason why the history of science pays equal attention to the work of those we currently think were right and those who seem to have been wrong; de Vries's theory produced much important science, even though much of it ultimately disproved his ideas.

Rather than a steady ascent towards the truth, science is perhaps better imagined as groping about in the dark, feeling the way and trying to imagine how to make sense of whatever fragments have been grasped. Even though today's scientists are doing their groping with computerized DNA sequencers and advanced genetic technologies, they are still doing what Aristotle was doing on the shores of the bay of Lesbos: looking, thinking, imagining, testing and wondering. With hindsight we might try to claim that when Darwin and Galton were looking for gemmules they were 'really' looking for genes, but that is simply replacing the commonsense of their time with the commonsense of our own, which is most unlikely to be the final word on the subject. For a historian, it is more appropriate to study Darwin and Galton in the context of the commonsense of their time, because that at least allows us to understand what they were doing and why they were doing it.

The history of science suggests that science does not deal in eternal, unchanging truths, only in provisional, short-lived ones. It makes sense to evaluate today's science in the same way we judge that of the past, accepting that today's scientific breakthroughs may eventually be discarded or, more likely, will simply become uninteresting because they have become part of the routine background knowledge that every scientists takes for granted.

So does that mean that modern science has not improved on older science, or that scientific accounts of the world are no better than other kinds of explanations? Does the history of science suggest, to take an example that is back in the news yet again, that Darwin's theory of evolution really is no better than Creationism? Obviously not, because although scientists can never be certain that they have arrived at the final truth, they do have good reasons for becoming increasingly confident that they are on the right track. As new experiments produce evidence that supports a theory, it becomes increasingly likely that the theory is substantially correct. Some details may need changing in the light of new evidence, but the major hypotheses underpinning the theory become increasingly well established.

As we have seen, science is a social activity, carried out by communities of researchers; it takes evidence to persuade your peers that your theory is right. At the same time, there are glittering careers to be made in demolishing established theories, so scientists are always on the look out for surprising new evidence that might challenge the prevailing wisdom. Nevertheless, the theory of evolution by natural selection has gained strength –

despite occasional setbacks – during the 150 years since Darwin first published it. There is now no serious opposition to any of its major premises within the scientific community, even though debates continue about many of its details. The controversies and discoveries I have been describing eventually put a great deal of flesh on the bare bones of Darwin's theory, but they did not alter its major tenets.

Does that mean Darwin can never be proved wrong? Of course not. Scientific truths are not dogmas founded on faith, but hypotheses founded on evidence. New evidence might always come along, but a mature, well-supported theory such as evolution by natural selection is most unlikely to be overturned. Scientists form a large, diverse community, with agreed standards of evidence and well-established mechanisms for testing and evaluating theories; if most scientists tell us there are genes, there are almost certainly genes. They could all be wrong, but the odds are strongly against it. And by the same token, if the vast majority of reputable scientists tell us that global warming is a real problem, we would be well advised to listen, even if a handful of scientists are telling us the opposite. Again, the tiny minority may be right, but it is most unlikely. Especially if they all work for oil companies.

However, let us consider a slightly more complex case. Is genetically modified food (GM) safe for humans to eat and safe for the environment? In the first case, the scientific consensus seems pretty strong (although there are a few dissenters): yes, it is. In the second, the picture is more complex. As we have seen, government-sponsored field trials of genetically modified crops in Britain have suggested that the kinds of modifications being made by the biotech companies are indeed causing problems. For example, some GM crops have had genes inserted that make them highly resistant to specific weedkillers, which as we have seen has been shown to seriously reduce the biodiversity in the fields where such crops are grown. Another potential problem with these crops is that, thanks to hybridization, the gene for herbicide resistance sometimes spreads to other plants, producing what are called 'superweeds' that are no longer vulnerable to the herbicide. If this were to happen often, clearly all the benefits of herbicide-resistant crops are going to be lost.[16]

Partly because of these concerns, the governments of Britain and many other European countries still do not allow GM crops to be grown commercially, but of course the decision has also been influenced by the widespread public opposition to eating GM food. Whether or not one regards this hostility as rational, there is clearly no market for GM food in Britain at present, so there is very little incentive to grow the crops. By contrast, in the United States most consumers and farmers seem to have decided – whether actively or passively – that any environmental risks are worth running to ensure the continued supply of cheap, plentiful food. In a country where most people are starving, the political equation would also look rather different; many Africans might, for example, decide that losing a few wildflowers and birds, or doing some extra crop-spraying, were risks worth running to avoid famine. Indeed, the hope of feeding the world is regularly cited as a main reason for creating genetically modified crops.

However, opponents of GM technology are cynical about such claims; one does not have to be an anti-capitalist to agree that any company exists to make a profit; if there were profits to be made by feeding the starving, critics argue, they would already be eating because most hunger is caused by people being too poor to buy food, not by an absolute shortage of it. Producing more food will not change that equation, unless it is much cheaper than existing food and it is difficult to see how biotechnology companies can increase their profits by driving the cost of food down.

The history of hybrid corn tends to support scepticism about the intentions of the biotechnology companies. Hybrid corn caught on partly because of its high yields and improved cropping, but also because of an alliance between government scientists and commercial seed companies. In one of his official reports, Edward East had noted that almost all the 12,000 bushels of seed corn which Connecticut farmers used each year had been saved from their previous year's crops, making them perhaps the thriftiest farmers in America. 'Yet', he added, 'this fact is more of a curse than a blessing' because most of the corn saved was of inferior quality and 'these varieties should be discarded because they produce smaller yields than other existing varieties'. He believed 'too many corn growers hold on to a poor variety through misplaced loyalty to some early ancestor who originated it'.[17] Scientists like East urged farmers to give up saving seed and buy new seed each year, but farmers were understandably reluctant to do so, suspicious that the college-educated science bureaucrats were once again putting fancy notions ahead of practical experience.

However, if the farmers were slow to embrace East's advice, the seed companies were more enthusiastic. They might not have known much about genetics, but they certainly recognized a profitable innovation. The beauty of hybrid corn from their perspective was that it forced farmers to buy new seed every year – saving seed from hybrids was no good, because whatever hybrid vigour was, it was only present in the first hybrid generation; as we have seen, yields declined rapidly in each subsequent generation. Few farmers had the time, money or expertise to hybridize corn for themselves, so – as long as hybrid yields were high enough – it made sense for them to buy new seed every year. The seed companies' cause was accidentally helped by the series of devastating droughts of the mid-1930s, which gradually turned the Great Plains from a breadbasket to a dust bowl. As the harvests withered and died, there was much less seed corn to save. Farmers were forced to buy and the seedsmen only sold them hybrids, which permanently locked them into a dependence on the seed companies.[18] While most maize biologists celebrated hybrid corn as one of the first great triumphs of twentieth-century genetics, others have argued that the real benefits went not to the farmers, but to the seed companies.[19]

Among the innovations of modern biotechnology companies is the so-called terminator gene, which makes seed from GM crops infertile so that it cannot be saved and resown in succeeding years, thus requiring farmers to buy new seed every season, just as they had to with hybrid corn. The biotech companies claim that

427

terminator-type technology helps make GM crops safer by ensuring that their genes do not escape into the wild, but of course it also protects their intellectual property, ensuring that they can continue to profit from their investments in scientific research. Meanwhile, their opponents claim that GM technology is just another way of ensuring that developing countries remain entirely dependent on the wealthy countries. The history of the hybrid corn revolution suggests that it is big agribusiness that benefits most from new agricultural technologies, not small-scale, poor farmers.

Brave new worlds?

Genetically modified food raises both ethical and political issues, as do all genetic technologies. However, to return to OncoMouse, genetically modified animals are, inevitably, the focus of the most intense anxieties. Twentieth-century science has shown us that what we can do to mice today, we can – and probably will – do to humans tomorrow. That is, after all, the principle on which many life-saving drugs have been developed and tested, but while many people do not oppose the genetic modification of mice, interfering with human genes is widely seen as something quite different. It seems as though the eugenicists' fantasies might finally become a reality: genetic engineering could allow the creation of 'better' humans, stronger and smarter – and the elimination of those deemed unfit. Possibilities that, of course, raise all the old dilemmas: who is to decide which characteristics are desirable or not? As one of Galton's contemporaries observed, when eugenics was first mooted, 'Who is to decide whether a man's issue is not likely to be well fitted "to play their part as citizens?"' Galton felt that the answer was self-evident; society needed more of the strongest, healthiest and above all, the most intelligent. But not everyone has accepted his answer. In Aldous Huxley's dystopian novel *Brave New World,* John Savage asks why, given the World State's ability to engineer people to order, they do not make everyone an Alpha, a brilliant intellectual, instead of creating huge cloned groups of Epsilon 'semi-morons' to do society's menial work. One of the world controllers explains that this has been tried: an experimental society composed entirely of Alphas was created, but soon disintegrated in arguments over whose turn it was to do the dirty work. A stable society, he concludes, has to be a stratified one: one where there is a place for everyone and everyone has been engineered to fit in – to be intelligent or stupid, to be a controller or to be subservient and infantilized. These objections might still be offered to those who think genetic engineering can make humans more intelligent; it could be argued, for example, that the most intelligent members of our species are not invariably the most moral or compassionate ones, and a case can certainly be made that such qualities are needed more urgently than raw intellectual horsepower. Who is to say?

However, even if eugenics in the broadest sense continues to be illegal, genetic technologies are already allowing us to practise it on a more modest scale. Genetic screening and testing already makes it possible to abort foetuses with serious genetic defects, which is controversial enough in many people's eyes. However, it is becoming possible to repair such defects by inserting a working copy of a gene into a human embryo to replace a faulty one. This kind of gene therapy is already being widely researched, and while there are considerable problems still to be overcome, its potential is already clear – as are the ethical dilemmas it raises. Once again, who is to decide which genes are 'normal', and which need to be 'repaired'?

It would be good to be able to avoid these questions, but ignoring them is not an option; the research continues and the technologies continue to develop. Those of us who are not scientists have little chance of understanding many of the technicalities of the science, but we need to have opinions; if we leave it to the experts, we abandon our duties and rights as citizens. One reason for writing this book was to learn more about a science that is already affecting me and is certain to affect my children and grandchildren even more profoundly. I have changed my mind about several issues in the course of my research, or at least found better evidence to support my opinions.

I have also become slightly less concerned by my inability to master the arcane details of modern genetics, since the science itself does not tell us much – if anything – about how to think about the ethical and political problems it raises.

Consider, for example, the possibility that homosexuality might have a genetic basis, an idea that has generated considerable discussion ever since researchers first raised it. What would the implications of a 'gene for homosexuality' be? For a start, geneticists tend to nominate the phrase 'the gene for . . .' as the single most annoying and misleading aspect of non-scientific discussions of these issues. As we saw in the case of fruit flies, genes affect each other; even the simplest characters can involve many genes. In addition every gene is affected by the environment of the cell it is in, from its temperature to the other genes present. And in many cases, such as that of cancer and oncogenes, the relevant gene only *increases* the chances of getting the disease, it does not actually cause it; some people with the gene will get ill while others will not and we still do not entirely understand why. All of which makes it most unlikely that there is anything we could reasonably call 'the gene for' a complex human behaviour (assuming that 'homosexuality' can even be meaningfully described as a single 'behaviour', which can be identified and defined as clearly as eye colour). Finally, there is an unspoken assumption in many debates about genetics that if something has a genetic basis, it is fixed and unalterable, as least not without some radical intervention, such as genetic engineering. But many forms of short-sightedness are inherited and it takes nothing more complicated than a pair of glasses to fix them; the fact that some feature of an organism is largely or wholly under genetic control does not necessarily have any implications as to how difficult it might be to alter that feature.

So, with all those caveats in mind, let us pretend for a moment that there is something clearly definable as homosexuality, which is largely determined by a single gene, and that it cannot be modified by any non-genetic means – all of which is utterly unlikely. What would that tell us about the morality of homosexual behaviour?

Precisely nothing.

Some gay rights campaigners believe that a gay gene, if there were one, would prove once and for all that homosexuality is normal, simply part of the rich genetic diversity of humanity. However, the genes for cancer, schizophrenia and Huntington's disease are also part of that diversity, and humanity would arguably be better off without such variations. So it would be entirely possible for the discovery of a 'gay gene' to be used by anti-gay campaigners to argue that the gene had finally provided a 'cure' for the disease of homosexuality – by using elective abortions, genetic therapy or preventing its 'carriers' from reproducing.

That the same scientific fact can be used to support entirely opposed moral and political positions suggests that such facts cannot be used to settle moral and political arguments. Philosophers have argued for at least 200 years that ethical guidance, help in knowing how we should act, cannot be derived from the way the world happens to be; you cannot get an *ought* from an *is*. In philosophical terms, attempting to do so is referred to as committing the naturalistic fallacy.

If the naturalistic fallacy is indeed a fallacy, it is rather reassuring. No matter what science discovers, it cannot tell us what to believe or how to behave. Racists who try to prove that black people have lower IQs than whites are – in more than one sense – wasting their time. Even if such a claim were shown to be true (and all the objections to the idea of a gay gene apply here with equal force), it would tell us nothing about how black and white people ought to treat one another. The same argument also offers comfort to those who look to their religious faith for moral guidance and are concerned that science may invalidate their faith – it cannot. Science is only a threat to those who insist on looking to religious dogma for answers to what are properly scientific questions, such as how old the earth is; to a philosopher, this is known as a category mistake. And it works both ways: science cannot prove whether or not there is a God, any more that it can tell us what ethical code we should live by. We need to consult our consciences and make these decisions for ourselves.

431

Avoiding the naturalistic fallacy also leads to the conclusion that when it comes to arguments about the implications of genetics, whether it is GM crops or any other new technology – we need to focus on the political and ethical debates, not just the scientific ones. Clearly there may well be scientific reasons not to pursue a technology, and an awareness of the history of science can help in understanding these. As several of the stories in this book have shown, we use animals and plants as models precisely because they are *not* like human beings – not as large, not as slow-breeding, not as recalcitrant, and we do not feel we need to seek their consent. Highlighting such differences should make us cautious when we start trying to apply what we have learnt from them to us. Next time the headlines proclaim a new cause or cure for cancer, homosexuality, or any other human phenomenon, check whether the research in question was done using fish, flies or flowers; that may help you assess how applicable the research may, or may not, be to humans.

Nevertheless, once the scientific community has reached a broad consensus about the safety or efficacy of a technology, the real debate has only just begun. The fact that we *can* do something does not mean that we have to. The issues raised by genetics are ultimately political ones: who controls the technology and who decides how it will be used. Personally, I do not have unlimited faith in market mechanisms to make such complex decisions; geneticists have convinced me that GM food is safe to eat, but I have decided not to – because I do not believe the biotechnology companies have either the environment or society's best interests at heart. That is not their fault – it would be unrealistic and unreasonable to expect companies to act altruistically; it is up to governments and international bodies to decide whether genetic technologies can be used in a way that benefits us all and makes the world a better place for our children to grow up in. And, since we elect those governments, that means it is up to us. Now that we have the knowledge to intervene so effectively in the engineering of living things, we need to ask whether we have the wisdom to use such power wisely.

Bibliography, sources and notes

Preface and acknowledgements

Notes

1. K.A. Rader, *Making Mice: Standardizing animals for American biomedical research, 1900–1955* (Princeton University Press, 2004); A.N.H. Creager, *The Life of a Virus: Tobacco Mosaic Virus as an Experimental Model, 1930–1965* (University of Chicago Press, 2002). For Latour and Callon's work, see: B. Latour, *Science in Action: How to follow scientists and engineers through society* (Harvard University Press, 1987); B. Latour, *The Pasteurization of France* (Harvard University Press, 1988); and M. Callon, 'Some Elements of a Sociology of Translation: Domestication of the Scallops and the Fishermen of St Brieuc Bay', in *Power, Action and Belief*, ed. J. Law (Routledge & Kegan Paul, 1985).

Chapter 1: *Equus quagga* and Lord Morton's mare

For **Ancient Greek science**, see: P.J. Bowler, *The Fontana History of the Environmental Sciences* (Fontana Press, 1992); D.C. Lindberg, *The Beginnings of Western Science* (University of Chicago Press, 1992); G.E.R. Lloyd, *Early Greek Science from Thales to Aristotle* (Chatto & Windus, 1970); A. Preus, 'Science and Philosophy in Aristotle's *Generation of Animals*', *Journal of the History of Biology*, 1970: 1–52.

The history of **breeding and generation** is described in: J. Diamond, *Guns, Germs and Steel: A Short History of Everybody for the Last 13,000 Years* (Vintage, 1998); H. Ritvo, 'Animal Planet', *Environmental History*, 2004, 9.2; N. Russell, *Like Engend'ring Like: Heredity and Animal Breeding in Early Modern England* (Cambridge University Press, 1986); C. Zirkle, 'The Early History of the Idea of the Inheritance of Acquired Characteristics and of Pangenesis', *Transactions of the American Philosophical Society*, 1946, Vol. XXXV: 91–150.

Notes

1. Earl of Morton, 'A Communication of a Singular Fact in Natural History', *Philosophical Transactions of the Royal Society of London* III (1821): 20.

2. J.F. Gmelin, *Ergänzungen zu Linnés Systema Naturae* (Göttingen); JE Gray, 'Revision of the Family *Equidae*', *Zoological Journal of London* 1,1825: 241–8; H. Lichtenstein, *Travels in Southern Africa in the Years 1803–1806*, trans. H. Plumptre (Henry Colbourn, London 1812; reprinted 1928, 1930, Van Riebeck Society, Cape Town); *Cassell's Popular Natural History* (London, undated), Vol.1: 220. All quoted in David Barnaby, *Quagga Quotations: A Quagga Bibliography* (Bartlett Society, 2001): 18, 32–3, 47.

3. Charles Darwin, *The Variation of Animals and Plants under Domestication, Volume 1*, ed. Harriet Ritvo, facsimile of 2nd edition (1875) (Johns Hopkins University Press, 1998): 435.

4. Quoted in G.E.R. Lloyd, *Early Greek Science from Thales to Aristotle* (Chatto & Windus, 1970): 115–16.

5. Quoted in ibid.: 116.

6. Plato, *Republic*.

7. Shakespeare, *King Lear* (1606), Act 1, Scene II.

8. G. Markham, *Cavalarice* (1607), quoted in N. Russell, *Like Engend'ring Like: Heredity and Animal Breeding in Early Modern England* (Cambridge University Press, 1986): 71–2.

9. N. Morgan, *The Horseman's Honour* (1620), quoted in Russell, *Like Engend'ring Like:* 78.

10. Russell, *Like Engend'ring Like:* 93–4.
11. Richard H. Drayton, *Nature's Government: Science, Imperial Britain and the 'Improvement' of the World* (Yale University Press, 2000): 85–7.
12. My analysis of improvement and its importance is heavily indebted to Richard Drayton's *Nature's Government.*

Chapter 2: *Passiflora gracilis*: Inside Darwin's greenhouse

Darwin's own works were my main source, especially: *The Descent of Man, and Selection in Relation to Sex* (1871; Princeton University Press, 1981); *The Effects of Cross- and Self-Fertilisation in the Vegetable Kingdom* (1878; New York University Press, 1989); *On the Movements and Habits of Climbing Plants* (Longman, Green, Longman, Roberts & Green, 1865); *On the Origin of Species by Means of Natural Selection: or the preservation of favoured races in the struggle for life* (1859; Penguin Books, 1964); *On the various contrivances by which British and foreign orchids are fertilised by insects, and on the good effects of intercrossing* (John Murray, 1862 and 1904).

For **Darwin's life**, see his autobiography: *The Autobiography of Charles Darwin* (Collins, 1958) and the comparative edition of the two versions of it, C. Darwin, *Autobiographies* (1887/1903; Penguin Books, 2003). The best biography of Darwin is Janet Browne's *Charles Darwin: Voyaging* (Jonathan Cape, 1995) and the second volume, *Charles Darwin: The Power of Place* (Jonathan Cape, 2002). Adrian Desmond and James Moore's *Darwin* (Michael Joseph, 1991) is also excellent. *The Correspondence of Charles Darwin* (Cambridge University Press) is an absolutely invaluable resource.

Surprisingly little has been written on **Darwin's botanical work** since Mea Allan's *Darwin and his Flowers: the key to Natural Selection* (Faber and Faber, 1977) apart from S.M. Walters and E.A. Stow, *Darwin's Mentor: John Stevens Henslow, 1796–1861* (Cambridge University Press, 2001) and an excellent but brief essay by David Kohn, 'Darwin's botanical research', in *Charles Darwin at Down House* (English Heritage 2003).

Darwin's research on reproduction and related topics is covered in: P.H. Barrett et al., *Charles Darwin's Notebooks, 1836–1844: Geology, Transmutation of Species, Metaphysical Enquiries* (Cambridge University Press, 1987); M.M. Bartley, 'Darwin and Domestication: Studies on Inheritance', *Journal of the History of Biology*, 1992: 307–33; J. Endersby, 'Darwin on generation, pangenesis and sexual selection', in *Cambridge Companion to Darwin* (Cambridge University Press); M. Hodge, 'Darwin as a lifelong generation theorist', in *The Darwinian Heritage* (Princeton University Press, 1985); J.A. Secord, 'Darwin and the Breeders: A Social History', in *The Darwinian Heritage* (Princeton University Press, 1985); R. Stott, *Darwin and the Barnacle* (Faber and Faber, 2003).

On **Victorian gardening** and the historical background to Darwin's period, the best place to start is Martin Hoyles, *The Story of Gardening* (Journeyman, 1991). Other useful sources are: J. Fisher, *The Origins of Garden Plants* (Constable, 1982); T. Carter, *The Victorian Garden* (Bell & Hyman, 1984); B. Elliott, *Victorian Gardens* (Batsford, 1986); E.C. Nelson and E.M. McCracken, *The Brightest Jewel: A history of the National Botanic Gardens, Glasnevin, Dublin* (Boethius Press, 1987); J. Morgan and A. Richards, *A Paradise out of a Common Field: the Pleasures and Plenty of the Victorian Garden* (Century, 1990).

For more on **passionflowers** see E.E. Kugler and L.A. King, 'A Brief History of the Passionflower', in *Passiflora: Passionflowers of the World* (Timber Press, 2004). The other essays in this book are also excellent for botanical and gardening information.

Notes

1. Both are quoted in E.E. Kugler and L.A. King, 'A Brief History of the Passionflower', in *Passiflora: Passionflowers of the World* (Timber Press, 2004): 17, 18.

2. The word 'stigmata' derives from the Greek, στιγμα, or stigma, which means a mark of shame, originally a tattoo indicating that someone was a slave or criminal. In the

sixteenth century, the receptive top of the female carpel had no consistent name. The modern term, 'stigma', comes from the same Greek root as stigmata, and was introduced by Linnaeus in the *Species Plantarum* (1753): he may have chosen the name because in many plants the stigma is cross-shaped when viewed from above. My thanks to Nick Jardine for this suggestion.

3. Kugler and King, 'A Brief History of the Passionflower'in, Ulmer and MacDougal, Passiflora-Passionflowers of the World (Timber Press 2004): 22.

4. [Lindley?], 'Editorial', *Gardener's Chronicle & Agricultural Gazette*, 1845: 114.

5. Quoted in Carter, *The Victorian Garden* (Bell & Hyman, 1984): 72–3.

6. Quoted in ibid: 67.

7. Quoted in ibid: 72–3.

8. [Lindley?], 'Editorial', *Gardener's Chronicle & Agricultural Gazette*, 1845: 114.

9. *Gardeners' Chronicle*, 1851: 707–8, quoted in Elliott, *Victorian Gardens* (Batsford, 1986): 107.

10. Quoted in Carter, *The Victorian Garden*: 72–3.

11. J.C. Loudon, 'Growing Ferns and Other Plants in Glass Cases', *Gardener's Magazine*, Vol. 10, April 1834: 162.

12. Stephen H. Ward, *Wardian Cases and their applications*, a lecture delivered to the Royal Institution, 17 March 1854 (Van Voorst, London, 1854). I am indebted to Mathew Underwood for bringing both this and the previous quote to my attention.

13. W.J. Hooker, *Botanical Magazine*, February 1838: plate 3635. The plant was also described in the *Gardener's Magazine*, 14 (96), 1838: 138.

14. *Botanical Register*, April 1838: 276. The species is now known as *Passiflora amethystina*.

15. My thanks to David Kohn for this suggestion.

16. J. D. Hooker to C. Darwin [12 December 1843–11 January 1844]: F. Burkhardt and S. Smith (eds.), *The Correspondence of*

Charles Darwin (Volume 2: 1837–1843) (Cambridge University Press, 1986).

17. In his first paper on Galápagos plants: J.D. Hooker, *Transactions of the Linnean Society of London*, 1847, 20: 222, 223. Hooker produced a fuller account of Darwin's Galápagos plants two years later: J.D. Hooker, 'An enumeration of the Plants of the Galapagos Islands', *Proceedings of the Linnean Society of London*, 1849, 1: 276–9.

18. Barrett, Gautrey, Herbert, Kohn and Smith, *Charles Darwin's Notebooks, 1836–1844: Geology, Transmutation of Species, Metaphysical Enquiries* (Cambridge University Press, 1987): 505.

19. D. Beaton, 'Greenhouse and Window Gardening', *The Cottage Gardener*, 1849: 37.

20. Jane Austen, *Northanger Abbey* (1818): Ch. 22. Greenhouses also crop up in *Persuasion* (Ch. 23) and *Sense and Sensibility* (Ch. 42).

21. Errington, 'Culture of the Passifloras for the Dessert', *The Cottage Gardener*, 1850: 342–3.

22. D. Beaton, 'Passion-flowers', *The Cottage Gardener*, 1850: 152–3.

23. Beaton considered himself an expert on hybridization, but in a letter to Hooker, Darwin observed: 'he strikes me as a clever, but d——d cock-sure man', C. Darwin to J.D. Hooker, 14 May [1861]: 127

24. T. Malthus, *An Essay on the Principle of Population* (6th edition, John Murray, 1826), Book IV, Chapter IV: 10.

25. C. Darwin, *Autobiographies* (1887/1903; Penguin Books, 2003): 72.

26. Darwin's oft-quoted list was first published in *The Autobiography of Charles Darwin* (Collins, 1958): 232–3.

27. Quoted in J. Browne, *Charles Darwin: The Power of Place* (Jonathan Cape, 2002): 276–82.

28. C. Darwin, *On the Origin of Species by Means of Natural Selection: or the preservation of favoured races in the struggle for life* (1859; Penguin Books, 1964): 96–7.

29. C. Darwin to J. D. Hooker, [15 February 1863]: F. Burkhardt et al., *The Correspondence of Charles Darwin (Volume 11: 1863)* (Cambridge University Press, 1999): 134.

30. C. Darwin to J. D. Hooker, [13 January 1863]: *The Correspondence of Charles Darwin (Volume 11: 1863)*: 36.

31. J. D. Hooker to C. Darwin, [15 January 1863]: *The Correspondence of Charles Darwin (Volume 11: 1863)*: 43.

32. C. Darwin to J. D. Hooker, [21 February 1863]: *The Correspondence of Charles Darwin (Volume 11: 1863)*: 161.

33. C. Darwin to J. D. Hooker, [5 March 1863]: *The Correspondence of Charles Darwin (Volume 11: 1863)*: 200.

34. C. Darwin, *The various contrivances by which British and foreign orchids are fertilised by insects, and on the good effects of intercrossing* (John Murray, 1904): 285–6.

35. C. Darwin to J. D. Hooker, [13 January 1863]: *The Correspondence of Charles Darwin (Volume 11: 1863)*: 36.

36. C. Darwin to W.E. Darwin, [25 July 1863]: *The Correspondence of Charles Darwin (Volume 11: 1863)*: 56.

37. C. Darwin, *On the Movements and Habits of Climbing Plants* (Longman, Green, Longman, Roberts & Green, 1865): 90. A grain is roughly 64 milligrams, so $^1/_{32}$ of a grain would be about $^2/_{1000}$ of a gram.

38. ibid.

39. ibid.: 89.

40. ibid.: 107–8.

41. C. Darwin, *Autobiography* (Penguin Books, 2002): 82

42. C. Darwin to J. Scott, [11 December 1862]: F. Burkhardt et al., *The Correspondence of Charles Darwin (Volume 10: 1862)* (Cambridge University Press, 1997): 594. *Melastoma* is a genus of evergreen tropical shrubs; they get their name from the black berries of some species which stain the mouth (the Greek for black is '*melas*' and for mouth '*stoma*').

43. J. Scott to C. Darwin, [11 November 1862]: *The Correspondence of Charles Darwin (Volume 10: 1862)*: 516.

44. C. Darwin to J. Scott, [12 November 1862]: *The Correspondence of Charles Darwin (Volume 10: 1862)*: 522.

45. C. Darwin to J. Scott, [19 November 1862]: *The Correspondence of Charles Darwin (Volume 10: 1862)*: 538.

46. J. Scott to C. Darwin, [20 November–2 December 1862]: *The Correspondence of Charles Darwin (Volume 10: 1862)*: 542.

47. Darwin, *On the Origin of Species by Means of Natural Selection: or the preservation of favoured races in the struggle for life* (1859: Penguin Books, 1964): 250–51.

48. J. Scott to C. Darwin, [3 March 1863]: *The Correspondence of Charles Darwin (Volume 11: 1863)*: 189. Darwin had mentioned *Passiflora* in passing in the *Origin of Species*: 250–1.

49. C. Darwin to J. Scott, [6 March 1863]: *The Correspondence of Charles Darwin (Volume 11: 1863)*: 213–14; J. Scott to C. Darwin, [21 March 1863]: *The Correspondence of Charles Darwin (Volume 11: 1863)*: 251–2; C. Darwin to J. Scott, [24 March 1863]: *The Correspondence of Charles Darwin (Volume 11: 1863)*: 262–3.

50. C. Darwin to J. Scott, [3 December 1862]: *The Correspondence of Charles Darwin (Volume 10: 1862)*: 582.

51. C. Darwin to J. Scott, [11 December 1862]: *The Correspondence of Charles Darwin (Volume 10: 1862)*: 594.

52. J. Scott to C. Darwin, [17 December 1862]: *The Correspondence of Charles Darwin (Volume 10: 1862)*: 607–8. Darwin later helped Scott emigrate to India and recruited Hooker's support to get Scott the curatorship of the Royal Botanic Garden, Calcutta.

53. Darwin, *The Effects of Cross-and Self-Fertilisation in the Vegetable Kingdom* (1878; New York University Press, 1989): 384.

54. Darwin, *On the various contrivances by which British and foreign orchids are fertilised by insects, and on the good effects of intercrossing* (John Murray, 1862): 286.

55. Darwin, *The Descent of Man, and Selection in Relation to Sex* (1871; Princeton University Press, 1981): II, 403.

56. Jonathan Smith has noted (*'Une Fleur du Mal?* Swinburne's "The Sundew" and Darwin's Insectivorous Plants') that Darwin's *Orchids* was reviewed with works on 'consanguineous' marriage as early as 1863 in [G.W. Child],

'Marriages of Consanguinity,' *Westminster Review*, 1863, 24 n.s. 88–109. Darwin's son, George, took up the questions extensively in 'On Beneficial Restrictions to Liberty of Marriage,' *Contemporary Review*, 1873, 22, 412–26, and 'Marriages Between First Cousins in England and Their Effects,' *Fortnightly Review* 18 n.s. (1875), 22–41. A.H. Huth, who incorporated the Darwins' work in his *Marriage of Near Kin* (1875; 2nd edition 1887), reviewed father and son together in 'Cross-Fertilisation of Plants, and Consanguineous Marriage,' *Westminster Review*, 1877, 52 n.s. 466–85.

57. Darwin, *The Descent of Man, and Selection in Relation to Sex*: II, 402–3.

Chapter 3: *Homo sapiens*: Francis Galton's fairground attraction

I have used **Galton's own publications** as a major source, most of which are available online at www.galton.org – an invaluable site. For the anthropometric laboratory, see in particular: 'The Anthropometric Laboratory', *Fortnightly Review*, 1882: 332–8; 'Blood-Relationship', *Proceedings of the Royal Society*, 1872: 394–402; *English Men of Science: Their Nature and Nurture* (Macmillan, 1874); 'Experiments in pangenesis, by breeding from rabbits of a pure variety, into whose circulation blood taken from other varieties had previously been largely transfused', *Proceedings of the Royal Society*, 1871: 393–410; *Hereditary Genius* (1869; Macmillan, 1892); 'Hereditary Improvement', *Fraser's Magazine*, 1873: 116–30; 'Hereditary talent and character', *Macmillan's Magazine*, 1865: 157–66, 318–27; *Memories of My Life* (Methuen, 1908); *The narrative of an explorer in tropical South Africa* (John Murray, 1853); 'On the Anthropometric Laboratory at the Late International Health Exhibition', *Journal of the Anthropological Institute*, 1884: 205–18; and 'Some Results of the Anthropometric Laboratory', *Journal of the Anthropological Institute*, 1884: 275–87.

For **Francis Galton's life**, see: D.W. Forrest, *Francis Galton: The Life and Work of a Victorian Genius* (Paul Elek, 1974); K. Pearson, *The life, letters and labours of Francis Galton* (Cambridge University Press, 1914–30); and, most useful of all, N.W. Gillham, *A Life of Sir Francis Galton: From African Exploration to the Birth of Eugenics* (Oxford University Press, 2001).

For **Darwin's theory of inheritance**, see the sources cited in Chapter 2, plus: G.L. Geison, 'Darwin and Heredity: the Evolution of His Hypothesis of Pangenesis', *Journal of the History of Medicine and Allied Sciences*, 1969: 375–411; D.L. Hull, *Darwin and his Critics: the Reception of Darwin's Theory of Evolution by the Scientific Community* (Harvard University Press, 1973); R.W. Burkhardt, 'Closing the Door on Lord Morton's Mare: The Rise and Fall of Telegony', *Studies in the History of Biology*, 1979: 1–21; P.H. Barrett, *The Collected Papers of Charles Darwin* (University of Chicago Press, 1980); L.J. Jordanova, *Lamarck* (Oxford University Press, 1984); M. Bulmer, 'Did Jenkin's swamping argument invalidate Darwin's theory of natural selection?', *British Journal for the History of Science*, 2004: 281–97.

Studies of **Galton's scientific work**, methods and its background include: C. Zirkle, 'The Early History of the Idea of the Inheritance of Acquired Characteristics and of Pangenesis', *Transactions of the American Philosophical Society*, 1946: 91–150; R.S. Cowan, 'Nature and Nurture: The Interplay of Biology and Politics in the Work of Francis Galton', *Studies in the History of Biology*, 1977: 133–208; J.A. Secord, 'Nature's Fancy: Charles Darwin and the Breeding of Pigeons', *Isis*, 1981: 163–86; R. Olby, *Origins of Mendelism* (University of Chicago Press, 1985); R.S. Cowan, *Sir Francis Galton and the study of heredity in the 19th century* (Garland, 1985); M. Hodge, 'Darwin as a lifelong generation theorist', in *The Darwinian Heritage*, (Princeton University Press, 1985); D.J. Kevles, *In the Name of Eugenics: Genetics and the uses of Human Heredity* (Penguin Books, 1986); E.A. Gökyigit, 'The reception of Francis Galton's *Hereditary Genius* in the Victorian periodical press', *Journal of the history of biology*, 1994: 215–40; R. Olby, 'The Emergence of Genetics', in *Companion to the History of Modern Science* (Routledge, 1996).

For **London in Galton's day**, see: F. Engels, 'The Condition of the Working Class in England', in *Literature and Science in the Nineteenth Century: An Anthology*, (Oxford University Press, 2002); J. Greenwood, *The Seven Curses of London* (1869; Blackwell 1981); S. Halliday, *The Great Stink of London: Sir Joseph Bazalgette and the Cleansing of the Victorian Metropolis* (Sutton Publishing, 2001); M. Daunton, 'London's 'Great Stink': The Sour Smell of Success' (British Broadcasting Corporation, n.d., www.bbc.co.uk/history/lj/victorian_britainlj/smell_of_success_1.shtml).

Notes
1. D. Galton, 'The International Health Exhibition', *The Art Journal*, 1884: 156
2. [G.A. Sala], 'The Health Exhibition: a look around', *Illustrated London News*, 1884: 94. Emphasis in original.
3. Anon., *Miscellaneous (Including: Return of Number of Visitors and Statistical Tables & Official Guide)* (Executive Council of the International Health Exhibition and the Council of the Society of Arts, 1884): 12–13.
4. My thanks to Judith Flanders for this information.
5. J. Greenwood, Ch. IX, 'The Thief Non-professional'. *The Seven Curses of London* (1869; Blackwell Publishers, 1981).
6. [G.A. Sala], 'Echoes of the Week [International Health Exhibition]', *Illustrated London News*, 1884: 439, 438.
7. Anon., 'International Health Exhibition', *Saturday Review: of Politics, Literature, Science and Art*, 1884: 634–5.
8. J.E. Ady, 'The International Health Exhibition', *Knowledge*, 1884: 387–8, 415–18, 434–5, 454–5, 476–7.
9. 'International Exhibitions', *Punch*, June 8, Vol. LXII, 1872: 240.
10. J.E. Ady, 'The International Health Exhibition': 388.
11. F. Engels, 'The Condition of the Working Class in England', in *Literature and Science in the Nineteenth Century: An Anthology* (Oxford University Press): 492.
12. D. Galton, 'The International Health Exhibition', *The Art*

Journal, 1884: 153–6, 161–4, 293–6: R. H. Vetch, 'Galton, Sir Douglas Strutt (1822–1899)', rev. David F. Channell, *Oxford Dictionary of National Biography*, Oxford University Press, 2004

13. Anon., *Miscellaneous (Including: Return of Number of Visitors and Statistical Tables & Official Guide)*: 8.

14. J.E. Ady, 'The International Health Exhibition': 434–5, 454–5, 476–7.

15. [J. Manley], 'The International Health Exhibition', *Journal of Science and Annals of Astronomy, Biology, Geology, Industrial Arts, Manufactures, and Technology*, 1884: 350–54, 412–16, 579–85: 583; Anon., *Miscellaneous (including Jury Awards and Official Catalogue)*: 59.

16. D. Galton, 'The International Health Exhibition': 155–6; [J. Manley], 'The International Health Exhibition': 413.

17. [G.A. Sala], 'The Health Exhibition: a look around': 94.

18. Anon., *Miscellaneous (Including: Return of Number of Visitors and Statistical Tables & Official Guide)*: 14.

19. ibid.

20. F. Galton, *Memories of My Life* (Methuen, 1908), 245–6.

21. F. Galton, 'On the Anthropometric Laboratory at the Late International Health Exhibition', *Journal of the Anthropological Institute*, 1884: 205–18: 206–7.

22. F. Galton, *Memories of My Life*, 249.

23. R.S. Cowan, *Sir Francis Galton and the study of heredity in the 19th century* (Garland, 1985): viii–ix; F. Galton, *The narrative of an explorer in tropical South Africa* (John Murray, 1853), 54; N.W. Gillham, *A Life of Sir Francis Galton: From African Exploration to the Birth of Eugenics* (Oxford University Press, 2001), 76. The nineteenth-century term Hottentot is now considered offensive and the indigenous people of Namibia are now usually known as the Khoikhoi.

24. F. Galton to C. Darwin, [9 December 1859]: F. Burkhardt and S. Smith (eds.), *The Correspondence of Charles Darwin (Volume 7: 1858–1859)* (Cambridge University Press, 1991): 417.

25. F. Galton, *Memories of My Life*, 288.

26. C. Darwin, *On the Origin of Species by Means of Natural Selection: or the preservation of favoured races in the struggle for life* (1859; Penguin Books, 1964), 490.

27. F. Galton to C. Darwin, [24 December 1869]: K. Pearson, *The life, letters and labours of Francis Galton: I. Birth 1822 to marriage 1853* (Cambridge University Press, 1914–30), 6–7.

28. C. Darwin, *On the Origin of Species*, 22–3, 84; J.A. Secord, 'Nature's Fancy: Charles Darwin and the Breeding of Pigeons', *Isis*, 1981, 163–86.

29. F. Galton, F, 'Hereditary talent and character', *Macmillan's Magazine*, 1865: 157–66, 318–27: 157.

30. ibid., 165–6.

31. R.S. Cowan, 'Nature and Nurture: The Interplay of Biology and Politics in the Work of Francis Galton', *Studies in the History of Biology*, 1977: 133–208: 163–4.

32. Viriculture first appeared in 'Hereditary Improvement', *Fraser's Magazine*, 1873, 7: 116–30; eugenics was coined in *Inquiries into Human Faculty and Its Development* (1883*)*.

33. *Guardian*, 4 April 1883: 1001. Quoted in N.W. Gillham, *A Life of Sir Francis Galton: From African Exploration to the Birth of Eugenics* (Oxford University Press, 2001): 207–8.

34. Quoted in E.A. Gökyigit, 'The reception of Francis Galton's *Hereditary Genius* in the Victorian periodical press', *Journal of the history of biology*, 1994: 215–40: 234; N.W. Gillham, *A Life of Sir Francis Galton*: 171.

35. My thanks to John Waller for clarifying this point for me.

36. F. Galton, *English Men of Science: Their Nature and Nurture* (Macmillan, 1874): 12. He had also used the nature/nurture pairing in the title of an address to the Royal Institution earlier that year: N.W. Gillham, *A Life of Sir Francis Galton*: 191–2.

37. F. Galton, *Hereditary Genius* (1892; Macmillan, 1869): 14.

38. *Guardian*, 4 April 1883: 1001. Quoted in N.W. Gillham, *A Life of Sir Francis Galton*: 207–8.

39. F. Galton, *Memories of My Life*: 288.

40. C. Darwin to Francis Galton, [23 December 1869]: F. Darwin and A.C. Seward. *More Letters of Charles Darwin*, Vol II (London: John Murray, 1903): 41.

41. C. Darwin to C. Kingsley, [10 June 1867]: in F.E. Kingsley (ed.), *Charles Kingsley: his letters and memories of his life* (1878, 2 vols. London), 2: 242.

42. D.L. Hull, *Darwin and his Critics: the Reception of Darwin's Theory of Evolution by the Scientific Community* (Harvard University Press, 1973): 315–6.

43. C. Darwin, *The Variation of Animals and Plants under Domestication, volume 2* (1875; Johns Hopkins University Press, 1998): 35–6.

44. R.W. Burkhardt, 'Closing the Door on Lord Morton's Mare: The Rise and Fall of Telegony', *Studies in the History of Biology*, 1979: 1–21: 3; M. Hodge, 'Darwin as a lifelong generation theorist', in *The Darwinian Heritage*, Princeton University Press: 224; C. Darwin, *The Variation of Animals and Plants under Domestication, volume 1* (1875; Johns Hopkins University Press, 1998): 435.

45. C. Darwin, *The Variation of Animals and Plants under Domestication, volume 2*: 370.

46. ibid.: 394–5.

47. ibid.: 346–7.

48. ibid.: 398–7.

49. F. Galton, *Hereditary Genius*: 370; N.W. Gillham, *A Life of Sir Francis Galton*: 174–5.

50. F. Galton to C. Darwin, [17 December 1870]; and [8 April 1870]: K. Pearson, *The life, letters and labours of Francis Galton: II. Researches of middle life* (Cambridge University Press, 1914–30): 158–9.

51. F. Galton, 'Experiments in pangenesis, by breeding from rabbits of a pure variety, into whose circulation blood taken from other varieties had previously been largely transfused', *Proceedings of the Royal Society*, 1871: 393–410: 404.

52. C. Darwin, *Nature*, 27 April 1871, in: P.H. Barrett, *The Collected Papers of Charles Darwin* (University of Chicago Press,

1980): 165–6.

53. F. Galton to C. Darwin, [25 April 1871]: K. Pearson, *The life, letters and labours of Francis Galton: II. Researches of middle life*: 162.

54. N.W. Gillham, *A Life of Sir Francis Galton*: 176–9; F. Galton to C. Darwin, [15 November l872]: K. Pearson, *The life, letters and labours of Francis Galton: II. Researches of middle life*: 175.

55. F. Galton, 'Blood-Relationship', *Proceedings of the Royal Society*, 1872: 394–402: 173–4.

56. ibid.: 175–6; R. Olby, *Origins of Mendelism* (University of Chicago Press, 1985): 55–63.

57. F. Galton, 'Blood-Relationship': 175.

58. Quoted in D.W. Forrest, *Francis Galton: The Life and Work of a Victorian Genius* (Paul Elek, 1974): 188; N.W. Gillham, *A Life of Sir Francis Galton*: 205.

59. F. Galton, *Hereditary Genius*: 332.

60. F. Galton, 'The Anthropometric Laboratory', *Fortnightly Review*, 1882: 332–8: 332–4, 37–8.

61. F. Galton, *Memories of My Life*: 246.

62. F. Galton, 'On the Anthropometric Laboratory at the Late International Health Exhibition': 211.

63. ibid.: 208, 209–10.

64. G.A. Sala, 'The Health Exhibition: a look around': 91.

65. F. Galton, 'On the Anthropometric Laboratory at the Late International Health Exhibition': 206–7.

66. F. Galton, 'Some Results of the Anthropometric Laboratory', *Journal of the Anthropological Institute*, 1884: 275–87: 278.

67. *Punch*, 15 April 1884, quoted in K. Pearson, *The life, letters and labours of Francis Galton: II. Researches of middle life*: 375.

68. F. Galton, 'Some Results of the Anthropometric Laboratory': 275.

69. F. Galton, 'On the Anthropometric Laboratory at the Late International Health Exhibition': 210.

70. Quoted in K. Pearson, *The life, letters and labours of Francis Galton: II. Researches of middle life*: 381–5.

71. F. Galton, 'Hereditary Improvement': 129.

Chapter 4: *Hieracium auricula*: **What Mendel did next**

Mendel published very little, and many of his papers were destroyed after his death; however, his main publications, 'Experiments on Plant Hybrids', 1865, and 'On Hieracium – Hybrids Obtained by Artificial Fertilisation', 1869, are readily available in *The Origin of Genetics: A Mendel Source Book*, (W.H. Freeman, 1966). L.K. Piternick and G. Piternick, 'Gregor Mendel's letters to Carl Nägeli, 1866–1873' (Electronic Scholarly Publishing, 1950, http://www.esp.org/foundations/genetics/classical/holdings/m/gm-let.pdf).

For **Mendel's life and scientific work**, its significance and its consistent misrepresentation by historians, the most important sources are: L.A. Callender, 'Gregor Mendel: An opponent of descent with modification', *History of Science*, 1988: 41–75; and, Robert Olby's work, particularly 'Mendel no Mendelian?' *History of Science*, 1979: 53–72; *Origins of Mendelism* (University of Chicago Press, 1985); and 'Mendel, Mendelism and Genetics' (1997, http://www.mendelweb.org/archive/MWolby.txt). I have also found Vitezslav Orel's work useful, especially *Mendel* (Oxford University Press, 1984); *Gregor Mendel: the first geneticist* (Oxford University Press, 1996); and 'Constant Hybrids in Mendel's Research', *History and Philosophy of the Life Sciences*, 1998: 291–9. Other works on Mendel consulted include: A.F. Corcos and F.V. Monaghan, 'Was Nägeli to Blame for Mendel's Choice to Work with Hawkweeds?', *Michigan Academician*, 1988: 221–33; W. George, 'The Mendel Enigma, the Farmer's Son: the key to Mendel's motivation', *Archives internationales d'histoire des sciences*, 1982: 177–83.

For the **history of botany** and the **background to Mendel's work**: J. Farley, *Gametes and Spores: Ideas about sexual reproduction, 1750–1914* (Johns Hopkins University Press, 1982); W.M. Montgomery, 'Germany', in *The comparative reception of Darwinism* (University of Chicago Press, 1988); E. Cittadino, *Nature as the Laboratory: Darwinian plant ecology in the German Empire, 1880–1900* (Cambridge University Press, 1990); P. Mazzarello, 'A unifying concept: the history of cell theory', *Nature Cell Biology*, 1999: 13–15;

R.H. Drayton, *Nature's Government: Science, Imperial Britain and the 'Improvement' of the World* (Yale University Press, 2000).

Notes

1. N. Culpeper, *The English physitian: or an astrologo-physical discourse of the vulgar herbs of this nation* (Peter Cole, 1652): 62.
2. Pliny, *Natural history: in ten volumes with an English translation by H. Rackham* (Heinemann, 1968): 36–9.
3. *American Journal of Pharmacy*, August 1881, Vol 53, #8.
4. P. Mraz, '*Hieracium alpinum* subsp. *augusti-bayeri* Zlatnik in the Muntii Rodnei Mts.: an interesting taxon in the flora of Romania', *Thaiszia*, 1999, 9(1): 27–30.
5. C. Darwin to J. D. Hooker, 1 August [1857]: F. Burkhardt and S. Smith(eds.), *The Correspondence of Charles Darwin (Volume 6: 1856–1857)* (Cambridge University Press, 1990): 438. The *OED* gives this as the first recorded use of the terms, though they were already clearly familiar enough for Hooker and Darwin to use without explanation.
6. J.D. Hooker and T. Thomson, *Introductory essay to the Flora Indica* (W. Pamplin, 1855): 43.
7. Linnaeus, *Critica Botanica* (1737), quoted in L.A. Callender, 'Gregor Mendel: An opponent of descent with modification', *History of Science*, 1988: 41–75: 43.
8. Linnaeus, *Somnus plantarum* (1755), quoted in L.A. Callender, 'Gregor Mendel: An opponent of descent with modification': 43.
9. Linnaeus, *Disquisitio de sexu plantarum* (1757), quoted in L.A. Callender, 'Gregor Mendel: An opponent of descent with modification': 43. The Goatsbeard was a cross between *Tragopogon pratense* and *T. porrifolius*.
10. Quoted in V. Orel, *Gregor Mendel: the first geneticist* (Oxford University Press, 1996): 18.
11. Quoted in ibid.: 17.
12. V. Orel, *Gregor Mendel: the first geneticist*: 78, 82–3.
13. R. Olby, *Origins of Mendelism* (University of Chicago Press, 1985): 16–18, 21.

14. G. Mendel, 'Experiments on Plant Hybrids', in *The Origin of Genetics: A Mendel Source Book* (W.H. Freeman, 1966): 8.

15. ibid.: 4.

16. ibid.

17. V. Orel, *Gregor Mendel: the first geneticist*: 97–9.

18. Quoted in R. Olby, *Origins of Mendelism*: 96. See also: V. Orel, *Mendel* (Oxford University Press, 1984): 39–40; R.C. Olby, 'Carl Wilhelm von Nägeli', in *Dictionary of scientific biography*, Scribner.

19. L.A. Callender, 'Gregor Mendel: An opponent of descent with modification': 67; A.F. Corcos and F.V. Monaghan, 'Was Nägeli to Blame for Mendel's Choice to Work with Hawkweeds?', *Michigan Academician*, 1988: 221–33: 221, 23; Barrett's story is in her collection *Ship Fever and Other Stories* (W.W. Norton, 1996).

20. G. Mendel to C. Nägeli, [31 December 1866]: L.K. Piternick and G. Piternick, 'Gregor Mendel's letters to Carl Nägeli. 1866–1873' (Electronic Scholarly Publishing, 1950, http://www.esp.org/foundations/genetics/classical/holdin gs/m/gm-let.pdf.

21. G. Mendel, 'On Hieracium-Hybrids Obtained by Artificial Fertilisation', in *The Origin of Genetics: A Mendel Source Book* (W.H. Freeman): 50–51.

22. ibid.: 52.

23. Mendel to Nägeli, [3 July 1870]. Quoted in L.A. Callender, 'Gregor Mendel: An opponent of descent with modification': 43.

24. G. Mendel, 'On Hieracium-Hybrids Obtained by Artificial Fertilisation': 49–50.

25. ibid.: 51.

26. Nägeli, 1866. Quoted in L.A. Callender, 'Gregor Mendel: An opponent of descent with modification': 60.

27. Ascertaining Mendel's view is difficult, not least because so many of his papers were deliberately destroyed after his death, and many Mendel experts would disagree with my characterization of it. However, I'm persuaded by the

arguments of L.A. Callender, in 'Gregor Mendel: An opponent of descent with modification'.

28. G. Mendel, 'On Hieracium-Hybrids Obtained by Artificial Fertilisation': 54–5.

Chapter 5: *Oenothera lamarckiana*: Hugo de Vries led up the primrose path

De Vries's own works include: 'The Evidence of Evolution', *Science*, 1904: 395–401; *Intracellular Pangenesis: Including a paper on Fertilization and Hybridization* (The Open Court Publishing Co., 1910); 'On the Origin of Species', *Popular Science Monthly*, 1903: 481–96; 'The Origin of Species by Mutation', *Science*, 1902: 721–9; *Species and varieties: their origin by mutation; lectures delivered at the University of California* (The Open Court Publishing Co., 1905).

Works by **supporters of the Mutation Theory** include: R.R. Gates, *The mutation factor in evolution, with particular reference to Oenothera* (Macmillan, 1915); D.T. MacDougal et al., *Mutations, Variations, and Relationships of the Oenotheras* (Carnegie Institution of Washington, 1907); D.T. MacDougal et al., *Mutants and Hybrids of the Oenotheras* (Carnegie Institution of Washington, 1905); and C.H. Merriam, 'Is Mutation a Factor in the Evolution of the Higher Vertebrates?', *Science*, 1906: 241–57.

For **de Vries's life and work**, my main sources were: P.W. Van der Pas, 'The correspondence of Hugo de Vries and Charles Darwin', *Janus*, 1970: 173–213; E. Zevenhuisen, 'Hugo de Vries: life and work', *Acta botanica neerlandica*, 1998: 409–17; I.H. Stamhuis, 'The reactions on Hugo de Vries's *Intracellular Pangenesis*; the Discussion with August Weismann', *Journal of the History of Biology*, 2003: 119–52.

For the **'rediscovery' of Mendelism**, see the Mendel references in Chapter 4, plus B. Theunissen, 'Closing the Door on Hugo de Vries's Mendelism', *Annals of Science*, 1994: 225–48.

De Vries's **Mutation Theory and its reception** around the world are analysed in: A.H. Sturtevant, *A History of Genetics* (Cold

Spring Harbor Laboratory Press/Electronic Scholarly Publishing Project, 1965, 2001); G.E. Allen, 'Hugo de Vries and the Reception of the "Mutation Theory"', *Journal of the History of Biology*, 1969: 55–87; S.E. Kingsland, 'The Battling Botanist: Daniel Trembly MacDougal, Mutation Theory, and the Rise of Experimental Evolutionary Biology in America, 1900–1912', *Isis*, 1991: 479–509.

For **Lyell, Darwinism and the age of the earth**, see: H.E. Gruber, *Darwin on Man: A Psychological Study of Scientific Creativity* (Wildwood House, 1974); C. Smith and M.N. Wise, *Energy and empire: a biographical study of Lord Kelvin* (Cambridge University Press, 1989); P.J. Bowler, *The Eclipse of Darwinism: Anti-Darwinian evolution theories in the decades around 1900* (Johns Hopkins University Press, 1992); J.A. Secord, 'Introduction to Lyell's *Principles of Geology*', in *Principles of Geology* (Penguin Books, 1997).

For twentieth-century biology and the **laboratory revolution**, I am indebted to Garland Allen's work, in particular: 'The introduction of *Drosophila* into the study of heredity and evolution, 1900–1910', *Isis*, 1975: 322–33; *Life Science in the Twentieth Century* (Cambridge University Press, 1978); and *Thomas Hunt Morgan: the man and his science* (Princeton University Press, 1978). Other useful sources included: S. Chadarevian, 'Instruments, Illustrations, Skills, and Laboratories in nineteenth century German Botany', in *Non-verbal communication in science prior to 1900* (Olschki, 1993); S. Chadarevian, 'Laboratory science versus country-house experiments. The controversy between Julius Sachs and Charles Darwin', *British Journal for the History of Science*, 1996: 17–41; L.K. Nyhart, 'Natural history and the 'new' biology', in *Cultures of Natural History* (Cambridge University Press, 1996).

Notes

1. F.V. Coville, 'Notes on the Plants used by the Klamath Indians of Oregon', *Contributions from the US National Herbarium* 1897, 5: 87–108. Southern Oregon Digital Archives, http://soda.sou.edu/
2. *OED.*

3. H. de Vries, 'The Origin of Species by Mutation', *Science*, 1902: 721–29: 722; H. de Vries, *Species and varieties: their origin by mutation; lectures delivered at the University of California* (The Open Court Publishing Company, 1905): 523–4.

4. H. de Vries, *Species and varieties: their origin by mutation; lectures delivered at the University of California*: 27.

5. C. Darwin, *On the Origin of Species by Means of Natural Selection: or the preservation of favoured races in the struggle for life* (1859; Penguin Books, 1964): 471.

6. ibid.: 95.

7. Darwin used the phrase in a letter to A. R. Wallace, [12 July 1871]: F. Darwin, *The Life and Letters of Charles Darwin (Volume III)* (John Murray, 1888): 146.

8. Quoted in K. Pearson, *The life, letters and labours of Francis Galton: II. Researches of middle life* (Cambridge University Press, 1914–30): 381–5.

9. N.W. Gillham, *A Life of Sir Francis Galton: From African Exploration to the Birth of Eugenics* (Oxford University Press, 2001): 281–3.

10. W.F.R. Weldon, 'Remarks on Variation in Animals and Plants', *Proceedings of the Royal Society of London*, 1894: 379–82: 381.

11. W. Bateson, 'Wm. Keith Brooks: A Sketch of His Life by Some of His Former Pupils and Associates', *Journal of Experimental Zoology*, 1910, 9: 1–52. Quoted in G.E. Allen, 'The introduction of *Drosophila* into the study of heredity and evolution, 1900–1910', *Isis*, 1975: 322–33: 323.

12. A.R. Wallace, 'The Method of Organic Evolution', *Fortnightly Review*, 1895: 211–24, 435–45: 216–17, 220, 437.

13. Anonymous, quoted by N. Mitchison, 'Beginnings', in *Haldane and modern biology* (Johns Hopkins University Press): 302.

14. Eberhardt Dennert, quoted in G.E. Allen, 'Hugo de Vries and the Reception of the "Mutation Theory', *Journal of the History of Biology*, 1969: 55–87: 56.

15. Quoted in G.E. Allen, 'Hugo de Vries and the Reception of the 'Mutation Theory'": 56–7.

16. E.B. Wilson, quoted in G.E. Allen, *Thomas Hunt Morgan: the man and his science* (Princeton University Press, 1978): 34–5.

17. M.J. Schleiden, *Principles of Scientific Botany: or Botany as an Inductive Science* (1849; Johnson Reprint Company, 1849 1969): 575, 80.

18. H. de Vries to C. Darwin, [15 October 1881]: P.W. Van der Pas, 'The correspondence of Hugo de Vries and Charles Darwin', *Janus*, 1970: 173–213: 200.

19. E. Zevenhuisen, 'The Hereditary Statistics of Hugo de Vries', *Acta botanica neerlandica*, 1998: 427–63.

20. H. de Vries, 'On the Origin of Species', *Popular Science Monthly*, 1903: 481–96: 491.

21. ibid.: 492.

22. H. de Vries, *Species and varieties: their origin by mutation*: 525–6.

23. H. de Vries, 'On the Origin of Species': 494.

24. H. de Vries, *Species and varieties: their origin by mutation*: 28–9.

25. ibid.: 549–50.

26. ibid.: frontis.

27. H. de Vries, *Mutation Theory*, I: 5–6, quoted in G.E. Allen, 'Hugo de Vries and the Reception of the "Mutation Theory"', *Journal of the History of Biology*, 1969: 55–87: 59–60.

28. H. de Vries, *Species and varieties: their origin by mutation*: 28–9; G.E. Allen, 'Hugo de Vries and the Reception of the 'Mutation Theory'": 62–3.

29. C.B. Davenport, '[review of] Species and Varieties, Their Origin by Mutation', *Science*, 1905: 369–72: 369.

30. K. Pearson, 'Mathematical Contributions to the Theory of Evolution. On the Law of Ancestral Heredity'. Proceedings of the Royal Society of London, Vol. 62 (1897–1898), pp 386–412

31. [Anon.], '*Oenothera* and Mutation', *Nature*, 19 August 1915.

32. H. de Vries, *Mutation Theory*, I: 5–6. Quoted in G.E. Allen, 'Hugo de Vries and the Reception of the "Mutation Theory"': 59–60.

33. Quoted in J. Sapp, *Genesis: the Evolution of Biology* (Oxford University Press, 2003): 132.

34. H. de Vries, 'The Evidence of Evolution', *Science*, 1904: 395–401: 400.

35. S.E. Kingsland, 'The Battling Botanist: Daniel Trembly MacDougal, Mutation Theory, and the Rise of Experimental Evolutionary Biology in America, 1900–1912', *Isis*, 1991: 479–509: 486–8.

36. C.H. Merriam, 'Is Mutation a Factor in the Evolution of the Higher Vertebrates?', *Science*, 1906: 241–57: 242.

37. ibid.: 243; G.E. Allen, 'Hugo de Vries and the Reception of the "Mutation Theory"': 68.

38. G.E. Allen, 'Hugo de Vries and the Reception of the "Mutation Theory"': 74–5.

39. H. de Vries to W. Bateson, [20 October 1901]. Quoted in B. Theunissen, 'Closing the Door on Hugo de Vries's Mendelism', *Annals of Science*, 1994: 225–48: 248.

40. [Anon.], *Athenaeum*, 28 August 1915.

41. 'F.L.', *Botanical Journal*, October 1915.

42. R.R. Gates, 'Review of "The Mutation Theory"', *American Naturalist*, 1911: 254–6: 255–6.

43. *The Times*, Monday 13 August 1962. Homosexuality was not legal in Britain until the passing of the 1967 Sexual Offences Act, which decriminalized sex in private between men over the age of twenty-one.

44. B. Theunissen, 'Closing the Door on Hugo de Vries's Mendelism': 248.

45. Morgan, 1909. Quoted in G. Allen, *Life Science in the Twentieth Century* (Cambridge University Press, 1978): 53–4.

Chapter 6: *Drosophila melanogaster*: Bananas, bottles and Bolsheviks

My main source for the **history of Drosophila** is Robert Kohler's work, particularly: 'Systems of production: Drosophila, neurospora, and biochemical genetics', *Historical Studies in the*

Physical and Biological Sciences, 1991: 87–130; and *Lords of the Fly: Drosophila Genetics and the Experimental Life* (University of Chicago Press, 1994). Garland Allen's work on Thomas Hunt Morgan and his career was also invaluable. Other sources included: E.A. Carlson, 'The 'Drosophila' group: The transition from Mendelian unit to individual gene', *Journal of the History of Biology*, 1974: 31–48; N. Roll-Hansen, 'Drosophila Genetics: A Reductionist Research Program', *Journal of the History of Biology*, 1978: 159–210; and S.G. Brush, 'How Theories became Knowledge: Morgan's Chromosome Theory of Heredity in America and Britain', *Journal of the History of Biology*, 2002: 471–535.

For the **Soviet Drosophila work and Chetverikov**, I have relied on Mark Adams's work, in particular: 'The Founding of Population Genetics: Contributions of the Chetverikov School 1924–1934', *Journal of the History of Biology*, 1968: 23–39; 'Towards a Synthesis: Population Concepts in Russian Evolutionary Thought', *Journal of the History of Biology*, 1970: 107–29; and 'Sergei Chetverikov, the Kol'tsov Institute, and the Evolutionary Synthesis', in *The Evolutionary Synthesis: Perspectives on the Unification of Biology* (Harvard University Press, 1998). I also used: S.S. Chetverikov, *On Certain Aspects of the Evolutionary Process from the Standpoint of Modern Genetics* (1926; Genetics Heritage Press, 1997); T. Dobzhansky, 'The Birth of the Genetic Theory of Evolution in the Soviet Union in the 1920s', in *The Evolutionary Synthesis: Perspectives on the Unification of Biology*, (Harvard University Press, 1998).

My thinking about this chapter and **experimental organisms in general** was greatly influenced by reading B.T. Clause, 'The Wistar Rat as a Right Choice: Establishing Mammalian Standards and the Ideal of a Standardized Animal', *Journal of the History of Biology*, 1993: 329–49.

For **chromosomes and classical genetics**, see: G.P. Rédei, *Genetics* (Macmillan, 1982); E.W. Crow and J.F. Crow, 'Walter Sutton and the Chromosome Theory of Heredity', *Genetics*, 2002: 1–4; S.R. Nelson and P.S. Nelson, 'Walter Sutton's chromosome theory of heredity: one hundred years later' (2002,

http://www.kumc.edu/research/medicine/anatomy/sutton/in
dex.html); and O.S. Harman, 'Cyril Dean Darlington: the man
who "invented" the chromosome', *Nature Reviews: Genetics*, 2005:
79–85.

Background material on **twentieth-century history,
including eugenics**, came from: N. Stepan, *The Idea of Race in
Science: Great Britain 1800–1960* (Macmillan, 1982); D.J. Kevles, *In
the Name of Eugenics: Genetics and the uses of Human Heredity* (Penguin
Books, 1986); W.H. Tucker, *The Science and Politics of Racial Research*
(University of Illinois Press, 1994); and S.J. Gould, *The Mismeasure
of Man* (Penguin Books, 1997).

The **banana details** are drawn from: N.W. Simmonds, *The
evolution of the bananas* (Longmans, 1962); N.S. Price, 'The origin
and development of banana and plantain cultivation', in *Bananas
and Plantains* (Chapman & Hall, 1995); and V.S. Jenkins, *Bananas:
an American history* (Smithsonian Institution Press, 2000).

Notes

1. C. Darwin to J. D. Hooker, [15 December 1876]:
   unpublished Darwin letter, Cambridge University Archives,
   DAR 95: 429.
2. Aristotle, *The History of Animals*, Book V: 19.
3. A.H. Sturtevant, *A History of Genetics* (1965; Cold Spring
   Harbor Laboratory Press / Electronic Scholarly Publishing
   Project, 2001): 43.
4. Kirby and Spence, *Introduction to entomology or elements of the
   natural history of insects* (1815).
5. [R. Chambers], *Vestiges of the Natural History of Creation: and
   other evolutionary writings* (1844; University of Chicago Press,
   1994): 183.
6. J.A. Secord, *Victorian Sensation: The Extraordinary Publication,
   Reception, and Secret Authorship of Vestiges of the Natural History of
   Creation* (University of Chicago Press, 2000); J.A. Secord,
   'Extraordinary experiment: Electricity and the creation of
   life in Victorian England', in *The uses of experiment: studies in the
   natural sciences* (Cambridge University Press); J.E. Strick,

Sparks of Life: Darwinism and the Victorian Debates over Spontaneous Generation (Harvard University Press, 2000).

7. A. Carnegie, *The Gospel of Wealth* (1900), quoted: in D.R. Oldroyd, *Darwinian Impacts: an introduction to the Darwinian Revolution* (University of New South Wales Press, 1980): 215.

8. William Graham Sumner, quoted in D.R. Oldroyd, *Darwinian Impacts: an introduction to the Darwinian Revolution*: 214.

9. Lord Roseberry, quoted in D. Trotter, 'Modernism and Empire: Reading The Waste Land', in *Futures for English* (Manchester University Press): 143–153: 150.

10. K. Pearson, *The Groundwork of Eugenics* (1909), quoted in W.H. Tucker, *The Science and Politics of Racial Research* (University of Illinois Press, 1994): 59.

11. Quoted in R.E. Kohler, *Lords of the Fly: Drosophila Genetics and the Experimental Life* (University of Chicago Press, 1994): 26.

12. Fernandus Payne to A.H. Sturtevant, [16 October 1947]. Sturtevant Papers. Quoted in G.E. Allen, 'The introduction of *Drosophila* into the study of heredity and evolution, 1900–1910', *Isis*, 1975: 322–33: 330.

13. R.E. Kohler, *Lords of the Fly: Drosophila Genetics and the Experimental Life*: 33–4.

14. W.S. Sutton, 'On the morphology of the chromosome group in *Bracystola magna*', *Biological Bulletin* 1902, 4: 24–39, and W.S. Sutton, 'The chromosomes in heredity', *Biological Bulletin*, 1903, 4,: 231–51. See E.W. Crow and J.F. Crow, 'Walter Sutton and the Chromosome Theory of Heredity', *Genetics*, 2002: 1–4: 1.

15. Morgan, *Evolution and Adaptation*, (London: Macmillan and Co. 1903): 286–287. Quoted in G.E. Allen, *Thomas Hunt Morgan: the man and his science* (Princeton University Press, 1978): 111.

16. X-rays were originally known as Röntgen rays in honour of their German discoverer, Wilhelm Conrad Röntgen.

17. G.E. Allen, *Thomas Hunt Morgan: the man and his science*: 152–3. Lilian Morgan's recollection was that it was *white* that

Morgan discussed so enthusiastically when their first child was born on 5 January 1910, but as Kohler has pointed out, she must have been mistaken: it could only have been *with* Morgan was talking about, as *white* did not turn up until May. R.E. Kohler, *Lords of the Fly: Drosophila Genetics and the Experimental Life*: 46.

18. F.W. Taylor, *The Principles of Scientific Management* (1911; Routledge/Thoemmes, 1993).

19. Quoted in B.T. Clause, 'The Wistar Rat as a Right Choice: Establishing Mammalian Standards and the Ideal of a Standardized Animal', *Journal of the History of Biology*, 1993: 329–49: 343.

20. T.H. Morgan, 'Random Segregation Versus Coupling in Mendelian Inheritance', *Science*, 1911: 384; G.E. Allen, *Thomas Hunt Morgan: the man and his science*: 160–1.

21. By J.B.S. Haldane, see: R.E. Kohler, *Lords of the Fly: Drosophila Genetics and the Experimental Life*: 47–8, 79–80.

22. G.E. Allen, *Thomas Hunt Morgan: the man and his science*: 191.

23. M.B. Adams, 'Sergei Chetverikov, the Kol'tsov Institute, and the Evolutionary Synthesis', in *The Evolutionary Synthesis: Perspectives on the Unification of Biology* (Harvard University Press): 262–4.

24. T.H. Morgan to H. de Vries, [5 January 1918]. Archives of the Biological Laboratory, Vrije Universsiteit, Netherlands. My thanks to Elliot Meyerowitz for showing me this letter and to Tom Gerats for permission to quote from it.

25. R.J. Greenspan, *Fly Pushing: The theory and practice of Drosophila genetics* (Cold Spring Harbor Press, 1997): 125.

26. T.H. Morgan, 'A critique of the theory of evolution', Princeton; Princeton University Press, 1916.

Chapter 7: *Cavia porcellus*: Mathematical guinea pigs

The **primary sources** were: C.É. Brown-Séquard, 'Hereditary Transmission of an Epileptiform Affection Accidentally

Produced', *Proceedings of the Royal Society of London*, 1859–60: 297–8; W.E. Castle and H. MacCurdy, *Selection and Cross-breeding in Relation to the Inheritance of Coat-pigments in Rats and Guinea-Pigs* (Carnegie Institution of Washington, 1907); W.E. Castle, 'An expedition to the home of the guinea-pig and some breeding experiments with material there obtained', in *Studies of Inheritance in Guinea-Pigs and Rats* (Carnegie Institution of Washington, 1916); and T. Dobzhansky, 'Genetics of natural populations. XIV. A response of certain gene arrangements in the third chromosome of *Drosophila pseudoobscura* to natural selection', *Genetics*, 1947: 142–60.

For the **history of the guinea pig**, see: C. Cumberland, *The Guinea Pig or Domestic Cavy for Food , Fur and Fancy* (L. Upcott Gill, 1897); B.J. Weir, 'Notes on the origin of the domestic guinea-pig', in *The Biology of Hystricomorph Rodents* (Zoological Society of London, 1974); B. Müller-Haye, 'Guinea-pig or cuy', in *Evolution of domesticated animals* (Longman, 1984); E. Morales, *The Guinea Pig: healing, food, and ritual in the Andes* (University of Arizona Press, 1995); S. Pritt, 'The history of the guinea pig (*Cavia porcellus*) in society and veterinary medicine', *Veterinary Heritage*, 1998: 12–16; J. Clutton-Brock, *A Natural History of Domesticated Mammals* (Cambridge University Press, 1999); and J.C. Castillo, 'Naming Difference: The Politics of Naming in Fernández de Oviedo's *Historia general y natural de las Indias*', *Science in Context*, 2003: 489–504.

For **vitamin C** and the guinea pig's role in its discovery: L.G. Wilson, 'The clinical definition of scurvy and the discovery of Vitamin C', *Journal of the History of Medicine and Allied Sciences*, 1975: 40–60; and K.J. Carpenter, *The History of Scurvy and Vitamin C* (Cambridge University Press, 1986).

For **J.B.S. Haldane's life and work**, I used: his archives at University College London; his own publications, especially: *Possible worlds and other essays* (Chatto & Windus, 1940); and *Science advances* (G. Allen & Unwin, 1947). Also invaluable were: R. Clark, *J.B.S.: The Life and Work of J.B.S. Haldane* (Hodder & Stoughton, 1968); A. Lacassagne, 'Recollections of Haldane', in *Haldane and modern biology* (Johns Hopkins University Press, 1968); N. Mitchison, 'Beginnings', in *Haldane and modern biology* (Johns

Hopkins University Press, 1968); N. Mitchison, 'The Haldanes: Personal notes and historical lessons', *Proceedings of the Royal Institution of Great Britain*, 1974: 1–21; D.B. Paul, 'A War on Two Fronts: J.B.S. Haldane and the Response to Lysenkoism in Britain', *Journal of the History of Biology*, 1983: 1–37; M.B. Adams, 'Last Judgment: The visionary biology of J.B.S. Haldane', *Journal of the History of Biology*, 2001: 457–91.

For **Sewall Wright's life and work**, I am greatly indebted to William Provine's publications: 'The Role of Mathematical Population Geneticists in the Evolutionary Synthesis of the 1930s and 1940s', *Studies in the History of Biology* (1978); *Sewall Wright and Evolutionary Biology* (University of Chicago Press, 1986). Provine has deposited all his research materials for his biography of Wright with the American Philosophical Society, whose archives I also consulted. I also made use of: J.F. Crow, 'Sewall Wright's place in twentieth-century biology', *Journal of the History of Biology*, 1990: 57–89.

In addition to the background material on **twentieth-century biology** listed for Chapters 5 and 6, I also used: P.G. Abir-Am, 'The Molecular Transformation of Twentieth-Century Biology', in *Companion to Science in the Twentieth Century*, (Routledge, 2003).

Notes
1. Oviedo, introduction to book 9, vol. 117, 278. Quoted in J.C. Castillo, 'Naming Difference: The Politics of Naming in Fernández de Oviedo's *Historia general y natural de las Indias*', *Science in Context*, 2003: 489–504: 491.
2. Power, *Experimental philosophy* (1664), I: 16.
3. G. Eliot, *Daniel Deronda* (Everyman, 1999): 383. See also: *Scenes of Clerical Life*: 15. 1873, US edition; *Silas Marner*: 423.
4. C. Cumberland, *The Guinea Pig or Domestic Cavy for Food, Fur and Fancy* (L. Upcott Gill, 1897): 2.
5. ibid.: 2–6.
6. ibid.: 7–8.
7. ibid.: 21–2, 34–44.
8. P. Gibier, 'Dr. Koch's Discovery', *The North American review*, 1890: 726–32: 728.

9. R. Wheatley, 'Hygeia in Manhattan', *Harper's New Monthly Magazine*, 1897: 384–401: 386.

10. C.É. Brown-Séquard, 'Hereditary Transmission of an Epileptiform Affection Accidentally Produced', *Proceedings of the Royal Society of London*, 1859–60: 297–8.

11. A.J. Leffingwell, 'Does Vivisection Pay?', *Scribners Monthly, an illustrated magazine for the people*, 1880: 391–9.

12. H.C. Wood, 'The Value of Vivisection', *Scribners Monthly, an illustrated magazine for the people*, 1880: 766–71: 768.

13. ibid.: 770.

14. E. Glasgow, 'A Point In Morals', *Harper's New Monthly Magazine*, 1895: 976–82.

15. Ouida, 'Some Fallacies of Science', *The North American Review*, 1886: 139–53: 151.

16. L.G. Wilson, 'The clinical definition of scurvy and the discovery of Vitamin C', *Journal of the History of Medicine and Allied Sciences*, 1975: 40–60: 47–51; C. Funk, *The Vitamines* (1st English edition, 1922).

17. E.V. McCollum. Quoted in H.E. Smith, et al., 'Architecture and science associated with the Dairy Barn at the University of Wisconsin-Madison' (Department of Landscape Architecture University of Wisconsin-Madison, 2000).

18. L.G. Wilson, 'The clinical definition of scurvy and the discovery of Vitamin C': 56–7.

19. W.E. Castle, 'The Mutation Theory of Organic Evolution, from the Standpoint of Animal Breeding', *Science*; 1905, 21 (536): 524.

20. W.E. Castle and H. MacCurdy, *Selection and Cross-breeding in Relation to the Inheritance of Coat-pigments in Rats and Guinea-Pigs* (Carnegie Institution of Washington, 1907): 3.

21. As we'll see in Chapter 9, the Danish botanist, Wilhelm Johannsen, had done the first pure line experiments with edible beans, Phaseolus.

22. W.E. Castle and H. MacCurdy, *Selection and Cross-breeding in Relation to the Inheritance of Coat-pigments in Rats and Guinea-Pigs*: 3–4.

23. ibid.: 34.

24. W.E. Castle. Quoted in W.B. Provine, *Sewall Wright and Evolutionary Biology* (University of Chicago Press, 1986): 53–4.

25. S. Wright, 'Birth and Family (Series III: Biographical and autobiographical materials)' (Sewall Wright Papers, American Philosophical Society, Philadelphia).

26. ibid.

27. ibid.

28. N. Mitchison, 'The Haldanes: Personal notes and historical lessons', *Proceedings of the Royal Institution of Great Britain*, 1974: 1–21: 3; R. Clark, *J.B.S.: The Life and Work of J.B.S. Haldane* (Hodder & Stoughton, 1968): 17.

29. N. Mitchison 'Beginnings', in *Haldane and modern biology* (Johns Hopkins University Press): 302–3; N. Mitchison, 'The Haldanes: Personal notes and historical lessons': 8–9; R. Clark, *J.B.S.: The Life and Work of J.B.S. Haldane*: 29–30.

30. N. Mitchison, 'Beginnings', in *Haldane and modern biology*: 303.

31. Cedric Davidson to J.B.S. Haldane, [20 April 1952]: J. Haldane, 'Box 20a: Letters from the public, General Correspondence' (Haldane Collection: Library Services, University College London).

32. J.B.S. Haldane to S. Wright, [5 July 1919]: S. Wright, 'Series II: Correspondence' (Sewall Wright Papers, American Philosophical Society, Philadelphia).

33. Quoted in K. Burns, *Jazz: A History of America's Music* (PBS television, http://www.pbs.org/jazz/index.htm).

34. Quoted in W.B. Provine, *Sewall Wright and Evolutionary Biology*: 110–11.

35. J.B.S. Haldane 'Box 17: Scientific Correspondence, A–D', 1940–52. Folder: Scientific correspondence, 1945–52, C–D.

36. [Name withheld] to J.B.S. Haldane, [24 September 1946]: J. Haldane, 'Box 18: Scientific Correspondence, E-K' (Haldane Collection: Library Services, University College London).

37. A. Lacassagne, 'Recollections of Haldane', in *Haldane and modern biology* (Johns Hopkins University Press): 308.

38. J.B.S. Haldane to S. Wright, [5 July 1919]: S. Wright, 'Series II: Correspondence'.

39. John Maynard Smith, quoted in M.B. Adams, 'Last Judgment: The visionary biology of J.B.S. Haldane', *Journal of the History of Biology*, 2001: 457–91: 477. Haldane's sister Naomi also remembered them both as 'clumsy and accident-prone': N. Mitchison, 'Beginnings', in *Haldane and modern biology*: 300.

40. S. Wright to J.B.S. Haldane, [31 March 1948]: J. Haldane, 'Box 20: Scientific Correspondence' (Haldane Collection: Library Services, University College, London).

41. S. Wright, 'Birth and Family (Series III: Biographical and autobiographical materials)'.

42. J. Cain, 'Interviews with Professor Robert E. Sloan', 1996 http://www.ucl.ac.uk/sts/cain/projects/sloan/

43. Dobzhansky, 1947. Quoted in G. Allen, *Life Science in the Twentieth Century* (Cambridge University Press, 1978): 142.

44. W.E. Castle and H. MacCurdy, *Selection and Cross-breeding in Relation to the Inheritance of Coat-pigments in Rats and Guinea-Pigs*: 3.

45. S. Wright to J.B.S. Haldane, [8 June 1934]: S. Wright, 'Series II: Correspondence'.

Chapter 8: Bacteriophage: The virus that revealed DNA

For **viruses** and their discovery, see: S.S. Hughes, *The virus: a history of the concept* (Heinemann Educational, 1977); A.P. Waterson and L. Wilkinson, *An introduction to the history of virology* (Cambridge University Press, 1978).

My account of **Felix d'Hérelle** and **the history of bacteriophage** relies on the work of William Summers, in particular: 'From culture as organism to organism as cell: Historical origins of bacterial genetics', *Journal of the History of Biology*, 1991: 171–90; 'How Bacteriophage Came to Be Used by the Phage Group', *Journal of the History of Biology*, 1993: 255–67;

and *Félix d'Herelle and the origins of molecular biology* (Yale University Press, 1999).

For more about ***Arrowsmith* and its influence**, see: P. de Kruif, *The Sweeping Wind, A Memoir* (Harcourt, Brace & World, Inc., 1962); W.C. Summers, 'On the origins of the science in "Arrowsmith": Paul de Kruif, Felix d'Hérelle and Phage', *Journal of the history of medicine and allied sciences*, 1991: 315–32.

Details about **Max Delbrück and the Phage Group** come from: T.F. Anderson, 'Electron Microscopy of Phages', in *Phage and the origins of molecular biology* (Cold Spring Harbor Laboratory of Quantitative Biology, 1966); E.L. Ellis, 'Bacteriophage: one-step growth', in *Phage and the origins of molecular biology* (Cold Spring Harbor Laboratory of Quantitative Biology, 1966); C. Harding, 'Max Delbrück (oral history interview, July 14–September 11, 1978)' (Archives, California Institute Of Technology, 1978, http://resolver.caltech.edu/CaltechOH:OH_Delbruck_M); W. Hayes, 'Max Delbrück and the birth of molecular biology', *Social Research*, 1984: 641–73; E.P. Fischer and C. Lipson, *Thinking about science: Max Delbrück and the origins of molecular biology* (W. W. Norton & Company, 1988); W. Beese, 'Max Delbrück: A physicist in biology', in *World views and scientific discipline formation* (Kluwer Academic Publishers, 1991); T. Helvoort, 'The controversy between John H. Northrop and Max Delbrück on the formation of bacteriophage: Bacterial synthesis or autonomous multiplication?', *Annals of Science*, 1992: 545–75; G. Bertani, 'Salvador Edward Luria', *Genetics*, 1992: 1–4.

My ideas about the **importance of phage** and related issues are particularly indebted to the work of Lily E. Kay, especially: 'Conceptual models and analytical tools: The biology of physicist Max Delbrück', *Journal of the History of Biology*, 1985: 207–46; 'Quanta of life: atomic physics and the reincarnation of phage', *History and philosophy of the life sciences*, 1992: 3–21; and *Who Wrote the Book of Life? A History of the Genetic Code* (Stanford University Press, 2000).

For the rise of **molecular biology**, see: R.C. Olby, *The path to the double helix* (Macmillan, 1974); P.G. Abir-Am, 'The discourse of physical power and biological knowledge in the 1930s: a reappraisal

465

of the Rockefeller Foundations "policy" in molecular biology', *Social Studies of Science*, 1982: 341–82; N. Rasmussen, *Picture Control: The Electron Microscope and the Transformation of Biology in America, 1940–1960* (Stanford University Press, 1997); P.G. Abir-Am, 'The Molecular Transformation of Twentieth-Century Biology', in *Companion to Science in the Twentieth Century* (Routledge, 2003).

Notes

1. S. Lewis, *Arrowsmith* (1925; Harcourt, Brace & World, Inc., 1952): 35–7.
2. ibid.: 166.
3. S.S. Hughes, *The virus: a history of the concept* (Heinemann Educational, 1977): 49, 109–12.
4. W.C. Summers, 'On the origins of the science in "Arrowsmith": Paul de Kruif, Felix d'Hérelle and Phage', *Journal of the history of medicine and allied sciences*, 1991: 315–32: 319; A.P. Waterson and L. Wilkinson, *An introduction to the history of virology* (Cambridge University Press, 1978): 87–8.
5. T. Helvoort, 'The controversy between John H. Northrop and Max Delbrück on the formation of bacteriophage: Bacterial synthesis or autonomous multiplication?', *Annals of Science*, 1992: 545–75; A.P. Waterson and L. Wilkinson, *An introduction to the history of virology*: 86.
6. A.P. Waterson and L. Wilkinson, *An introduction to the history of virology*: 91.
7. L.E. Kay, 'Quanta of life: atomic physics and the reincarnation of phage', *History and philosophy of the life sciences*, 1992: 3–21: 9–10.
8. P. de Kruif, *The Sweeping Wind, A Memoir* (Harcourt, Brace & World, Inc., 1962): 13–14.
9. ibid.: 60–61.
10. ibid.: 93; W.C. Summers, 'On the origins of the science in "Arrowsmith": Paul de Kruif, Felix d'Hérelle and Phage': 317; B.G. Spayd, 'Introduction', in S. Lewis, *Arrowsmith*: xv.
11. P. de Kruif, *The Sweeping Wind, A Memoir*: 96; W.C. Summers, 'On the origins of the science in "Arrowsmith": Paul de

Kruif, Felix d'Hérelle and Phage': 317; B.G. Spayd, 'Introduction', in S. Lewis, *Arrowsmith*: xv.

12. T.J. LeBlanc, '"Arrowsmith" (review)', *Science*, 1925: 632–4.

13. *Lancet*, 1 August 1925, 234/2.

14. J.E. Greaves, 'Do Bacteria have Disease?', *Scientific Monthly*, 1926: 123–5: 124.

15. E.L. Ellis, 'Bacteriophage: one-step growth', in *Phage and the origins of molecular biology*, Cold Spring Harbor Laboratory of Quantitative Biology: 55.

16. W. Hayes, 'Max Delbrück and the birth of molecular biology', *Social Research*, 1984: 641–73: 648; E.L. Ellis, 'Bacteriophage: one-step growth': 55; W.C. Summers, 'How Bacteriophage Came to Be Used by the Phage Group', *Journal of the History of Biology*, 1993: 255–67: 258–9.

17. Delbrück interview: C. Harding, 'Max Delbrück (oral history interview, July 14–September 11, 1978)' (Archives, California Institute of Technology, 1978, http://resolver.caltech.edu/CaltechOH:OH_Delbruck_M.

18. ibid.; L.E. Kay, 'Quanta of life: atomic physics and the reincarnation of phage': 4.

19. Quoted in P.G. Abir-Am, 'The discourse of physical power and biological knowledge in the 1930s: a reappraisal of the Rockefeller Foundation's "policy" in molecular biology', *Social Studies of Science*, 1982: 341–82: 350.

20. H.J. Muller, 'Variations due to change in the individual gene', *American Naturalist*, 1922, 56: 48–9. Quoted in W.C. Summers, 'How Bacteriophage Came to Be Used by the Phage Group': 262.

21. Delbrück interview: C. Harding, 'Max Delbrück (oral history interview, July 14–September 11, 1978)'.

22. ibid.

23. ibid.

24. Quoted in L.E. Kay, 'Quanta of life: atomic physics and the reincarnation of phage': 11.

25. W. Hayes, 'Max Delbrück and the birth of molecular biology': 653–4.

26. J. Robert Oppenheimer, Lecture at MIT, 25 November 1947; 'Physics in the Contemporary World', *Bulletin of the Atomic Scientists*, Vol. IV, No. 3, March 1948: 66.

27. W. Hayes, 'Max Delbrück and the birth of molecular biology': 654.

28. Quoted in E.M. Witkin, 'Chances and Choices: Cold Spring Harbor, 1944–1955', *Annual Review of Microbiology*, 2002: 1–15: 14.

29. W.C. Summers, *Félix d'Herelle and the origins of molecular biology* (Yale University Press, 1999): 167–8.

Chapter 9: *Zea mays*: Incorrigible corn

Some of the **original scientific papers** referred to in this chapter include: G.H. Shull, 'A study in heredity (review of Johannsen, W., 1903)', *Botanical Gazette*, 1904: 314–15; W.E. Castle and H. MacCurdy, *Selection and Cross-breeding in Relation to the Inheritance of Coat-pigments in Rats and Guinea-Pigs* (Carnegie Institution of Washington, 1907); G. H. Shull, 'The Composition of a field of Maize', in *Report of the meeting held at Washington, D.C., January 28–30, 1908, and for the year ending January 12, 1908* (Kohn & Polloock, 1908); E.M. East, 'A Mendelian Interpretation of Variation that is Apparently Continuous', *American Naturalist*, 1910: 65–82; B. McClintock, 'Introduction to *The Discovery and Characterization of Transposable Elements*', in *The dynamic genome: Barbara McClintock's ideas in the century of genetics* (Cold Spring Harbor Laboratory Press, 1992).

For the **history of corn**, see: P.C. Mangelsdorf, *Corn: its origin, evolution, and improvement* (Harvard University Press, 1974); N.P. Hardeman, *Shucks, shocks, and hominy blocks: corn as a way of life in pioneer America* (Louisiana State University Press, 1981); J.B. Longone, *Mother maize and king corn: the persistence of corn in the American ethos* (William L. Clements Library University of Michigan, 1986); B. Fussell, 'Translating maize into corn: The transformation of America's native grain'. *Social Research*, 1999: 41–65.

For **Kellogg and cornflakes**, see: J.H. Kellogg, 'Plain facts for old and young: embracing the natural history and hygiene of organic life' (Electronic Text Center, University of Virginia Library, 1877 (1999), http://etext.lib.virginia.edu); J. Money, *The Destroying Angel: Sex, fitness & food in the legacy of degeneracy theory, Graham Crackers, Kellogg's Corn Flakes & American Health History* (Prometheus Books, 1985).

The story of **hybrid corn** and of **early maize genetics** is told in: P. de Kruif, *Hunger fighters* (Jonathan Cape, 1929); H.P. Riley, 'George Harrison Shull', *Bulletin of the Torrey Botanical Club*, 1955: 243–8; D.B. Paul and B.A. Kimmelman, 'Mendel in America: Theory and Practice, 1900–1919', in *The American Development of Biology* (University of Pennsylvania Press, 1988); D.K. Fitzgerald, *Business of Breeding: Hybrid Corn in Illinois, 1890–1940* (Cornell University Press, 1990); B. Kimmelman, 'Organisms and Interests in Scientific Research: R.A. Emerson's claims for the unique contributions of agricultural genetics', in *The Right tools for the job: at work in twentieth-century life sciences* (Princeton University Press, 1992); M.H. Rhoades, 'The Early Years of Maize Genetics', in *The dynamic genome: Barbara McClintock's ideas in the century of genetics* (Cold Spring Harbor Laboratory Press, 1992).

The main sources for **Barbara McClintock's life and work** are: E.F. Keller, *A feeling for the organism: the life and work of Barbara McClintock* (W.H. Freeman, 1983); H.B. Creighton, 'Recollections of Barbara McClintock's Cornell Years', in *The dynamic genome: Barbara McClintock's ideas in the century of genetics* (Cold Spring Harbor Laboratory Press, 1992); N. Fedoroff, 'Maize Transposable Elements: A Story in Four Parts', in *The dynamic genome: Barbara McClintock's ideas in the century of genetics* (Cold Spring Harbor Laboratory Press, 1992); N.C. Comfort, 'The real point is control: the reception of Barbara McClintock's controlling elements', *Journal of the history of biology*, 1999: 133–62; N.C. Comfort, *The Tangled Field: Barbara McClintock's Search for the Patterns of Genetic Control* (Harvard University Press, 2001).

Notes

1. J.B. Longone, *Mother maize and king corn: the persistence of corn in the American ethos* (William L. Clements Library University of Michigan, 1986): 6–8.

2. 'An Excersion [*sic*] to Cape Cod', from *Bradford's and Winslow's Journal* ('A Diary of Occurrences') by William Bradford (1622).

3. J.B. Longone, *Mother maize and king corn: the persistence of corn in the American ethos*: 10–11.

4. 'Homespun' [Benjamin Franklin], letter to *The Gazetteer and New Daily Advertiser* (2 January 1766), in *The Writings of Benjamin Franklin, Volume III*: London, 1757–75.

5. J.H. Kellogg, 'Plain facts for old and young: embracing the natural history and hygiene of organic life' (Electronic Text Center, University of Virginia Library, 1877 (1999), http://etext.lib.virginia.edu/etcbin/toccer-new2?id= KelPlai.sgm&images=images/modeng&data=/texts/englis h/modeng/parsed&tag=public&part=all.

6. ibid.

7. ibid.

8. ibid. The modern edition of the book notes that 'The so-called "facts" and dogma in this text are over 100 years old. Many are obsolete practices as well as inhuman. Do NOT accept the medical information contained herein as a contemporary practice!'

9. ibid.

10. ibid.

11. ibid.

12. 'J. H. Kellogg Dies; Health Expert, 91', *New York Times*, 16 December 1943.

13. E.M. Thomas, 'Mondamin', *Atlantic Monthly*, 1885: 364–9: 365.

14. ibid.: 369.

15. P. de Kruif, *Hunger fighters* (Jonathan Cape, 1929: 184).

16. C. Darwin to J. Scott, [11 December 1862]: F. Burkhardt et al., *The Correspondence of Charles Darwin (Volume 10: 1862)*

(Cambridge University Press, 1997): 594. Scott took the hint and later published two papers on maize: J. Scott, 'Remarks on the sexual changes in the inflorescence of *Zea Mays*', *Edinburgh New Philosophical Journal*, 1864, n.s. 19: 213–20; and 'Remarks on the sexual changes in the inflorescence of *Zea Mays*', *Transactions of the Botanical Society of Edinburgh*, 1866 8,: 55–62.

17. Quoted in D.K. Fitzgerald, *Business of Breeding: Hybrid Corn in Illinois, 1890–1940* (Cornell University Press, 1990): 43.

18. Quoted in P.C. Mangelsdorf, *Corn: its origin, evolution, and improvement* (Harvard University Press, 1974): 5. I have modernized Lyte's spelling for clarity's sake.

19. H.P. Riley, 'George Harrison Shull', *Bulletin of the Torrey Botanical Club*, 1955: 243–8: 245.

20. P. de Kruif, *Hunger fighters*: 192.

21. ibid.

22. G.H. Shull, 'A study in heredity (review of Johannsen, W., 1903)', *Botanical Gazette*, 1904: 314–15: 314.

23. George McCleur, 1892. Quoted in D.K. Fitzgerald, *Business of Breeding: Hybrid Corn in Illinois, 1890–1940* (Cornell University Press, 1990): 14–15.

24. P. de Kruif, *Hunger fighters*: 182.

25. G.H. Shull, 'The Composition of a field of Maize', in *Report of the meeting held at Washington, D.C., January 28–30, 1908, and for the year ending January 12, 1908*, Kohn & Polloock: 299; D.K. Fitzgerald, *Business of Breeding: Hybrid Corn in Illinois, 1890–1940*: 36–7.

26. P. de Kruif, *Hunger fighters*: 199.

27. Cyril G. Hopkins, quoted in D.K. Fitzgerald, *Business of Breeding: Hybrid Corn in Illinois, 1890–1940*: 21–2.

28. E.M. East, 'A Mendelian Interpretation of Variation that is Apparently Continuous', *American Naturalist*, 1910: 65–82: 401–3.

29. ibid.: 403–4.

30. ibid.: 422.

31. Both letters are quoted in D.K. Fitzgerald, *Business of Breeding: Hybrid Corn in Illinois, 1890–1940*: 38–9.

32. R.A. Emerson and E.M. East 1913; 'The inheritance of quantitative characters in maize', *Research Bulletin of the Nebraska Agricultural Experimental Station*, 1913, 2.

33. R.A. Emerson, quoted in B. Kimmelman, 'Organisms and Interests in Scientific Research: R.A. Emerson's claims for the unique contributions of agricultural genetics', in *The Right tools for the job: at work in twentieth-century life sciences* (Princeton University Press, 1992): 211–12.

34. M.H. Rhoades, 'The Early Years of Maize Genetics', in *The dynamic genome: Barbara McClintock's ideas in the century of genetics*, Cold Spring Harbor Laboratory Press, 1992: 45.

35. H.B. Creighton, 'Recollections of Barbara McClintock's Cornell Years', in *The dynamic genome: Barbara McClintock's ideas in the century of genetics*: 15.

36. John Evelyn, *Kalendarium hortense, or Gardn'ers Almanac* (J. Walthoe, 1664): 77.

37. Emerson, in *American Naturalist* (1914). Quoted in N. Fedoroff, 'Maize Transposable Elements: A Story in Four Parts', in *The dynamic genome: Barbara McClintock's ideas in the century of genetics*: 390; N.C. Comfort, *The Tangled Field: Barbara McClintock's Search for the Patterns of Genetic Control* (Harvard University Press, 2001): 69–70.

38. N. Comfort, personal communication (via email, [29 August 2005].

39. B. McClintock, 'Introduction to *The Discovery and Characterization of Transposable Elements*', in *The dynamic genome: Barbara McClintock's ideas in the century of genetics*: x–xi.

40. See N.C. Comfort, '"The real point is control": the reception of Barbara McClintock's controlling elements', *Journal of the history of biology*, 1999; and *The Tangled Field: Barbara McClintock's Search for the Patterns of Genetic Control* (Harvard University Press, 2001).

41. N.C. Comfort, '"The real point is control": the reception of Barbara McClintock's controlling elements': 152.

42. Quoted in G.R. Fink, 'Barbara McClintock (1902–1992)', *Nature*, 1992: 272.

Chapter 10: *Arabidopsis thaliana*: A fruit fly for the botanists

My **main sources** were interviews and emails with Caroline Dean, Ian Furner, Maarten Koornneef, Eliot Meyerowitz and Chris Somerville. The history of Arabidopsis research is also described in: G.P. Rédei, '*Arabidopsis* as a genetic tool', *Annual Review of Genetics*, 1975, 9: 111–27; E.M. Meyerowitz, '*Arabidopsis*: a Useful Weed', *Cell*, 1989, 56: 263–9; G.R. Fink, 'Anatomy of a revolution', *Genetics*, 1998, 149(2): 473–7; E. Pennisi, '*Arabidopsis* comes of age', *Science*, 2000, 290(5489): 32–5; E.M. Meyerowitz, 'Prehistory and history of *Arabidopsis* research', *Plant Physiology*, 2001, 125: 15–19; and C. Somerville and M. Koornneef, 'A fortunate choice: the history of *Arabidopsis* as a plant model', *Nature Reviews: Genetics*, 2002, 3: 883–9.

Background material on the **Manhattan Project** and the rise of **Big Science** came from: D.B. Paul, 'H.J. Muller, Communism, and the Cold War', *Genetics*, 1988, 119: 223–5; J. Beatty, 'Genetics in the Atomic Age: the Atomic Bomb Casualty Commission, 1947–56', in K.R. Benson, J. Maienschein and R. Rainger, *The Expansion of American biology* (Rutgers University Press, 1991); J. Hughes, *The Manhattan Project: Big Science and the Atom Bomb* (Icon Books, 2002). For the rise of molecular biology see: S.S. Hughes, 'Making Dollars out of DNA: The First Major Patent in Biotechnology and the Commercialization of Molecular Biology, 1974–1980', *Isis*, 2001, 95(3): 541–75; L.E. Kay, *Who Wrote the Book of Life? A History of the Genetic Code* (Stanford University Press, 2000). For Sydney Brenner and *C. elegans*, see: S. de Chadarevian, 'Of worms and programs: *Caenorhabditis elegans* and the study of development', *Studies in the History and Philosophy of the Biological and Biomedical Sciences*, 1998, 29(1): 81–105. For Seymour Benzer and the return of Drosophila see: J. Weiner, *Time, Love, Memory: a great biologist and his quest for the origins of behaviour* (Faber and Faber, 2000).

Notes

1. Quoted in W. Blunt and W.T. Stearn, *The Art of Botanical Illustration* (Antique Collector's Club, 1994).

2. 'Muller, Biologist, wins Nobel Prize', *New York Times*, 1 November 1946.

3. E.M. Meyerowitz, 'Prehistory and history of *Arabidopsis* research', *Plant Physiology*, 2001, 125: 15–19.

4. Quoted in J. Beatty, 'Genetics in the Atomic Age: the Atomic Bomb Casualty Commission, 1947–56'. K.R. Benson, J. Maienschein and R. Rainger, in *The Expansion of American biology* (New Brunswick, Rutgers University Press, 1991): 284–324.

5. C. Somerville, 'Interview about *Arabidopsis* history', conducted by Jim Endersby, Carnegie Institution, 260 Panama St, Stanford University, 14 April 2005. All quotes from Chris Somerville are from this interview unless otherwise stated.

6. D.L. Meadows, *The Limits to Growth; a report for the Club of Rome on the Predicament of Mankind.* (London, 1972): 23

7. E.M. Meyerowitz (2005), 'Interview about *Arabidopsis* history', conducted by Jim Endersby, Kerckhoff laboratory, California Institute of Technology, Pasadena, California, 13 April 2005. All quotes from Elliott Meyerowitz are from this interview unless otherwise stated.

8. E.M. Meyerowitz, personal communication via email [25 September 2005]. For Meinke and Sussex's work, see: D.W. Meinke, 'Embryo lethal mutants of *Arabidopsis thaliana*', *Genetics*, 1978, 88 (4): s67–s68; D.W. Meinke and I.M. Sussex, 'Embryo-lethal mutants of *Arabidopsis thaliana*', *Developmental Biology*, 1979, 72 (1): 50–61.

9. For a more detailed account of this – very complex – story, see W.L. Ogren, 'Affixing the O to Rubisco: discovering the source of photorespiratory glycolate and its regulation', *Photosynthesis Research*, 2003, 76: 53–63.

10. A process biologists refer to as cloning, but which is completely distinct from the cloning of whole animals, such as Dolly the sheep.

11. Bennett and Smith, *Phil. Trans. Roy. Soc. Lond. B.* 1976, 274, 227–74. Table 8, p. 248.

12. L.S. Leutwiler, B.R. Hough-Evans and E.M. Meyerowitz, 'The DNA of *Arabidopsis thaliana*', *Molecular and General Genetics*, 1984, 194: 15.

13. I. Furner, 'Interview about *Arabidopsis* history', conducted by Jim Endersby, Furner Lab, Department of Genetics, Cambridge University, 7 March 2005. All quotes from Ian Furner are from this interview unless otherwise stated.

14. M. Koornneef, 'Interview about *Arabidopsis* history', conducted by Jim Endersby, via email, 19 September 2005. All quotes from Maarten Koornneef are from this interview unless otherwise stated.

15. M.A. Estelle and C.R. Somerville, 'The mutants of *Arabidopsis*', *Trends in Genetics*, 1986, 2: 89–93.

16. C. Somerville and M. Koornneef, 'A fortunate choice: the history of *Arabidopsis* as a plant model', *Nature Reviews: Genetics*, 2002, 3: 883–9.

17. M. Koornneef, M. Personal communication with Jim Endersby via email, 19 September 2005.

18. The term had first been used in the 1960s, but it was only in the mid-1970s – as the possibility of genetic engineering became a reality – that the word became widely used.

19. The case, *Diamond vs. Chakrabarty*, involved a bacterium that was supposed to 'eat' oil spilt in accidents, and the Court decided that the bacteria was in effect an invention and therefore patentable. The bacterium was not genetically engineered, but the decision that it could be patented made the biotechnology era possible.

20. Genetic transformation was possible before the Agrobacterium technique, but cumbersome and slow.

21. C. Dean, 'Interview about *Arabidopsis* history', conducted by Jim Endersby, 21 March 2005, John Innes Institute, Norwich. All quotes from Caroline Dean are from this interview unless otherwise stated.

22. Strictly speaking, what happens in tulips is not true vernalization, because they are bulbs, not seeds, but the genetic mechanism controlling it is almost certainly the same.

23. The British effort was taken over by Mike Bevan at John Innes, who suggested that Caroline Dean should do less administrative work and concentrate on her flowering-time research.

24. I. Furner, personal communication (via email), 23 Setember 2005.

25. G.R. Fink, 'Anatomy of a revolution', *Genetics*, 1998, 149(2): 473–7.

26. G.R. Fink, 'Anatomy of a revolution'.

27. Among the sequencing labs was Cereon Genomics, a subsidiary of Monsanto based in Cambridge, Mass.

28. I. Furner, Personal communication (via email), 23 September 2005.

29. J. Durant and N. Lindsey (2000), *The Great GM Food Debate: a survey of media coverage in the first half of 1999*, London, Parliamentary Office of Science and Technology, 2000; MORI opinion poll, reported in H. Gibson, 'Who is Afraid of GM Food?', *TIME* magazine, 1 March 1999.

30. H. Gibson, 'Who is Afraid of GM Food?'.

31. Paul Kelso, 'Greenpeace wins key GM case', *Guardian*, 21 September 2000.

Chapter 11: *Danio rerio*: Seeing through zebrafish

My **main sources** were interviews and emails with Kate Lewis, Derek Stemple, Will Talbot and Steve Wilson. The history of zebrafish research is also described in J. Bradbury, 'Small Fish, Big Science', *PLoS Biology* 2.5, 2004: 568–72; D.J. Grunwald and J.S. Eisen, 'Headwaters of the Zebrafish – Emergence of a New Model Vertebrate', *Nature Reviews: Genetics* 3 (2002): 717–24. K. Lewis, 'The Emergence of the Zebrafish as a Model Vertebrate Organism for Developmental Genetics' (Unpublished essay, 1994)

For **George Streisinger's life** and the history of the **Oregon fish research**, I have relied heavily on: L. Streisinger, *From the Sidelines: A Personal History of the Institute of Molecular Biology at the University of Oregon*, 1st edn, University of Oregon Press, 2004; F.W. Stahl, 'George Streisinger' *Biographical Memoirs, National Academy of Sciences*, 1996: 68, 352–61.

Some of the more **accessible scientific papers** on zebrafish include: D. Marcey and C. Nüsslein-Volhard, 'Embryology goes fishing', *Nature*, 321, 1986: 380–381; C.B. Kimmel, 'Genetics and Early Development of Zebrafish', *Trends in Genetics*, 1989: 5, 283–8; and P. Kahn, 'Zebrafish hit the big time', *Science*, 1994: 264, 904–5.

In addition to the interviewees' own websites (which you can find by searching), **internet sites dealing with zebrafish** include: *zebrafish.stanford.edu* as well as *zfin.org* and *zebra.biol.sc.edu*

Notes

1. Hamilton had been called Francis Buchanan until 1818, when he succeeded to his mother's estate and took her name of Hamilton.

2. P.H. Gosse, *The Aquarium* (1854). Quoted in B. Brunner, *The Ocean at Home: An Illustrated History of the Aquarium* (New York: Princeton Architectural Press, 2003): 41. While Gosse is usually credited with inventing the aquarium, that honour should be shared with the chemist Robert Warington and Anna Thynne, wife of the dean of Westminster Abbey. Thynne was the one who discovered the oxygenating effect of plants, which kept the sea water fresh, which allowed marine creatures to be brought home by train, carefully housed in buckets or jars, and then transferred to an aquarium; R. Stott, *Theatres of Glass: The Woman Who Brought the Sea to the City* (Short Books, 2003).

3. H.D. Butler, *The Family Aquarium, or Aqua Vivarium* (N.Y., 1858). Quoted in Brunner, *The Ocean at Home*: 69.

4. *The New York Aquarium Journal* (1876). Quoted in Brunner, *The Ocean at Home*: 81.

5. Quoted in J.M. Oppenheimer, 'Historical Introduction to the Study of Teleostean Development,' *Osiris*,1936, 2.

6. E. Leitholf, '*Danio Analipunctatus*', *Aquatic Life*, Philadelphia, August 1917: 161.

7. C. W. Creaser, 'The Technic of Handling the Zebra Fish (*Brachydanio Rerio*) for the Production of Eggs Which Are Favorable for Embryological Research and Are Available at Any Specified Time Throughout the Year,' *Copeia*, 1934, 4. See also R. Riehl and H.A. Baensch, *Aquarium Atlas* (Baensch Press, Hong Kong, 1987): 408.

8. Creaser, 'The Technic of Handling the Zebra Fish': 160.

9. ibid.: 161.

10. L. Streisinger, *From the Sidelines: A Personal History of the Institute of Molecular Biology at the University of Oregon*, 1st edn. (University of Oregon Press, 2004): 53.

11. If you are a fish-keeper and would like to try it, it's made from liver chopped up and mixed with a precooked baby cereal such as Pablum. Blend the two together with a high-speed blender, spoon the mix into small glass containers, such as baby food jars, and place the open jars in a pan of boiling water for a few minutes – the poaching helps stop the liver rotting.

12. K.D. Kallman, 'How the *Xiphophorus* Problem Arrived in San Marcos, Texas,' *Marine Biotechnology*, 2001, 3, Supplement 1,: S6–8, S11–13.

13. Quoted in L. Streisinger, *From the Sidelines*: 24.

14. Quoted in L. Streisinger, *From the Sidelines*: 36–7.

15. David Junah Grunwald and Judith S. Eisen. 'Headwaters of the Zebrafish – Emergence of a New Model Vertebrate! *Nature Renews Genetics 3* (2003): 717–24

16. Streisinger, *From the Sidelines*: 61.

17. Information from the Oregon-based Zebrafish Information Network, ZFIN, http://zfin.org.

18. William Talbot, 'Interview About Zebrafish History,' 2005, conducted by Jim Endersby, Stanford University School of Medicine, Stanford, 12 April 2005. All quotes from Will Talbot are from this interview unless otherwise stated.

19. Stephen Wilson, 'Interview About Zebrafish History,' 2005, conducted by Jim Endersby, University College London, 9 March 2005. All quotes from Steve Wilson are from this interview unless otherwise stated.

20. You can experience this for yourself, by visiting one of the websites listed in the sources for this chapter: several have fish movies available online.

21. J.M. Schindler, 'Zebrafish: *Drosophila* with a spine (but can they fly?)', *The New Biologist*, 1991, 3: 47–9.

22. Information from the US National Centre for Biotechnology Information's PubMed database, http://www.ncbi.nlm.nih.gov/entrez/.

23. Quoted in J. Bradbury, 'Small Fish, Big Science,' *PLoS Biology*, 2004, 2.5 2004: 569.

24. Quoted in ibid.

Chapter 12: *OncoMouse*®: Engineering organisms

My **main sources** were: Sarah Blaffer Hrdy, 'Empathy, Polyandry and the Myth of the Coy Female', in Ruth Bleier (ed.), *Feminist approaches to science*, Pergamon, 1986: 119–146; Adele E. Clarke and Joan H. Fujimura (eds.) *The Right tools for the job: at work in twentieth-century life sciences* (Princeton University Press, 1992); Richard Burian, 'How the choice of experimental organism matters: Epistemological reflections on an aspect of biological practice', *Journal of the History of Biology*, 1993, 26: 351–67; Susan Aldridge, *The Thread of Life: The story of genes and genetic engineering* (Cambridge University Press, 1996); G.J.V. Nossal and R.L. Coppel, *Reshaping Life: key issues in genetic engineering* (Cambridge University Press, 2002); W. Faulkner and E.A. Kerr, 'On Seeing Brockenspectres: Sex and Gender in Twentieth-Century

479

Science', in John Krige and Dominique Pestre (eds.), *Companion to Science in the Twentieth Century*, Routledge, 2003: 43–60; and Karen A. Rader, *Making Mice: Standardizing animals for American biomedical research, 1900–1955* (Princeton University Press, 2004).

Notes
1. Rader, *Making Mice*.
2. Following Chakrabarty's successful patent in March 1980 (Chapter 10), which allowed living bacteria to be patented, a further case in 1987 – known as *Ex parte Allen* – extended the principle to nonhuman multicellular organisms; editorial, *Nature Biotechnology*, 2003, 21, 341 (2003).
3. See S.J. Gould, 'The Misnamed, Mistreated and Misunderstood Irish Elk', in *Ever Since Darwin: Reflections in Natural History* (1977, Penguin Books, 1987).
4. C. Darwin, *The Descent of Man, and Selection in Relation to Sex* (1871, Princeton University Press, 1981).
5. ibid.
6. ibid.
7. C.W. Creaser, 'The Technic of Handling the Zebra Fish': 160.
8. M.L. Wayne, 'Walking a tightrope: the feminist life of a Drosophila biologist,' *NWSA Journal*, 2000, 12(3): 142.
9. A.J. Bateman, 'Intra-sexual selection in Drosophila', *Heredity*, 1948, 2: 360.
10. ibid.: 364–5.
11. ibid.: 365.
12. S.B. Hrdy, 'Empathy, Polyandry and the Myth of the Coy Female,' in *Feminist approaches to science*, ed. R. Bleier Pergamon, 1986): 121–3.
13. ibid.: 123–4.
14. M.L. Wayne, 'Walking a tightrope': 141–2.
15. Quoted in Creager, *Life of a Virus*.
16. P. Brown, 'GM crops created superweed, say scientists', *Guardian*, 25 July 2005: 3.
17. E.M. East, *Report of the Agronomist: being Part VII of the biennial*

report of 1907–8 (New Haven: Connecticut Agricultural Experiment Station, 1908).

18. D. Fitzgerald, 'Farmers Deskilled: Hybrid Corn and Farmers' Work', *Technology and Culture*, 1993, 34: 340–41.

19. R. Lewontin and J.-P. Berlan, 'The political economy of hybrid corn', *The Monthly Review*, 1986: 38.

Additional sources

As well as the various books and articles used above, I have used **general reference works,** especially the *Oxford English Dictionary*, the *Oxford Dictionary of National Biography* and the *Oxford American National Biography*. Overall **histories of biology** include: G. Allen, *Life Science in the Twentieth Century* (Cambridge University Press, 1978); J. Farley, *Gametes and Spores: Ideas about sexual reproduction, 1750–1914* (Johns Hopkins University Press, 1982); P.J. Bowler, *Evolution: the History of an Idea* (University of California Press, 1989); P.J. Bowler, *The Fontana History of the Environmental Sciences* (Fontana, 1992); and most recently, J. Sapp, *Genesis: the Evolution of Biology* (Oxford University Press, 2003). I have used **environmental history** to help tell some of the organisms' stories, in particular: J. Diamond, *Guns, Germs and Steel* (Vintage, 1998); and W. Cronon, *Changes in the Land: Indians, colonists, and the ecology of New England* (Hill & Wang, 2003). For **technological history**, see: T.P. Hughes, *American Genesis: A Century of Invention and Technological Enthusiasm, 1870–1970* (University of Chicago Press, 2004). And for general **material on modern genetics**, I have relied on: S. Aldridge, *The Thread of Life: The story of genes and genetic engineering* (Cambridge University Press, 1996); and G.J.V. Nossal and R.L. Coppel, *Reshaping Life: key issues in genetic engineering* (Cambridge University Press, 2002).

Index

Index

Index

Index

Index

Index

Index

Index